Reviews of
117 Physiology Biochemistry and Pharmacology

Editors
M.P. Blaustein, Baltimore · O. Creutzfeldt, Göttingen
H. Grunicke, Innsbruck · E. Habermann, Gießen
H. Neurath, Seattle · S. Numa, Kyoto
D. Pette, Konstanz · B . Sakmann, Heidelberg
M. Schweiger, Innsbruck · U. Trendelenburg, Würzburg
K.J. Ullrich, Frankfurt/M · E.M. Wright, Los Angeles

Roland Seifert and Günter Schultz

The Superoxide-Forming NADPH Oxidase of Phagocytes

An Enzyme System Regulated by Multiple Mechanisms

With 18 Tables

Springer-Verlag Berlin Heidelberg GmbH

Dr. med. ROLAND SEIFERT

Professor Dr. med. GÜNTER SCHULTZ

Institut für Pharmakologie
Universitätsklinikum Rudolf Virchow
Freie Universität Berlin
Thielallee 69/63
1000 Berlin 33
Federal Republic of Germany

ISBN 978-3-662-31156-1   ISBN 978-3-540-46943-8 (eBook)
DOI 10.1007/978-3-540-46943-8

Library of Congress-Catalog-Card Number 74–3674

© Springer-Verlag Berlin Heidelberg 1991

Originally published by Springer-Verlag Berlin Heidelberg New York in 1991.
Softcover reprint of the hardcover 1st edition 1991

The use of registered names, trademarks, etc. in this publication does not imply, even in the absence of a specific statement, that such names are exempt from the relevant protective laws and regulations and therefore free for general use.

Product Liability: The publisher can give no guarantee for information about drug dosage and application thereof contained in this book. In every individual case the respective user must check its accuracy by consulting other pharmaceutical literature.

Typesetting: International Typesetters Inc., Manila, Philippines
27/3130–543210 — Printed on acid-free paper

Contents

1 Introduction 1

2 NADPH Oxidase:
   A Superoxide-Forming Enzyme System 5

2.1   Catalytical Properties 5
2.2   Cofactor Requirements 6
2.3   Some Effects of Various Stimuli
         on the Catalytic Properties of
         NADPH Oxidase 7
2.4   Structural Components 9
2.4.1   NADPH-Binding Component, Flavoproteins,
         and Low Molecular Mass Components 9
2.4.2   Quinones 14
2.4.3   Cytochrome b_{-245} 15
2.5   Cellular Localization of NADPH
         Oxidase 17

3 Activation of NADPH Oxidase 19

3.1   Some General Mechanisms Involved
         in the Activation of NADPH Oxidase 19
3.1.1   Protein Kinase C 19
3.1.1.1 Phorbol Esters 20
3.1.1.2 Diacylglycerols 25
3.1.1.3 Other Agents 27
3.1.1.4 The Role of Ca^{2+}/Phospholipid-Independent
         Protein Kinase C 28
3.1.2   Fatty Acids 29
3.1.2.1 Lipid Specificity 29
3.1.2.2 Mechanistic Aspects 30
3.2   Mechanisms Involved in the Activation of
         NADPH Oxidase by Receptor Agonists 31

3.2.1   Guanine Nucleotide Binding Proteins
          (G-Proteins) 31
3.2.1.1   Interaction of Plasma Membrane Receptors
          with G-Proteins in Phagocytes 32
3.2.1.2   Low Molecular Mass GTP-Binding Proteins . . . 39
3.2.1.3   NaF 39
3.2.2   Protein Kinase C 41
3.2.2.1   Correlations and Dissociations Between
          Activation of Phospholipases C and D
          and Protein Kinase C, on one Hand,
          and NADPH Oxidase, on the Other 42
3.2.2.2   Priming by Activators of Protein Kinase C 45
3.2.2.3   Studies with Protein Kinase C Inhibitors 46
3.2.2.4   Studies with Inhibitors of Diacylglycerol
          Kinase and Diacylglycerol Lipase 53
3.2.2.5   The Inhibitory Role of Protein Kinase C
          in Receptor Agonist-Induced
          Cell Activation 54
3.2.3   Calcium and Calmodulin 55
3.2.3.1   Correlations and Dissociations Between
          Activation of Ca^{2+}-Dependent Processes
          and NADPH Oxidase 56
3.2.4   Phospholipase A_2 and Arachidonic Acid 59
3.2.4.1   Correlations and Dissociations Between
          Activation of Arachidonic Acid Release
          and NADPH Oxidase 59
3.2.5   Cytoskeleton 61
3.2.5.1   Cytochalasins 61
3.2.5.2   Botulinum C2 Toxin 63
3.2.5.3   Miscellaneous Agents 63
3.2.6   Cyclic Nucleotides 64
3.2.6.1   The Role of cAMP 64
3.2.6.2   The Role of cGMP 65
3.2.7   Protein Tyrosine Phosphorylation
          and Protein Phosphatases 67
3.2.8   Proteases 69
3.3   Activation of NADPH Oxidase
          by Various Classes of Stimuli 70
3.3.1   Agents Presumably Acting Through
          Plasma Membrane Receptors and G-Proteins . . 82
3.3.1.1   Formyl Peptides 82
3.3.1.2   Miscellaneous Peptides 86

3.3.1.3   Cytokines   91
3.3.1.4   Matrix Proteins   101
3.3.1.5   Complement Components
          and Immunoglobulins   102
3.3.1.6   Platelet-Activating Factor   107
3.3.1.7   Products of the Lipoxygenase Pathway   109
3.3.1.8   Purine and Pyrimidine Nucleotides   111
3.3.2   Miscellaneous Stimulatory Agents   115
3.3.2.1   Lectins   115
3.3.2.2   Lipopolysaccharides   116
3.3.2.3   Muramyl Peptides   119
3.3.2.4   Retinoids   120
3.3.2.5   Digitonin   120
3.3.2.6   Hexachlorocyclohexanes   121
3.3.2.7   Alcohols   122
3.3.2.8   Thymol   122
3.3.2.9   Bleomycin   122
3.3.2.10 Neuraminidase   123
3.3.2.11 1,25-Dihydroxyvitamin D_3   123
3.3.2.12 Particulate Stimuli   123
3.4   Miscellaneous Aspects of
          NADPH Oxidase Activation   126
3.4.1   Relation of the Respiratory Burst
          to the Synthesis of Reactive Nitrogen Oxide
          Intermediates: Role of Arginine   126
3.4.2   Relation of the Respiratory Burst
          to Ion Fluxes   128
3.4.2.1 Na^+/H^+ Exchange   128
3.4.2.2   Membrane Depolarization   129
3.4.2.3   Anion Transport   131
3.4.3   Adherence   131
3.4.4   Properties of the Respiratory Burst
          in some Specialized Cell Types   133
3.4.4.1   Phagocytes   133
3.4.4.2   Other Cell Types   140

4   Inhibition of NADPH Oxidase   147

4.1   cAMP-Increasing Agents   147
4.1.1   Receptor Agonists   159
4.1.1.1   Prostaglandins Including Prostacyclin   159
4.1.1.2   β-Adrenergic Agonists   160

4.1.1.3   Histamine   161
4.1.1.4   Adenosine   162
4.1.2   Other cAMP-Increasing Agents   163
4.1.3   Mechanistic Aspects   164
4.1.4   cAMP-Decreasing Agents   166
4.2   Anti-inflammatory Drugs   167
4.2.1   Glucocorticoids and Cyclosporin A   167
4.2.2   Nonsteroidal Anti-inflammatory Drugs   169
4.2.3   Cloroquine, Mepacrine,
          and Gold Compounds   170
4.3   Miscellaneous Inhibitory Agents   171
4.3.1   Protozoal, Fungal,
          and Bacterial Products   171
4.3.2   Endogenous Proteins   173
4.3.3   SH Reagents   175

5 Reconstitution and Regulation of NADPH Oxidase
   Activity in Cell-free Systems   177

5.1   Reconstitution and Regulation of NADPH
          Oxidase Activity by Fatty Acids
          and Sodium Dodecyl Sulfate   178
5.1.1   Historical Remarks   178
5.1.2   Some General Aspects   178
5.1.3   Activation by Fatty Acids   179
5.1.3.1   Lipid Specificity   179
5.1.3.2   Fatty Acids and the Role of
          Protein Kinase C   180
5.1.3.3   The Role of Calmodulin   182
5.1.3.4   Other Mechanistic Aspects   182
5.1.3.5   Physiological Relevance of
          Fatty Acid-Induced Activation of
          NADPH Oxidase in Cell-free Systems   183
5.1.4   The Role of G-Proteins   184
5.1.4.1   Guanine Nucleotides   184
5.1.4.2   NaF   187
5.1.4.3   Nucleoside Diphosphate Kinase   187
5.1.4.4   Some Open Questions   188
5.1.5   Cytosolic Activation Factors   189
5.1.5.1   Some General Properties   190
5.1.5.2   Involvement of Multiple Cytosolic Activation
          Factors in the Regulation of NADPH Oxidase .   191

5.2   Reconstitution and Regulation of NADPH
        Oxidase Activity by Phorbol Esters 200
5.3   Reconstitution and Regulation of NADPH
        Oxidase Activity Phosphatidic Acid 201

6 Pathology of NADPH Oxidase 203

6.1   Chronic Granulomatous Disease 203
6.1.1   Defect of Cytochrome b_{-245} 203
6.1.2   Defect of Cytosolic Activation Factors 205
6.1.3   Variant Chronic Granulomatous
        Disease 206
6.2   Other Pathological States 206
6.2.1   Hematological Disorders 206
6.2.2   Infections 211
6.2.3   Pulmonary and Allergic Disorders 212
6.2.4   Essential Hypertension 213
6.2.5   Myotonic Dystrophy 214
6.2.6   Diabetes Mellitus 214
6.2.7   Renal Disorders 215
6.2.8   Osteopetrosis 216
6.2.9   Glycogen Storage Disease 217

7 Age- and Sex-Related Alterations of the Activity
   of NADPH Oxidase 219

8 Concluding Remarks 221

References 225

Subject Index 333

Indexed in Current Contents

1 Introduction

Upon exposure to a multitude of stimuli, phagocytes, i.e., neutrophils, eosinophils, monocytes, and macrophages, undergo the so-called respiratory burst, which is characterized by activation of the hexose monophosphate shunt, increased oxygen consumption, and activation of the plasma membrane-bound NADPH oxidase. This enzyme catalyzes the formation of superoxide anions ($O_2^{\bullet-}$) at the expense of oxygen and NADPH. $O_2^{\bullet-}$ may enzymatically or spontaneously dismute to H_2O_2 and may subsequently be converted to a variety of active oxygen species such as singlet oxygen and hydroxyl radical (Babior 1984; Rossi 1986; DiGregorio et al. 1989; Cadenas 1989).

The respiratory burst in phagocytes was discovered by Baldrige and Gerard (1933), who observed increased oxygen consumption in phagocytosing canine neutrophils. More than two decades later, Stähelin et al. (1957) reported that glucose metabolism via the hexose monophosphate shunt is activated during phagocytosis of guinea pig neutrophils. Sbarra and Karnovsky (1959) showed that the stimulated oxygen consumption during phagocytosis is not prevented by inhibitors of mitochondrial respiration, suggesting its nonmitochondrial origin. In 1961, Iyer et al. presented indirect evidence for the assumption that H_2O_2 is produced during the respiratory burst. Subsequent studies suggested that a particulate NADPH oxidase is involved in H_2O_2 formation and increased oxygen consumption during phagocytosis (Rossi et al. 1969, 1972; Romeo et al. 1971; Patriarca et al. 1971). In 1973, Babior et al. reported that $O_2^{\bullet-}$ is generated in phagocytosing neutrophils.

It should be emphasized that NADPH oxidase is not the only source of $O_2^{\bullet-}$ in cellular systems. For example, $O_2^{\bullet-}$ is generated in the cyclooxygenase pathway of arachidonic acid, in xanthine oxidase reactions, and in the electron transport chain of mitochondria (Slater 1984; Cadenas 1989; Koner et al. 1989).

Reactive oxygen intermediates are assumed to play a role in the killing of bacteria, fungi, parasites, and tumor cells and in the pathogenesis of myocardial ischemia reperfusion injury, adult respiratory distress syndrome, cerebral vascular damage, tumorigenesis, and noninfectious acute and chronic inflammatory processes (Nathan et al. 1980; Nathan 1982; Dallegri et al. 1989; Hammond et al. 1985; Weitzman et al. 1985; Malech and Gallin 1987; Johnston 1988; Blake et al. 1987; Lunec et al. 1987; Halliwell et al. 1988; Weiss 1989; Kloner et al. 1989). Thus, as reactive oxygen species potentially possess both beneficial and deleterious effects, the production of the primary radical, $O_2^{\cdot-}$, would be expected to be carefully regulated.

The present review does not deal with the chemistry or biochemistry of oxygen radicals, as reviews on this topic are available (Klebanoff 1980; Roos 1980; Badwey and Karnovsky 1980; Halliwell and Gutteridge 1984; Slater 1984; Fridovich 1986, 1989; Naqui et al 1986; Britigan et al. 1987; Cadenas 1989).

Certain aspects of signal transduction processes in phagocytes have been reviewed by Omann et al. (1987a), Hamilton and Adams (1987), Sandborg and Smolen (1988), Dillon et al. (1988), Sha'afi and Molski (1988), Riches et al. (1988), and Adams (1989). Becker (1990) reviewed the history of the research on signal transduction in neutrophils. Various aspects concerning the structure and regulation of NADPH oxidase have been reviewed by Babior (1978a,b, 1984), Badwey et al. (1979), Roos (1980), Badwey and Karnovsky (1980, 1986), Baggiolini (1984), McPhail and Snyderman (1984), Forman and Thomas (1986), Rossi (1986), Tauber (1987), Sandborg and Smolen (1988), Lambeth (1988), Jesaitis and Allen (1988), Dillon et al. (1988), and Segal (1989a,b).

The methods available to measure the respiratory burst have been reviewed (Absolom 1986). A commonly employed method to measure the respiratory burst is based on the superoxide dismutase-inhibitable reduction of ferricytochrome c or derivatives of ferricytochrome c by $O_2^{\cdot-}$ (Drath and Karnovsky 1975; Butler et al. 1982; Nasrallah et al. 1983; Bellavite et al. 1983; Markert et al. 1984; Rajkovic and Williams 1985a; Pick 1986; Turrens and McCord 1988; Morel et al. 1988). H_2O_2 production is also a widely used measure to assess the respiratory burst (Rajkovic and Williams 1985a; Pick 1986). Another method to assess the activity of NADPH oxidase is the measurement of oxygen consumption that is insensitive to inhibitors of mitochondrial respiration, e.g., to NaN_3, KCN, and antimycin (Absolom 1986). In addition, various chemiluminescence methods have been described to measure oxygen radical formation (Sagone et al. 1976; Trush et al. 1978; Roos 1980; Welch et al. 1980; Schopf et al. 1984; Halstensen et al. 1986; Roberts et al. 1987; Wymann et al. 1987a; Lock et al. 1988; Johansson and Dahlgren 1989). Moreover, $O_2^{\cdot-}$-dependent reduction of nitro blue tetrazolium (NBT) to insoluble formazan (Baehner et al. 1975; Schopf et al. 1984; Absolom 1986; Pick 1986; DiGregorio et al. 1989) and measurement of hexose monophosphate shunt activity (Borregaard et al. 1984) reflect activation of the respiratory burst.

We review the literature on the regulation of NADPH oxidase with special respect to relevant publications from the past 4 years. The literature search for this review was concluded in January 1991. While we deal with a broad spectrum of aspects, we are aware of the fact that we cannot consider all literature available on this subject.

As is outlined below, the literature on NADPH oxidase regulation is controversial in many areas. Unfortunately, it is often difficult to analyze the reasons for conflicting reports. Some discrepancies are explained by cell type- and species-specific properties of the respiratory burst, by differences in the experimental procedures, by differences in the stimuli and inhibitors used, and by differences in their concentrations. Some aspects of the species specificity of the respiratory burst have been reviewed by Styrt (1989).

It is known that activation of NADPH oxidase in intact phagocytes under in vitro conditions is affected by various factors during cell isolation such as temperature, centrifugal forces, presence of serum, dextran, Ficoll-Hypaque, and buffer constituents (Berkow et al. 1984; English et al. 1988; Tennenberg et al. 1988). The storage time and storage conditions after cell isolation influence O_2^- formation as well (English et al. 1981a, 1988; J.J. Zimmerman et al. 1985, 1989; Dahlgren et al. 1987). The assay conditions during measurement of O_2^- formation such as time of contact with stimulus and the state of adherence are also of importance (Dahinden et al. 1983b; English et al. 1988). Moreover, the density of adherent and suspended neutrophils strongly affects the amount of O_2^- and H_2O_2 generated per cell (Mege et al. 1986; Peters et al. 1990). Finally, the density of adherent neutrophils determines the requirement for extracellular Ca^{2+} of O_2^- formation (Ishihara et al. 1990).

Another difficulty is certainly the fact that there are substantial inter- and intraindividual differences in the magnitude of the respiratory burst activated by various stimuli in neutrophils (J.J. Zimmerman et al. 1985, 1989; Pontremoli et al. 1988; Seifert et al. 1991a). It is also not known exactly how far the in vitro assay conditions for measurement of the respiratory burst activity reflect the in vivo conditions. For example, in vivo, phagocytes are exposed simultaneously to plasma proteins and/or extracellular matrix proteins, lipids, cytokines, and other intercellular signal molecules, and the phagocytes may interact with endothelial cells, platelets, lymphocytes, and/or their secretory products. Most in vitro studies are performed with purified cell populations, and the phagocytes are suspended in artificial, defined buffer solutions which are devoid of many components normally present in the blood or in the extravascular fluid. Discussion of all these aspects is, of course, beyond the scope of this review.

2 NADPH Oxidase:
A Superoxide-Forming Enzyme System

2.1 Catalytical Properties

NADPH oxidase (EC 1.6.99.6) catalyzes the univalent reduction of O_2 to $O_2^{\cdot-}$ according to the reaction: $2\ O_2 + NADPH \rightarrow 2\ O_2^{\cdot-} + NADP^+ + H^+$. NADPH oxidase has also been shown to catalyze the divalent reduction of O_2 to H_2O_2 under certain experimental conditions (Green and Wu 1986; Green and Pratt 1987). The K_m value for NADPH of NADPH oxidase amounts to 30–80 μM, that for NADH to 0.4–0.9 mM, and that for oxygen to about 10 μM (Babior et al. 1975, 1976; Gabig and Babior 1979; Cohen et al. 1980a; Lew et al. 1981; Chaudhry et al. 1982; Wakeyama et al. 1982; Yamaguchi et al. 1983; Suzuki et al. 1985; Tamura et al. 1988). NADPH oxidase shows a pH optimum at about 7.0 and is inhibited by various SH reagents (see also Sect. 4.3.3) but not by inhibitors of the respiratory chain (Babior et al. 1975, 1976; McPhail et al. 1976; Iverson et al. 1977; Tauber and Goetzl 1979; Cohen et al. 1980a; Green and Schaefer 1981; Gabig et al. 1981).

In the case of rabbit peritoneal neutrophils, NADPH oxidase not only catalyzes the formation of $O_2^{\cdot-}$ but also possesses NADPH diaphorase activity (EC 1.6.99.1) as measured by superoxide dismutase-insensitive reduction of ferricytochrome c (Laporte et al. 1990). Transition of NADPH oxidase from the diaphorase to the $O_2^{\cdot-}$-forming enzyme is enhanced by arachidonic acid and stable guanine nucleotides and is accompanied by a substantial increase in the K_m for NADPH. Iodonium compounds reduce $O_2^{\cdot-}$ formation but not diaphorase activity, and arachidonic acid at high concentrations or Triton X-100 induces reappearance of diaphorase activity.

The thermolability and the sensitivity to inhibition by salts have been obstacles for the purification of this enzyme (Babior and Peters 1981; Green and Pratt 1988; Wakeyama et al. 1982; Sakane et al. 1987a). Glutaraldehyde, glycerol, ethylene glycol, and dimethyl sulfoxide have been reported to stabilize NADPH oxidase (Tauber and Goetzl 1979; Sakane et al. 1987). The stabilizing effect of glutaraldehyde is apparently due to its protein cross-linking ability, as its monovalent analogue, butyraldehyde, is without stabilizing effect (Sakane et al. 1987).

The particulate NADPH oxidase from stimulated human neutrophils has been solubilized with various detergents (Gabig et al. 1982; Gabig and Babior 1979; Tauber and Goetzl 1979; Light et al. 1981; Tamura et al. 1988). Early studies showed that the solubilized NADPH oxidase passes through membrane filters retaining species with molecular masses over 300 kDa (Gabig et al. 1978). Using a gel filtration technique, Tauber and Goetzl (1979) reported that entities with apparent molecular masses of 150 kDa and of over 300 kDa show NADPH oxidase activity, and that the solubilized NADPH oxidase possesses at pI of 7.6–8.3. In agreement with the data obtained for the particulate enzyme, the K_m for NADPH of the Triton X-100-solubilized NADPH oxidase amounts to 33 μM and that for NADH to 930 μM (Gabig and Babior 1979). Solubilization of NADPH oxidase with deoxycholate plus Tween 20 results in a twofold increase in enzyme activity and in the K_m for NADPH (Tamura et al. 1988). Similar to the particulate enzyme, the solubilized NADPH oxidase shows maximal activity at pH 7.0–7.5 (Gabig and Babior 1979; Tauber and Goetzl 1979).

2.2 Cofactor Requirements

The activity of solubilized NADPH oxidase is modulated by phospholipids. Phosphatidylethanolamine but not phosphatidylcholine or phosphatidylserine were found to enhance the activity of the Triton X-100-solubilized NADPH oxidase from human neutrophils (Gabig and Babior 1979). The activity of the human neutrophil NADPH oxidase solubilized by deoxycholate plus Tween 20 is substantially augmented by phospholipids, in the order of effectiveness phosphatidylserine > cardiolipin > phosphatidylethanolamine > phosphatidylinositol, whereas phosphatidylcholine is inactive (Tamura et al. 1988). NADPH oxidase activity shows a quadratic dependence on the protein concentration of the solubilized preparation, and this property results in lower activity than expected with low protein concentrations (Gabig et al. 1978; Tamura et al. 1988). The addition of phospholipids to the solubilized enzyme restores the relation between protein concentration and enzyme activity to a linear function (Gabig et al. 1978; Gabig and Babior 1979; Tamura et al. 1988).

NADPH oxidase is regulated by divalent cations, but the results obtained by various groups are not consistent. Ca^{2+} at micromolar concentrations has been shown to inhibit O_2^- formation in phagocytic vesicles from rabbit alveolar macrophages, and preincubation of phagocytic vesicles with ethylene glycol tetraacetate (EGTA) desensitizes NADPH oxidase to inhibition by Ca^{2+} (Lew and Stossel 1981). The activity of NADPH oxidase from myristic acid-stimulated guinea pig neutrophils has been reported to be enhanced by Mg^{2+} but not by Ca^{2+} (Yamaguchi et al. 1983). Mg^{2+} enhances V_{max} and reduces the K_m for NADPH, and the effect of Mg^{2+} is maximal at 40–50 μM (Yamaguchi et al. 1983). NADPH oxidase from phorbol myristate acetate (PMA) treated human neutrophils and monocytes is inhibited by ethylenediaminetetraacetate (EDTA) and is stimulated by both Mg^{2+} and Ca^{2+} (Green et al. 1983; Suzuki et al. 1985). In addition, Ca^{2+} and Mg^{2+} protect the enzyme against thermal inactivation (Suzuki et al. 1985). These authors suggested that divalent cations do not interact with NADPH or modulate its binding to NADPH oxidase, but rather that they bind to a structural or regulatory component of the enzyme (Suzuki et al. 1985; see also Sect. 5.1).

The particulate NADPH oxidase is regulated by adenine nucleotides. Early studies suggested that NADPH oxidase is inhibited by AMP, ADP, ATP, and the nucleotide analogue, 2′,5′-ADP (Badwey and Karnovsky 1979; Babior and Peters 1981). Some years later, Melloni et al. (1986b) reported that ATP increases particulate NADPH oxidase activity from PMA-stimulated human neutrophils. This stimulation of O_2^- formation is accompanied by the release of a neutral serine protease from the membrane into the incubation medium and by the phosphorylation of membrane proteins (Melloni et al. 1986b; see also Sect. 3.1.1.4). The stimulatory effects of ATP have been suggested to involve protein kinase C-mediated phosphorylation reactions (Melloni et al. 1986; see also Sects. 3.3.1.8, 5.1.4.3).

2.3 Some Effects of Various Stimuli on the Catalytic Properties of NADPH Oxidase

Exposure of phagocytes to different substances which per se at the concentrations applied are not sufficient to activate the respiratory burst, may result in enhanced O_2^- formation upon exposure to a second stimulus, and this process is referred to as priming. Various aspects and examples of priming processes are described in Sects. 3.2.2.2 and 3.3.

Resident macrophages possess a lower capacity to undergo a respiratory burst upon exposure to PMA than macrophages which have been primed, i.e., "activated" by infection of the host with bacteria or "elicited" by intraperitoneal injection of various substances (Bellavite et al. 1981; Johnston and Kitagawa 1985). The priming of macrophages is associated with an increase in NADPH oxidase activity (Bellavite et al. 1981).

Primed murine peritoneal macrophages generate larger amounts of O_2^- upon stimulation with PMA than resident macrophages, and the K_m for NADPH of NADPH oxidase is lower in the former cells (Bellavite et al. 1981; Sasada et al. 1983; Berton et al. 1985a). NADPH oxidase from lipopolysaccharide (LPS) treated macrophages shows a higher V_{max} and a lower K_m for NADPH than the enzyme from resident cells, and LPS-treated macrophages contain larger amounts of NADPH than control cells (Sasada et al. 1983). The content of cytochrome b_{-245} apparently does not increase in primed macrophages, but these macrophages may utilize a higher number of cytochrome b_{-245} molecules than the nonprimed cells (Berton et al. 1986a). Tsunawaki and Nathan (1984) did not find substantial differences in the V_{max} of NADPH oxidase and the cellular content of cytochrome b_{-245} in primed and resident macrophages. In contrast, the K_m for NADPH of NADPH oxidase was found to be inversely related to the macrophages' ability to generate H_2O_2 (Tsunawaki and Nathan 1984). These data suggest that an increase in the affinity of NADPH oxidase for NADPH contributes to priming of macrophages for an augmented respiratory burst.

In neutrophils, various chemoattractants activate NADPH oxidase in the presence of cytochalasin B (CB) in a time- and concentration-dependent manner (McPhail and Snyderman 1983; see also Sect. 3.3.1). The temporal pattern of NADPH oxidase activation caused by the Ca^{2+} ionophore A 23187 or by PMA is different from that induced by chemoattractants, and activation of the enzyme by the latter two agents does not depend on the presence of CB (see also Sect. 3.2.5). The analysis of the K_m values for NADPH and NADH of NADPH oxidase suggests that the same oxidase is activated by various stimuli acting through different signal transduction pathways (McPhail and Snyderman 1983). Activation of the respiratory burst in human neutrophils by formyl peptides is associated with a decrease in the K_m of NADPH oxidase for oxygen (Edwards et al. 1983). In human neutrophils, priming for an enhanced respiratory burst by cell-permeant diacylglycerols, the Ca^{2+} ionophore, ionomycin, LPS and exsudation is accompanied by an enhanced capacity of NADPH oxidase for divalent reduction of oxygen (Follin and Dahlgren 1990; see also Sect. 2.1). Chemoattractants plus PMA interact synergistically to activate NADPH oxidase in human neutrophils (Bender et al. 1983). The effects of chemoattractants are concentration- and time-dependent and result in an increase in V_{max} of NADPH oxidase but not in an alteration of the K_m for NADPH (Bender et al. 1983).

2.4 Structural Components

NADPH oxidase is an enzyme system which apparently consists of multiple components. These components are localized in the plasma membrane, in specific granules, and in the cytosol of phagocytes. A number of components has been suggested to be involved in the redox chain of NADPH oxidase, including Flavine adenine dinucleotide (FAD), quinones, and a phagocyte-specific cytochrome referred to here as cytochrome b_{-245}. The electron flow may proceed as follows: NADPH → FAD protein → quinones (?) → cytochrome b_{-245} → O_2. The identity and functional organization of the redox components of NADPH oxidase is still a matter of intense debate. Table 1 summarizes some properties of putative plasma membrane-associated components of NADPH oxidase and of purified preparations of NADPH oxidase, and Table 16 summarizes some properties of putative cytosolic components of the enzyme system, in the following operationally referred to as cytosolic activation factors (see also Sect. 5.1.5).

When this review was in the final stages of preparation, a standardized nomenclature was suggested for these components of NADPH oxidase which are widely accepted to be involved in its activation (Clark 1990). The α-subunit of cytochrome b_{-245} is referred to as p22-*phox* (p, protein; 22, apparent molecular mass by sodium dodecyl sulfate (SDS)-polyacrylamide gel electrophoresis; *phox*, phagocyte oxidase). The β-subunit of cytochrome b_{-245} is designated gp91-*phox* (gp, glycoprotein). By analogy, the cytosolic 47-kDa and 67-kDa proteins are referred to as p47-*phox* and p67-*phox* (see also Sect. 5.1.5.2).

2.4.1 NADPH-Binding Component, Flavoproteins, and Low Molecular Mass Components

NADPH binds to a proximal component of the redox chain, possibly to a flavoprotein. The dye Cibacron blue inhibits O_2^- formation in membranes from PMA-activated guinea pig neutrophils, which effect may be due to the dye's ability to interact with the NADPH-binding component of NADPH oxidase (Yamaguchi and Kakinuma 1982). Affinity-labeling techniques have been used to identify the NADPH-binding component of NADPH oxidase. The 2'3'-dialdehyde derivative of NADPH serves both as an electron donor and as a competitive antagonist for NADPH (Umei et al. 1986). Inhibition of O_2^- formation by the NADPH analogue is prevented by NADPH, and NADPH dialdehyde has been shown to label a protein with an apparent molecular mass of 65–67 kDa, in the following referred to as 66-kDa protein (Umei et al. 1986). Pretreatment of the NADPH oxidase

Table 1. Some properties of putative plasma membrane-associated components of NADPH oxidase

Component	Properties	Selected references
Flavoprotein	Contains FAD. Redox midpoint potential = -280 mV; neutral semiquinone intermediates; reduction by NADPH. Flavoprotein component: 51 kDa protein?; NADPH cytochrome c reductase (87 kDa)? This reductase reduces cytochrome b_{-245} and is phosphorylated by protein kinase C in parallel with O_2^- formation. Antibody raised against purified porcine neutrophil flavoprotein recognizes 70 ± 72, 28 ± 32, and 16 ± 18-kDa peptides. Some dissociations between NADPH oxidase activity and FAD content. Diphenylene iodonium inhibits reduction of FAD.	Gabig and Lefker (1984a,b; Lutter et al. (1984), Bellavite et al. (1984); Sakane et al. (1984, 1987b); Berton et al. (1985b, Kakinuma et al. (1986), Parkinson and Gabig (1988), Fukuhara et al. (1988) Green and Pratt (1988), Ellis et al. (1989), Tamoto et al. (1989)
Quinones	Role as component of NADPH oxidase is controversial.	Cunningham et al. (1982), Crawford and Schneider (1982, 1983), Cross et al. (1983), Lutter et al. (1984), Gabig and Lefker (1985), M. Murakami et al. (1986)
Cytochrome b_{-245}	Correlation between O_2^- formation and cytochrome reduction, reduction by NADPH. Imidazole inhibits cytochrome reduction. Redox midpoint potential = -245 mV; hemoprotein, α, β, and Soret maxima at 558, 528, and 426 nm. Heterodimer: α-subunit (20-23 kDa), heme-containing moiety; β-subunit (76-92 kDa) glycosylated; "anchor function?" Cytochrome b_{-245} is missing in X-chromosomal CGD. Assumed to be the terminal oxidase for the univalent reduction of O_2 to O_2^-. Both subunits are phosphorylated by protein kinase C (not in parallel with O_2^- formation). Some dissociations between NADPH oxidase activity and cytochrome content. The GTP-binding protein, $rap1$, is associated with the cytochrome.	Morel and Vignais (1984), Pember et al. (1984), Cross et al. (1984, 1985), Iizuka et al. (1985a), Doussiere and Vignais (1985), Kakinuma et al. (1987), Edwards and Lloyd (1888), Garcia and Segal (1988); Parkos et al. (1988a,b), Ellis et al. (1989), Quinn et al. (1989), Yamaguchi et al. (1989)
31.5-kDa protein (pig neutrophils)	Phosphorylated in intact cells and in vitro by protein kinase C. Role in the redox chain not known exactly.	Papini et al. (1985)

Table 1. (continued)

Component	Properties	Selected references
14- to 18-kDa proteins (bovine and pig neutrophils)	Identified by immunoblotting with antibodies raised against purified NADPH oxidase. Function and identity not known (catalytic components? Flavoprotein components? Proteolytic degradation products?).	Fukuhara et al. (1988), Doussiere and Vignais (1988), Berton et al. (1989)
45-kDa protein (neutrophils and macrophages)	Labeled by diphenylene iodonium, inhibits O_2^- formation in intact cells and in cell-free systems. Binds FAD. Flavoprotein component?	Cross and Jones (1986), Hancock and Jones (1987), Cross et al. (1988), Ellis et al. (1988, 1989), Yea et al. (1990)
Purified NADPH oxidase (human neutrophils)	Red Sepharose dye affinity chromatography; holoenzyme = 150 kDa; 67-, 48-, and 32-kDa subunits; contains FAD but no quinone; content of cytochrome b_{-245} controversial.	Markert et al. (1985), Bellavite et al. (1986), Glass et al. (1986)
Purified NADPH oxidase (bovine neutrophils)	Ion exchange chromatography, gel filtration, isoelectric focusing; 65-kDa protein; pI 5.0; devoid of FAD and cytochrome b_{-245}. Further components: 16-, 18-, and 54-kDa proteins.	Doussiere and Vignais (1985, 1988)
Purified NADPH oxidase (pig neutrophils)	By isoelectric focusing, a pI 5.0, 67-kDa protein, contains FAD but no heme.	Kakinuma et al. (1987)

preparation with p-chloromercuribenzoate or with NADPH at excess prevents affinity labeling of the protein (Umei et al. 1986). An arylazido analogue of NADP inhibits NADPH oxidase from PMA-activated bovine neutrophils and affinity-labels a 66-kDa protein as well (Doussiere et al. 1986). A protein with similar apparent molecular mass is enriched during the purification of NADPH oxidase (Doussiere and Vignais 1985; see also below). These data suggest that a 66-kDa protein carries the NADPH-binding site and acts as NADPH dehydrogenase. Unexpectedly, the results of recent studies suggested that the NADPH-binding component of NADPH oxidase is one of the cytosolic activation factors (Sha'ag and Pick 1988; Volpp et al. 1988; Nunoi et al. 1988; Smith et al. 1989a,b; Takasugi et al. 1989; see also Sect. 5.1.5).

Solubilized NADPH oxidase from stimulated bovine neutrophils has been partially purified by ion-exchange chromatography, gel filtration, and isoelectric focusing (Doussiere and Vignais 1985). The apparent molecular mass of the purified NADPH oxidase is 65 kDa, with a pI of 5.0. With respect to the sensitivity to inhibitors, pH optimum, and K_m, the purified enzyme possesses similar properties as the crude enzyme, but the purified enzyme has been reported to contain neither FAD nor cytochrome b_{-245}. An antibody raised against purified NADPH oxidase obtained after isoelectric focusing inhibits $O_2^{\cdot-}$ generation in intact neutrophils and the activity of the particulate NADPH oxidase (Doussiere and Vignais 1988). This antibody recognizes four membrane-associated antigens with apparent molecular masses of 16, 18, 54, and 65 kDa. Another antibody raised against the 65-kDa protein excised from SDS polyacrylamide gels does not inhibit $O_2^{\cdot-}$ formation in intact neutrophils, but it inhibits the particulate enzyme. In neutrophils, this antiserum recognizes proteins with apparent molecular masses of 54 and 65 kDa (Doussiere and Vignais 1988). These data suggest that the 16- and 18-kDa proteins which are present in low amounts in the active NADPH oxidase preparation following isoelectric focusing are catalytic components of the respiratory burst enzyme.

A monoclonal antibody against purified NADPH oxidase from pig neutrophils reduces V_{max} but does not affect its K_m (Berton et al. 1989). The antibody apparently does not interfere with the NADPH-binding component but recognizes a heterodimer with apparent molecular masses of 14 and 16–18 kDa (Berton et al. 1989). How far these components are related to those described by Doussiere and Vignais (1988), is not yet known. Interestingly, a monoclonal antibody raised against a particulate preparation of PMA-stimulated guinea pig neutrophils, induces a respiratory burst in intact neutrophils and recognizes a 10-kDa antigen which may also be associated with NADPH oxidase (Berton et al. 1986b). Furthermore, antineutrophil cytoplasmic antibodies which are present in the plasma of patients with certain forms of necrotizing vasculitis and glomerulonephritis induce a respiratory burst in human neutrophils (Falk et al. 1990; see also Sect. 6.2). Finally, monoclonal antibodies against the sialoglycoprotein sialophorin show stimulatory effects on the respiratory burst in human monocytes; this process is accompanied by activation of phosphoinositide turnover, increase in cytoplasmic Ca^{2+}, and activation of protein kinase C (Wong et al. 1990).

A flavoprotein has been postulated to be involved in the redox chain of NADPH oxidase (Babior and Kipnes 1977; Light et al. 1981; Wakeyama et al. 1982). This assumption is supported by the findings that FAD and the FAD analogues 8-F-FAD, 8-phenyl-FAD, and 8-S-FAD are cofactors for $O_2^{\cdot-}$ formation, whereas the 2e⁻ donor 5-carba-deaza-FAD is inhibitory (Light et al. 1981; Parkinson and Gabig 1988). Membranes of human neutrophils contain FAD and cytochrome b_{-245} at similar concentrations (Lutter et al. 1984; Gabig and Lefker 1984b, 1985; Green and Pratt 1988). FAD is reduced by NADPH under anaerobic conditions, and the flavoprotein is resolved from cytochrome b_{-245} with concommittant loss of enzymatic activity (Lutter et al. 1984; Gabig and Lefker 1984b, 1985; Green and Pratt 1988). The redox intermediate of the flavoprotein may be a neutral semiquinone (Kakinuma et al. 1986). The putative flavoprotein of NADPH oxidase possesses characteristics of the dehydrogenase/electron transferase class and may have an apparent molecular mass of 51 kDa, as revealed by gel filtration (Parkinson and Gabig 1988; Green and Pratt 1988).

Solubilized NADPH oxidase from activated human neutrophils has been partially purified by red Sepharose dye affinity chromatography (Markert et al. 1985; Glass et al. 1986). The K_m for NADPH and turnover number of the purified NADPH oxidase are in agreement with the data obtained for the crude enzyme, and the purified preparation contains FAD (Glass et al. 1986). This purified NADPH oxidase has been suggested to consist of subunits with apparent molecular masses of 32, 48, and 67 kDa (Glass et al. 1986). In contrast, other preparations of partially purified NADPH oxidase have been reported to contain only very small amounts of FAD or no FAD (Serra et al. 1984; Bellavite et al. 1984, 1986; Berton et al. 1985b).

NADPH oxidase from fatty acid-activated pig neutrophils has been enriched by isoelectric focusing (Kakinuma et al. 1987). NADPH oxidase activity focuses at pI 5.0, exhibits a molecular mass of 67 kDa, and contains FAD. An antibody raised against the purified flavoprotein partially inhibits NADPH oxidase (Fukuhara et al. 1988). The antibody recognizes a major antigen with a molecular mass of 70–72 kDa and minor antigens with molecular masses of 16–18 and 28–32 kDa.

NADPH cytochrome c reductase has been suggested to represent the flavoprotein component of NADPH oxidase, at least in the case of guinea pig phagocytes (Sakane et al. 1983, 1984, 1987b; Tamoto et al. 1989). This reductase contains FAD, has an apparent molecular mass of 80 kDa, a pH optimum at 7.0–7.4, shows a much lower K_m for NADPH than for NADH, and is not inhibited by NaN_3. In the presence of phospholipids, purified NADPH cytochrome c reductase oxidizes NADPH and generates O_2^- upon addition of partially/purified cytochrome b_{-245} (Sakane et al. 1987). In addition, NADPH cytochrome c reductase reduces cytochrome b_{-245} under anaerobic conditions, and the reductase is phosphorylated upon stimulation of phagocytes with PMA (Sakane et al. 1987; Tamoto et al. 1989; see also Sect. 3.1.1.1).

The 45-kDa protein that is labeled with iodonium compounds such as diphenylene iodonium has been suggested to be a component of NADPH oxidase as well (Cross and Jones 1986; Ellis et al. 1988). Diphenylene iodonium prevents NADPH-dependent reduction of FAD and of cytochrome b_{-245} and inhibits O_2^- formation in mononuclear phagocytes and in human neutrophils presumably by interfering with the flavoprotein (Cross and Jones 1986; Hancock and Jones 1987; Ellis et al. 1988). In contrast, diphenylene iodonium does not inhibit phagocytosis, chemotaxis, or motility of neutrophils. These data point to the specificity of the effects of diphenylene iodonium on NADPH oxidase and suggest that this compound is of potential value as an anti-inflammatory agent (Cross 1987). Recently, the 45-kDa diphenylene iodonium binding protein has been purified by affinity chromatography (Yea et al. 1990). The purified protein binds FAD and has an isoelectric point of 4.0, and polyclonal antibodies against this protein inhibit the activity of the solubilized NADPH oxidase as well as O_2^- formation in the cell-free system (Yea et al. 1990; see also Sect. 5.1). The above results support the concept that the 45-kDa protein is the FAD-carrying redox component of NADPH oxidase.

2.4.2 Quinones

The role of quinones as further components of the redox chain of NADPH oxidase is very controversial, and the experimental data available do not convincingly support a role of quinones as part of the redox cascade. Some authors reported that neutrophils contain ubiquinone-10 and ubiquinone-50 and suggested that these redox carriers link the flavoprotein to cytochrome b_{-245} (Cunningham et al. 1982; Crawford and Schneider 1982, 1983; Gabig and Lefker 1985). This assumption is supported by the finding that certain quinones may show stimulatory effects on the respiratory burst

(Crawford and Schneider 1982). In addition, the redox state of ubiquinone-50 has been suggested to change according to the functional state of the neutrophils, and NADPH oxidase activity and ubiquinone have been reported to be enriched in phagolysosomes (Crawford and Schneider 1982, 1983).

In contrast, other authors reported that ubiquinone-1 does not accept electron equivalents from intermediate redox components of NADPH oxidase (M. Murakami et al. 1986). Cross et al. (1983) showed that ubiquinone is present only in mitochondria, not in particulate fractions enriched in NADPH oxidase activity and cytochrome b_{-245}, suggesting that quinones do not play a role in the regulation of O_2^- formation. Moreover, neutrophils and neutrophil "cytoplasts," i.e., neutrophils depleted of intracellular organelles and consisting only of the plasma membrane and cytosol, do not contain ubiquinone-50 (Lutter et al. 1984). Finally, various preparations of partially purified NADPH oxidase were found to contain no or only very low amounts of quinones (Markert et al. 1985; Glass et al. 1986; Bellavite et al. 1986).

2.4.3 Cytochrome b_{-245}

Cytochrome b_{-245} is a hemoprotein with absorption maxima at 426, 528, and 558 nm and is also referred to as cytochrome b_{558} (Pember et al. 1984; Iizuka et al. 1985a; Lutter et al. 1985; Yamaguchi et al. 1989). Cytochrome b_{-245} is a heterodimer consisting of a 20- to 23-kDa α-subunit and a glycosylated β-subunit. The molecular mass of latter component amounts to 74–115 kDa, depending on the type of phagocyte (Harper et al. 1984; Dinauer et al. 1987; Teahan et al. 1987; Parkos et al. 1987, 1988a,b; Yamaguchi et al. 1989; Nugent et al. 1989; Kleinberg et al. 1989). This large variation in apparent molecular mass of the β-subunit of cytochrome b_{-245} in various cell types is explained by the fact that the glycosylation pattern of this protein shows cell type specificity (Kleinberg et al. 1989). The deglycosylated β-subunit shows a molecular mass of 58 kDa (Kleinberg et al. 1989). The inhibition of glycosylation of proteins by tunicamycin in HL-60 cells does not abolish the cells' capacity to generate O_2^-, indicating that glycosylation of cytochrome b_{-245} is not obligatorily required for activation of the respiratory burst (Kleinberg et al. 1989).

The α-subunit of the cytochrome is assumed to carry heme, and the β-subunit is supposed to play a role in the functional assembly of the dimer (Yamaguchi et al. 1989; Verhoeven et al. 1989; Heyworth et al. 1989a; Nugent et al. 1989). The amino acid sequence deduced from the cDNA encoding the α-subunit of cytochrome b_{-245} shows no apparent homology to other known cytochromes, but it contains certain structural motifs common to other heme-carrying proteins (Parkos et al. 1988b). The gene for the β-subunit of cytochrome b_{-245} has also been cloned, and its identity has been confirmed by comparison with the amino acid sequence of the purified cytochrome and by immunological studies (Royer-Pokora et al. 1986; Teahan et al. 1987; Dinauer et al. 1987). The gene for the β-subunit is defective in most cases of X-chromosomal chronic granulomatous disease (CGD), and neither the α- nor the β-subunit is expressed in phagocytes of these patients (Royer-Pokora et al. 1986; Teahan et al. 1987; Dinauer et al. 1987; Parkos et al. 1989) (see also Sect. 6.1). Interestingly, RNA for the β-subunit is found only in phagocytes, whereas RNA encoding the α-subunit occurs also in other cell types (Parkos et al. 1988b). As the α-subunit is expressed only in phagocytes, these data support the view that the β-subunit is of importance for the stability of the dimer (Parkos et al. 1988b). Recently, cytochrome b_{-245} has been shown to be associated with the 22-kDa GTP-binding protein, *rap1* (Quinn et al. 1989) (see also Sects. 3.2.1, 5.1.4).

There is substantial experimental evidence for the assumption that cytochrome b_{-245} is the terminal redox component of NADPH oxidase. Cytochrome b_{-245} has been identified in various types of phagocytes including neutrophils, eosinophils, HL-60 cells, and mononuclear cells (Segal et al. 1981). Cytochrome b_{-245} possesses a midpoint redox potential of -245 mV which renders the cytochrome capable of catalyzing the univalent reduction of molecular oxygen to O_2^- at the expense of electrons delivered from the putative flavoprotein (Segal and Jones 1978; Segal et al. 1981; Cross et al. 1984, 1985; Gabig and Lefker 1984b; Pember et al. 1984; Lutter et al. 1985; Aviram and Sharabani 1986). Certain heterocyclic bases bind to heme iron in cytochromes and inhibit O_2^- formation, which process is accompanied by inhibition of reduction of cytochrome b_{-245} (Iizuka et al. 1985b; Ellis et al. 1989). The oxygen affinity for O_2^- formation in intact neutrophils and the oxygen tension at which cytochrome b_{-245} is oxidized to 50% of its aerobic steady-state level are similar, and there is a correlation between O_2^- formation and reduction of cytochrome b_{-245} (Morel and Vignais 1984; Edwards and Lloyd 1988; Ellis et al. 1989). Under anaerobic conditions, cytochrome b_{-245} is reduced by NADPH (Cross et al. 1984). The K_m values for NADPH of NADPH oxidase with respect to O_2^- formation and for cytochrome b reduction are similar, and the calculated aerobic rate of cytochrome b reduction correlates with the rate of O_2^- formation under various experimental conditions (Cross et al. 1985). Interestingly, an antibody raised against hepatic cytochrome P_{450} from guinea pig inhibits particulate NADPH oxidase obtained from PMA-stimulated guinea pig neutrophils in a concentration-dependent manner, suggesting that cytochrome b_{-245} and cytochrome P_{450} share some common epitopes (Takayama et al. 1984). Finally, NADPH oxidase activity has been reported to copurify with cytochrome b_{-245} (Serra et al. 1984; Bellavite et al. 1984, 1986; Berton et al. 1985b). In contrast to the above data, Markert et al. (1985), Glass et al. (1986), Doussiere and Vignais (1985) and Kakinuma et al. (1987) did not obtain positive evidence for the presence of cytochrome b_{-245} in partially purified NADPH oxidase preparations.

Recent functional studies support a role of cytochrome b_{-245} as the redox component of NADPH oxidase. An antibody raised against the cytoplasmic carboxy-terminal domain of the β-subunit of the cytochrome inhibits fatty acid-induced $O_2^{\cdot-}$ formation in a cell-free system (Rotrosen et al. 1990) (see also Sect. 5.1). Additionally, synthetic peptides corresponding to the carboxy-terminus of the β-subunit inhibit activation of $O_2^{\cdot-}$ formation when added to the assay mixture prior to but not after arachidonic acid. Moreover, these peptides have inhibitory effects on activation of NADPH by phorbol esters and chemotactic peptides in electropermeabilized human neutrophils. These data indicate that a cytoplasmic domain in the β-subunit of cytochrome b_{-245} mediates interactions with other components involved in the activation of NADPH oxidase.

2.5 Cellular Localization of NADPH Oxidase

The cellular localization of NADPH oxidase is also a subject of discussion. Early cytochemical and functional studies have shown that the formation of reactive oxygen intermediates in neutrophils occurs within the phagosomes and at the plasma membrane (Briggs et al. 1975; Goldstein et al. 1977; Dewald et al. 1979; Babior et al. 1981; Tsunawaki et al. 1983).

Cytochrome b_{-245} and the putative flavoprotein are localized both in the plasma membrane and in the specific granules of human neutrophils (Segal and Jones 1979, 1980; Millard et al. 1979; Clark et al. 1987; see also Sect. 5.1.2). In addition, specific granules contain substantial amounts of α-subunits of guanine nucleotide-binding proteins (G-proteins; Rotrosen et al. 1988; see also Sect. 3.2.1.1). Granule-associated cytochrome b_{-245} has been reported to be translocated to the plasma membrane upon stimulation of phagocytes with PMA or a Ca^{2+} ionophore (Borregaard et al. 1983; Higson et al. 1985). As the time courses of PMA-induced cytochrome b translocation and $O_2^{\cdot-}$ formation are in parallel, these findings suggest that translocation plays a role in the activation of NADPH oxidase (Borregaard et al. 1983; Higson et al. 1985). Translocation of the flavoprotein from intracellular granules to the plasma membrane has also been observed in stimulated neutrophils (Borregaard and Tauber 1984). Moreover, human neutrophils with a congenital defect in specific granules and neutrophil cytoplasts have been reported to generate lower amounts of O_2^- than control neutrophils upon stimulation with PMA, chemoattractants, or a Ca^{2+} ionophore (Ohno et al. 1985; see also Sect. 6.2).

Correlations between translocation of putative components of NADPH oxidase and the extent of O_2^- generation have not been observed in other experiments. For example, it has been reported that there is no correlation between the amount of cytochrome b_{-245} translocated to the plasma membrane and the amount of O_2^- generated (Ohno et al. 1985). Parkos et al. (1985) reported that PMA induces translocation of cytochrome b_{-245} but not of flavin components from specific granules to the plasma membrane. In addition, the concentrations of these redox components in the plasma membrane have been shown not to change considerably upon stimulation of neutrophils with chemoattractants (Parkos et al. 1985). Furthermore, neutrophil cytoplasts and intact neutrophils have been reported to generate O_2^- to similar extents upon stimulation with soluble and particulate stimuli (Roos et al. 1983). Moreover, membranes from resting and PMA-stimulated neutrophil cytoplasts contain similar amounts of FAD and cytochrome b_{-245} (Lutter et al. 1984). Finally, the Ca^{2+} ionophore- but not the chemoattractant-induced O_2^- formation depends on the presence of intracellular granule components (Dahlgren et al. 1989). These data suggest that the structural components of NADPH oxidase are present in a functioning state in the plasma membrane and in the cytosol, and that translocation of redox components is not necessarily required for enzyme activation (Roos et al. 1983; Parkos et al. 1985; see also Sect. 5.1).

The subcellular distribution of the α-subunit of cytochrome b_{-245} in human neutrophils, eosinophils, and monocytes was studied by im-munogold labeling with a monoclonal antibody (Ginsel et al. 1990). This technique has the advantage of avoiding cross-contamination of cell fractions isolated by differential centrifugation techniques. In neutrophils, the α-subunit is present on the cytosolic surface of the membrane of specific granules and the plasma membrane. In eosinophils and monocytes, the α-subunit also shows a dual localization in intracellular compartments and in the plasma membrane.

3 Activation of NADPH Oxidase

3.1 Some General Mechanisms Involved in the Activation of NADPH Oxidase

3.1.1 Protein Kinase C

In many cell types, stimulation of plasma membrane receptors with intercellular signal molecules results in the activation of phospholipase C, which catalyzes phosphoinositide degradation to hydrophilic inositolphosphates and to lipophilic diacylglycerol. A number of reviews on this topic are available (Michell 1975; Berridge 1984, 1987, 1989; Hokin 1985; Majerus et al. 1986; Abdel-Latif 1986).

Protein kinase C plays a crucial role in the signal transduction pathways activated by numerous intercellular signal molecules (Nishizuka 1984, 1986, 1988, 1989). In the presence of phospholipids, diacylglycerol activates protein kinase C by increasing the apparent affinity of the enzyme for Ca^{2+} (Nishizuka 1984, 1986, 1988, 1989). The importance of protein kinase C in signal transduction processes is supported by the findings that tumor-promoting phorbol esters, e.g., PMA, and cell-permeant diacylglycerols activate protein kinase C and mimic receptor agonist-induced cell activation in some instances (Castagna et al. 1982; Ashendel 1985; Nishizuka 1984, 1986, 1988, 1989).

Protein kinase C is a family consisting of several isoenzymes, which are assumed to play different functional roles (Coussens et al. 1986; Ohno et al. 1987; Nishizuka 1988, 1989). This assumption is supported by the fact that protein kinase C isoenzymes are differentially distributed among various cell types and in compartments of a given cell type (Nishizuka 1989). In addition, isoenzymes of protein kinase C show quantitative differences with respect to the activation by lipids. For example, the γ-isoenzyme is activated by arachidonic acid at lower concentrations than the α- and β-isoenzymes (Sekiguchi et al. 1987; Nishizuka 1989; see also Sects. 3.1.2, 3.2.4, 5.1.3).

Table 2 summarizes some properties of various substances which may directly or indirectly activate the respiratory burst through activation of protein kinase C.

3.1.1.1 Phorbol Esters

Neutrophils possess large amounts of protein kinase C, and the purified enzyme shows regulatory properties similar to protein kinase C in other cell types (Nishizuka 1984, 1986; Huang and Oshana 1986; Christiansen and Juhl 1986). Human myeloid cells, i.e., undifferentiated HL-60 cells, have been reported to contain the α-, β-, and γ-isoenzyme of protein kinase C, and the α-isoenzyme is the most abundant form (Makowske et al. 1988). In human neutrophils, the β- and α-isoenzyme of protein kinase C comprise 60% and 35% of the total protein kinase C activity, and no γ-isoenzyme is present (Pontremoli et al. 1990). Interestingly, human neutrophils also contain a Ca^{2+}- and phospholipid-dependent protein kinase which utilizes GTP instead of ATP as substrate (Stoehr and Smolen 1990). Myeloid differentiation is associated with an increase in amount of the α- and β-isoenzyme or of all three isoforms (Makowske et al. 1988). However, it is not yet known whether isoenzymes of protein kinase C play different roles in the regulation of NADPH oxidase.

It has been known for several years that neutrophils and mononuclear phagocytes of various species, including man, undergo a respiratory burst upon exposure to phorbol esters, and PMA is one of the most potent and effective activators of NADPH oxidase known so far (DeChatelet et al. 1976; Bass et al. 1978, 1983; Suzuki and Lehrer 1980; Badwey et al. 1980; Weiss et al. 1980; Hafeman et al. 1982). Primed peritoneal macrophages show higher capacities than resident cells to generate O_2^{-} upon stimulation with PMA (Bryant et al. 1982; Weinberg and Misukonis 1983; Badwey et al. 1983; Chung and Kim 1988; see also Sect. 2.3). Unlike resident murine peritoneal macrophages, bone marrow derived murine macrophages generate O_2^{-} upon exposure to zymosan but not to PMA (Philips and Hamilton 1989; see also Sects. 3.2.2, 3.3.1.5). The ability of bone marrow-derived macrophages to respond to PMA is restored, at least in part, by treatment with LPS or with the cytokines, granulocyte-macrophage colony-stimulating factor (GM-CSF), tumor necrosis factor-α (TNF-α), interferon-γ (IFN-γ), or interleukin-1α (IL-1α; Phillips and Hamilton 1989; see also Sect. 3.3.1.3). Thus, the responsiveness of resident peritoneal macrophages to PMA may be the result of in vivo exposure to cytokines (Phillips and Hamilton 1989). With respect to the rat, peritoneal macrophages show a greater capacity than alveolar macrophages to generate O_2^{-} upon stimulation with PMA (Peters-Golden et al. 1990).

Table 2. Activation of the respiratory burst by tumor promoters, activators of protein kinase C, and agents which lead to increased protein kinase C activation

Agent	Effect (cell type)	Mechanisms discussed	Selected references
Phorbol esters (e.g., PMA)	Potentiation (priming) and activation of $O_2^{\bullet-}$ formation (intact neutrophils and macrophages and cell-free system)	Translocation of protein kinase C and calpain to the plasma membrane, proteolytic activation of protein kinase C, phosphorylation of several proteins (e.g., the 47-kDa protein), accumulation of diacylglycerol and alkylacylglycerol. Protein kinase C-independent mechanisms?	Cox et al. (1985), Myers et al. (1985), Wolfson et al. (1985), Papini et al. (1985), Melloni et al. (1985, 1986a,b), Ohtsuka et al. (1986), Gennaro et al. (1986), Ohsaka et al. (1988), Rider et al. (1988), Garcia and Segal (1988), Tyagi et al. (1988), Badwey et al. (1989a), Sha'afi (1989)
Cell-permeant diacylglycerols (e.g., dioctanoylglycerol, OAG)	Potentiation and activation of $O_2^{\bullet-}$ formation (neutrophils)	Similar to phorbol esters, but no equivalence with respect to various parameters, e.g., phosphorylation pattern, protein kinase C translocation, kinetics, modulation by PGE_1. Protein kinase C-independent mechanisms?	Dale and Penfield (1985), Cox et al. (1986), Kiss and Luo (1986), Tsusaki et al. (1986), Wong and Chew (1986), Bass et al. (1988), Smith et al. (1988), Badwey et al. (1989c)
Alkylacylglycerols	Controversial; potentiation or inhibition of $O_2^{\bullet-}$ formation (neutrophils)	Modulation of protein kinase C activity, stimulation of arachidonic acid release. Mechanisms different from those of diacylglycerol?	Bauldry et al. (1988), Bass et al. (1988), Ford et al. (1989)
Phosphatidic acid	Activation of $O_2^{\bullet-}$ formation (intact neutrophils and cell-free system)	Ca^{2+}-independent processes, phosphorylation of the 47-kDa protein. Direct activation of NADPH oxidase?	Bellavite et al. (1988), Ohtsuka et al. (1989)
Benoxaprofen	Activation of $O_2^{\bullet-}$ formation and chemiluminescence (neutrophils)	Activation of protein kinase C.	Anderson and Eftychis (1986), Lukey et al. (1988)

Table 2. (continued)

Agent	Effect (cell type)	Mechanisms discussed	Selected references
Bryostatin	Activation of $O_2^{\bullet-}$ formation (neutrophils)	Activation of protein kinase C.	Berkow and Kraft (1985), Kraft et al. (1986), Wender et al. (1988)
Mezerein	Activation of $O_2^{\bullet-}$ formation (neutrophils)	Activation of protein kinase C. Protein kinase C-independent mechanisms?	Miyake et al. (1984), O'Flaherty et al. (1985b), Balazovich et al. (1986a)
R 59022	Potentiation of $O_2^{\bullet-}$ formation (neutrophils)	Inhibition of diacylglycerol kinase, activation of phospholipases C and D?	De Chaffoy de Courcelles et al. (1985), Gomez-Cambronero et al. (1987), Mege et al. (1988a), Mahadevappa (1988)
Indomethacin	Potentiation of OAG-induced $O_2^{\bullet-}$ formation (neutrophils)	Inhibition of diacylglycerol kinase and diacylglycerol lipase.	Dale and Penfield (1985, 1987)
RHC 80267	Potentiation of OAG-induced $O_2^{\bullet-}$ formation (neutrophils)	Inhibition of diacylglycerol lipase.	Dale and Penfield (1987)
Propranolol	Potentiation of fMet-Leu-Phe-induced respiratory burst (neutrophils)	Inhibition of phosphatidic acid phosphohydrolase, enhanced accumulation of phosphatidic acid	Billah et al. (1989), Rossi et al. (1990)
Inhalation anesthetics (e.g., halothane, chloroform)	Controversial, inhibition or activation of $O_2^{\bullet-}$ formation	Membrane perturbation, activation of protein kinase C	Welch and Zaccari (1982), Welch (1984), Nakagawara et al. (1986), Roghani et al. (1987), Tsuchiya et al. (1988)
Palytoxin, thapsigargin	Activation of $O_2^{\bullet-}$ formation (neutrophils)	Different from that of phorbol esters?	Kano et al. (1987)

Activation of $O_2^{\bullet-}$ formation by PMA does not depend on the presence of extracellular Ca^{2+} or on an increase in cytoplasmic Ca^{2+} (Sha'afi et al. 1983; Bass et al. 1983; Kiyotaki and Bloom 1984; Di Virgilio et al. 1984; Nasmith and Grinstein 1987a). However, under certain experimental conditions, i.e., when human neutrophils are seeded at low densities in microtiter plates, PMA-induced $O_2^{\bullet-}$ formation is dependent, at least in part, on the presence of extracellular Ca^{2+} (Ishihara et al. 1990). The PMA-induced respiratory burst is characterized by a lag time, requires temperatures above 17°C, has a pH optimum at 7.0, and is long lasting (Newburger et al. 1980a; Lehrer and Cohen 1981; Manara and Schneider 1985; J.J. Zimmerman et al. 1985). The lag time and the rate of PMA-induced $O_2^{\bullet-}$ formation in neutrophils are differently affected by temperature, pH, SH reagents, the concentration of PMA, and by other parameters (Newburger et al. 1980a). Interestingly, the magnitude of the PMA-induced respiratory burst in rat alveolar macrophages varies by about two- to threefold among individual phagocytes (DiGregorio et al. 1987).

There is a close correlation between the ability of various phorbol esters to activate protein kinase C and to induce $O_2^{\bullet-}$ formation (Robinson et al. 1985). In addition, the occupancy of phorbol ester binding sites with agonists is in parallel with the rate of $O_2^{\bullet-}$ formation (Tauber et al. 1982). There are, however, certain dissimilarities between the effects of various phorbol esters on the respiratory burst, pointing to their functional nonequivalence (Gaudry et al. 1990). Upon stimulation with PMA, protein kinase C is translocated from the cytosol to the phagocyte plasma membrane, and this process precedes $O_2^{\bullet-}$ formation and may explain the observed lag time (Kraft and Anderson 1983; Myers et al. 1985; Wolfson et al. 1985; Gennaro et al. 1986). A monoclonal antibody against an unidentified macrophage antigen recognizes a 90-kDa protein in neutrophil membranes and delays PMA-induced protein phosphorylation and $O_2^{\bullet-}$ formation, and phosphorylation occurs prior to the respiratory burst (Pontremoli et al. 1986d).

PMA induces the phosphorylation of numerous proteins in human myeloid cells (Andrews and Babior 1983; Helfman et al. 1983; Kiyotaki and Bloom 1984; Irita et al. 1984a,b; Feuerstein and Cooper 1984; Gennaro et al. 1985; Ohtsuka et al. 1986; Gaut and Carchman 1987). At the time being, most attention focuses on a 44- to 49-kDa protein or group of proteins (in the following referred to as 47-kDa protein) which is phosphorylated in neutrophils of healthy subjects but not in those of patients with autosomal-recessive CGD and is one of the cytosolic activation factors (see also Sects. 5.1.5, 6.1). There is a close correlation between phosphorylation of the 47-kDa protein and $O_2^{\bullet-}$ formation induced by PMA or a cell-permeable diacylglycerol, and the phosphorylated 47-kDa protein is apparently associated with cytochrome b_{-245} (Badwey et al. 1989a; Heyworth et al. 1989a).

Subsequent to phosphorylation, the 47-kDa protein is rapidly dephosphorylated and its continuous phosphorylation is required to maintain O_2^- formation in PMA-stimulated neutrophils (Heyworth and Badwey 1990).

Although a large body of data points to the central role of the 47-kDa protein as target for PMA, a role of additional protein kinase C-dependent and/or -independent mechanisms cannot be excluded. For example, PMA has been reported to induce phosphorylation of the α- and β-subunit of cytochrome b_{-245}, but the time courses of cytochrome b phosphorylation and O_2^- formation do not correlate (Garcia and Segal 1988). PMA has also been shown to induce phosphorylation of a 31.5-kDa protein which may be associated with cytochrome b_{-245} (Papini et al. 1985; see Table 1). In addition, PMA induces phosphorylation of NADPH cytochrome c reductase of guinea pig phagocytes, and this covalent modification correlates with the activation of O_2^- formation (Tamoto et al. 1989; see also Sect. 2.4.1).

PMA and cell-permeable diacylglycerols induce the accumulation of diradylglycerol, i.e. diacylglycerol and alkylacylglycerol, and diacylglycerol may amplify activation of protein kinase C (Rider et al. 1988; Reibman et al. 1988). Alkylacylglycerol has been shown to modulate diacylglycerol- and chemoattractant-induced O_2^- formation in a complex manner (Bauldry et al. 1988; Rider et al. 1988; Bass et al. 1989; see also Sect. 3.2.2.2). With respect to monocytic differentiation of HL-60 cells, PMA has been suggested to act by protein kinase C-independent mechanisms (Morin et al. 1987). PMA also induces rapid and substantial alterations in the plasma membrane fluidity of neutrophils, and alterations in the activity of phospholipase A_2 may contribute to PMA-induced activation of NADPH oxidase as well (Stocker et al. 1982; Stocker and Richter 1982; Henderson et al. 1989). Quite recently, it has been shown that various inhibitors of phospholipase A_2 blunt PMA-induced O_2^- formation in neutrophil cytoplasts, whereas exogenous arachidonic acid restores the respiratory burst (Henderson et al. 1989; see also Sects. 3.1.2, 3.2.4). Moreover, PMA-induced reduction of the intracellular ATP concentration has been claimed to be involved in the regulation of NADPH oxidase (Schinetti and Lazarino 1986). Finally, the potentiating effect of PMA on chemoattractant-induced O_2^- formation is discussed to involve protein kinase C-independent mechanisms (Sha'afi 1989; see also Sect. 3.2.2.2).

It is well known that activators of protein kinase C and Ca^{2+} ionophores can interact synergistically to activate various cell functions (Berridge 1984, 1987; Nishizuka 1984, 1986, 1989; Abdel-Latif 1986). In phagocytes, the Ca^{2+} ionophore A 23187 potentiates the PMA-induced respiratory burst, resulting in a reduction of the lag time and in an enhanced rate of O_2^- formation (Dale and Penfield 1984; Robinson et al. 1984; Strnad and Wong 1985a).

This synergism requires extracellular Ca^{2+} (Robinson et al. 1984). Ca^{2+} ionophores reduce the EC_{50} for PMA-induced $O_2^{\cdot-}$ formation. This synergism between phorbol esters and Ca^{2+} ionophores is explained, at least in part, by the fact that an increase in cytoplasmic Ca^{2+} enhances the affinity of protein kinase C for phorbol esters without altering the number of binding sites (Dougherty and Niedel 1986; French et al. 1987). Synergistic activation of $O_2^{\cdot-}$ formation by activators of protein kinase C and Ca^{2+}-mobilizing agents has been reported to be accompanied by increased phosphorylation of the 47-kDa protein (Heyworth et al. 1989b).

Exposure of murine and human peritoneal macrophages and human neutrophils to PMA desensitizes them to undergo a second respiratory burst upon subsequent stimulation with phorbol esters (Berton and Gordon 1983c; Gbarah et al. 1989). Desensitization of the respiratory burst to PMA in macrophages is associated with a reversible reduction in the number of phorbol ester binding sites (Berton and Gordon 1983c). In contrast, Kitagawa and Johnston (1986) did not find a decrease in the number or affinity of phorbol ester binding sites during deactivation of the respiratory burst. Interestingly, neutrophils have been reported to contain endogenous inhibitors of protein kinase C which may play an inhibitory role in the regulation of NADPH oxidase (Balazovich et al. 1986b; Huang and Oshana 1986).

3.1.1.2 Diacylglycerols

Like the physiologically relevant diacylglycerols which bear long-chain saturated and unsaturated fatty acids, cell-permeant diacylglycerols activate protein kinase C. Cell-permeant diacylglycerols are widely used to study the role of protein kinase C in signal transduction processes (Nishizuka 1984, 1986; Kreutter et al. 1985; Morin et al. 1987).

The diacylglycerols substituted with two short-chain saturated fatty acids, dihexanoylglycerol and dioctanoylglycerol, and the diacylglycerol substituted with one long-chain unsaturated fatty acid and one short-chain saturated fatty acid, 1-oleoyl-2-acetylglycerol (OAG), are effective activators of $O_2^{\cdot-}$ formation in neutrophils, but they are several orders of magnitude less potent than PMA (Fujita et al. 1984; Cox et al. 1986; Seifert et al. 1991a). Similar to phorbol esters, activation of NADPH oxidase by cell-permeant diacylglycerols does not depend on the presence of extracellular Ca^{2+} (Fujita et al. 1984). The effectiveness of OAG to induce $O_2^{\cdot-}$ formation is substantially enhanced by its incorporation into multilamellar liposomes (Tsusaki et al. 1986). In analogy to PMA, OAG and A 23187 interact synergistically to activate the respiratory burst (Penfield and Dale 1984). Cell-permeant diacylglycerols and phorbol esteres induce similar patterns of protein phosphorylation including the 47-kDa protein (Fujita et al. 1984; Badwey et al. 1989c). Interestingly, didecanoylglycerol has been reported to induce exocytosis but not O_2^- formation, suggesting that different species of protein kinase C are involved in the activation of different cell functions (Cox et al. 1986; see also Sect. 3.1.1.1).

There are, however, several differences between the effects of cell-permeant diacylglycerols and phorbol esters. Unlike in human neutrophils, OAG is a much less effective activator of O_2^- formation than PMA in HL-60 cells (Wong and Chew 1986; Cox et al. 1986; Bonser et al. 1986). Membranes of PMA-treated human neutrophils possess a higher NADPH oxidase activity than membranes of OAG-stimulated cells. In addition, the kinetics of the PMA- and OAG-induced respiratory burst are different (Wong and Chew 1986). These differences are explained, at least in part, by the fact that diacylglycerols but not phorbol esters can be rapidly metabolized (Wong and Chew 1986). Activation of the respiratory burst by phorbol esters and diacylglycerol also differs with respect to the requirement for cytoplasmic Ca^{2+}, i.e., cytoplasmic Ca^{2+} above 1 μM enhances OAG- but not PMA-induced O_2^- formation (Christiansen et al. 1988a). Prostaglandin E_1 (PGE_1) has been reported to inhibit the respiratory burst triggered by diacylglycerols but not that induced by phorbol esters (Dale and Penfield 1985; see also Sect. 4.1), whereas we did not observe stimulatory or inhibitory effects of various cAMP-increasing agents on dioctanoylglycerol-induced O_2^- formation in human neutrophils (unpublished results). With respect to the phosphorylation of membrane proteins, diacylglycerol and PMA are not equivalent as well (Kiss and Luo 1986). Under certain experimental conditions, the inhibitors of protein kinase C, staurosporine, and 1-(5-isoquinolinesulfonyl)-2-methylpiperazine (H-7) inhibit diacylglycerol-induced phosphorylation of the 47-kDa protein but not O_2^- formation, whereas the effects of phorbol esters on both parameters are inhibited in parallel by the protein kinase C inhibitors (Badwey et al. 1989c; see also Sect. 3.2.2.3). In agreement with these data, staurosporine does not completely inhibit O_2^- formation in human neutrophils induced by dioctanoylglycerol at 100 μM, whereas the effect of PMA at a maximally effective concentration is abolished by staurosporine (Seifert, unpublished results). Finally, PMA but not OAG has been reported to induce monocytic differentiation of HL-60 cells (Kreutter et al. 1985; Morin et al. 1987). These data indicate that the mechanisms by which PMA and diacylglycerol activate protein kinase C in myeloid cells are similar but not identical, and that a protein kinase C-independent pathway may be involved in the activation of O_2^- formation by diacylglycerols.

3.1.1.3 Other Agents

In addition to phorbol esters and diacylglycerols, various other agents have been suggested to activate NADPH oxidase via protein kinase C (see Table 2). The widely used inhalation anesthetic halothane has been shown to enhance PMA-induced O_2^- formation in guinea pig neutrophils (Tsuchiya et al. 1988). This effect of halothane is counteracted by H-7, and halothane activates purified protein kinase C and induces phosphorylation of a 47-kDa protein in these phagocytes (Tsuchiya et al. 1988). The inhalation anesthetic and tumor promoter chloroform activates O_2^- formation and also purified protein kinase C (Roghani et al. 1987; Tsuchiya et al. 1988). In contrast, other authors reported that halothane at therapeutically relevant concentrations inhibits the respiratory burst induced by various stimuli, possibly by interfering with Ca^{2+} mobilization, by causing membrane perturbation or by directly interfering with components of NADPH oxidase (Welch and Zaccari 1982; Welch 1984; Nakagawara et al. 1986; Tsuchiya et al. 1988; Lieners et al. 1989). Inhibition of the respiratory burst by halothane has been suggested to contribute to the reduced bactericidal activity of neutrophils following exposure to these agents (Welch and Zaccari 1982; Welch 1984; Nakagawara et al. 1986; Lieners et al. 1989). At least in the case of α_2-adrenergic synapses in rat brain, halothane may interfere with the interaction of receptors with G-proteins (Baumgartner et al. 1990).

Bryostatin is a macrocyclic lactone from the marine bryozoan *Bugula neritina* and is structurally dissimilar to phorbol esters (Kraft et al. 1986). Unlike PMA, bryostatin does not induce monocytic differentiation of HL-60 cells (Kraft et al. 1986). Bryostatin binds to the phorbol ester binding sites in human neutrophils and HL-60 cells, activates purified protein kinase C, induces protein phosphorylation patterns similar to PMA, and induces O_2^- formation in human neutrophils (Berkow and Kraft 1985; Kraft et al. 1986; Wender et al. 1988).

The tumor promoter mezerein is also an activator of O_2^- formation in human neutrophils (Balazovich et al. 1986a). Mezerein competitively inhibits binding of phorbol esters to purified protein kinase C and has been reported to stimulate protein kinase C by a mechanism similar to that of diacylglycerol and PMA (Miyake et al. 1984; O'Flaherty et al. 1985b). In contrast, Balazovich et al. (1986a) reported on a lack of stimulatory effect of mezerein on protein kinase C in human neutrophils. Mezerein has been reported not to induce protein kinase C translocation and to induce a protein phosphorylation pattern different from that induced by PMA (Balazovich et al. 1986a). These data suggest that protein kinase C-dependent and -independent pathways are involved in the activation of NADPH oxidase by mezerein. Finally, the non-PMA type tumor promoters palytoxin and thapsigargin have been reported to activate O_2^- formation in human neutrophils by a mechanism independent of protein kinase C (Kano et al. 1987).

The diterpene resiniferatoxin, which is an irritant but not a tumor promoter, activates the respiratory burst in elicited murine peritoneal macrophages (Evans et al. 1990). Resiniferatoxin is only a poor activator of protein kinase C but is more potent than PMA in activating a Ca^{2+}-inhibited kinase referred to as resiniferatoxin kinase. This kinase may play a part in the activation of NADPH oxidase by zymosan (see also Sect. 3.3.1.5.5).

Benoxaprofen had been used as a nonsteroidal anti-inflammatory drug until it was withdrawn because of severe phototoxicity (Allen 1983). Benoxaprofen has been reported to induce chemiluminescence and O_2^- formation in human neutrophils (Anderson and Eftychis 1986; Lukey et al. 1988). The effect of benoxaprofen is synergistically enhanced by UV radiation and is inhibited by H-7. In addition, benoxaprofen activates purified protein kinase C in a concentration-dependent manner. These data suggest that the benoxaprofen-induced dermatotoxicity is attributable, at least in part, to protein kinase C-mediated activation of NADPH oxidase in human neutrophils (Anderson and Eftychis 1986; Lukey et al. 1988).

Exogenous phospholipase C derived from bacteria such as *Clostridium perfringens* and *Bacillus cereus* activates the respiratory burst in macrophages and in neutrophils (Pick and Keisari 1981; Grzeskowiak et al. 1985; Styrt et al. 1989). Exogenous phospholipase C and NaF have been reported synergistically to stimulate the respiratory burst, whereas phospholipase C blunts activation by latex beads (Styrt et al. 1989; see also Sect. 3.3.2.12.2). Exogenous phospholipase C induces the hydrolysis of various classes of plasma membrane phospholipids, among other effects resulting in the formation of diacylglycerol (Grzeskowiak et al. 1985). Unlike fMet-Leu-Phe and ionomycin induce similar increases in cytoplasmic Ca^{2+}, whereas only the chemotactic peptide is an effective activator of O_2^- formation. Unlike formyl peptide induced activation of NADPH oxidase, that induced by exogenous phospholipase C apparently does not involve phosphoinositide degradation and Ca^{2+} mobilization. Phospholipase C-mediated activation of NADPH oxidase is qualitatively similar to that induced by PMA, suggesting that the effects of phospholipase C are mediated through the release of diacylglycerol and activation of protein kinase C (Grzeskowiak et al. 1985).

3.1.1.4 The Role of Ca^{2+}/Phospholipid-Independent Protein Kinase C

In addition to native protein kinase C, proteolytically modified protein kinase C has been suggested to play a role in the regulation of NADPH oxidase. Upon stimulation of neutrophils with PMA, the neutral Ca^{2+}-dependent protease calpain is translocated to the plasma membrane (Melloni et al. 1985). Calpain cleaves protein kinase C into an active catalytic and into an inactive fragment (Melloni et al. 1986a; Pontremoli et al. 1986b; Kishimoto et al. 1989). Proteolytically activated protein kinase C is no longer regulated by phospholipids and Ca^{2+}. Native protein kinase C has been suggested to mediate phosphorylation of membrane proteins, and the proteolytically activated kinase is assumed to phosphorylate predominantly cytosolic proteins (Pontremoli et al. 1986c). A monoclonal antibody against calpain inhibits the protease in intact human neutrophils and prevents appearance of Ca^{2+}/phospholipid-independent form of protein kinase C (Pontremoli et al. 1988). The inhibition of calpain is accompanied by a prolonged association of protein kinase C with the plasma membrane, enhanced phosphorylation of membrane proteins and an augmented respiratory burst (Pontremoli et al. 1988).

The regulatory domain of protein kinase C is inactivated by the product of the respiratory burst, H_2O_2, resulting in an increase in the activity of Ca^{2+}/phospholipid-independent protein kinase C (Gopalakrishna and Anderson 1989). The catalytic domain of protein kinase C is apparently less sensitive to inhibition by H_2O_2 (Gopalakrishna and Anderson 1989). This dual activation and inactivation of protein kinase C upon exposure to H_2O_2 may provide an effective on/off signal mechanism, but its role in the termination of the respiratory burst is not yet known (see also Sects. 3.3.1.1.3, 5.2).

3.1.2 Fatty Acids

Fatty acids have for several years been well known as activators of the respiratory burst. The role of fatty acids in the activation of NADPH oxidase in cell-free systems is dealt with in Sect. 5.1. Unsaturated fatty acids, especially arachidonic acid, have been suggested to play a role as intracellular signal molecules for activation of NADPH oxidase (Bromberg and Pick 1983; see also Sect. 3.2.4). This section describes some results concerning activation of NADPH oxidase by exogenous fatty acids in intact cells.

3.1.2.1 Lipid Specificity

Saturated and unsaturated fatty acids and the detergent SDS induce a respiratory burst in guinea pig neutrophils (Kakinuma 1974; Kakinuma and Minakami 1978; Washida et al. 1980). Activation of the respiratory burst correlates with hydrophobic binding of fatty acids to neutrophil plasma membranes (Kakinuma 1974; Kakinuma and Minakami 1978; Washida et al. 1980). In human neutrophils, *cis*-unsaturated fatty acids activate O_2^- formation in the presence of extracellular Ca^{2+} (Badwey et al. 1981, 1984; H.J. Cohen et al. 1986; Morimoto et al. 1986). In contrast, *trans*-unsaturated and saturated fatty acids have been reported to activate NADPH oxidase in intact human neutrophils in the absence of extracellular Ca^{2+}, whereas Ca^{2+} is inhibitory (Yamaguchi et al. 1986; Tanaka et al. 1987). Thus, ionic interactions between Ca^{2+} and fatty acids may determine their ability to activate the respiratory burst (Yamaguchi et al. 1986). Fatty acid-induced activation of NADPH oxidase is accompanied by changes in cell morphology (Badwey et al. 1984). Arachidonic acid is a similarly effective activator of NADPH oxidase in human neutrophils and differentiated HL-60 cells, as are PMA and dioctanoylgylcerol (Seifert et al. 1989c, 1991). Although there are certain similarities between fatty acid- and chemoattractant-induced O_2^- formation, it must be emphasized that arachidonic acid at concentrations which activate the respiratory burst is cytotoxic (Badwey et al. 1984; H.J. Cohen et al. 1986; Jesaitis et al. 1986; Tsunawaki and Nathan 1986).

3.1.2.2 Mechanistic Aspects

Although much work has been done in this field, the mechanism by which fatty acids activate NADPH oxidase is still incompletely understood. The stimulatory effects of fatty acids are apparently independent of their oxygenation products, as *cis*-mono-unsaturated and *trans*-unsaturated fatty acids are no substrates for lipoxygenases and cyclooxygenase, but they are activators of O_2^- formation (Kinsella et al. 1981; Needleman et al. 1986). In addition, inhibitors of lipoxygenases and cyclooxygenase do not inhibit fatty acid-induced O_2^- formation in guinea pig macrophages (Bromberg and Pick 1983). *Cis*-unsaturated fatty acids have been suggested to activate O_2^- formation by increasing the membrane fluidity (Klausner et al. 1980; Badwey et al. 1984). In contrast, activation of NADPH oxidase by saturated and *trans*-unsaturated fatty acids cannot be explained by this mechanism, as these lipids are known not to increase membrane fluidity (Klausner et al. 1980). The involvement of Ca^{2+}, protein kinase C and proteases in the activation of NADPH oxidase by fatty acids has been suggested by the findings that fatty acids increase cytoplasmic Ca^{2+}, and that inhibitors of calmodulin, protein kinase C, and proteases inhibit O_2^- formation (Curnutte et al. 1984; Morimoto et al. 1986). In contrast, staurosporine shows only a very moderate inhibitory effect on arachidonic acid-induced O_2^- formation in human neutrophils (Seifert, unpublished results; see also Sects.

3.2.2.3, 3.2.3.1, 3.2.8). Fatty acids and cell-permeant diacylglycerols interact synergistically to activate O_2^- formation (Ozawa et al. 1989).

cis-unsaturated but not saturated fatty acids activate protein kinase C (McPhail et al. 1984b; Murakami and Routtenberg 1985; Hansson et al. 1986; Linden et al. 1986; Murakami K et al. 1986, 1987; Sekiguchi et al. 1987; Seifert et al. 1988c). The physiological relevance of this protein kinase C activation is, however, not generally accepted (Dell and Severson 1989). The acetylenic analogue of arachidonic acid, eicosatetraynoic acid (ETYA), activates protein kinase C but not the respiratory burst (Badwey et al. 1981, 1984; Seifert et al. 1988c). Conversely, the *trans*-stereoisomer of oleic acid, elaidic acid, activates the respiratory burst but not protein kinase C (Murakami and Routtenberg 1985; Yamaguchi et al. 1986; Tanaka et al. 1987; Seifert et al. 1988c). These dissociations suggest that fatty acids do not activate NADPH oxidase through direct stimulation of protein kinase C. Another mechanistic possibility is that fatty acids act by stimulation of phospholipases C and A_2 (Irvine et al. 1979; Takenawa and Nagai 1981; Maridonneau-Parini and Tauber 1986; see also Sect. 5.1.3).

Positively charged alkylamines, e.g., cetyltrimethylammonium bromide, inhibit O_2^- formation in intact neutrophils induced by various agents, including PMA, A 23187, arachidonic acid, and chemotactic peptides (Miyahara et al. 1987, 1988). In addition, alkylamines inhibit the activity of the particulate NADPH oxidase and fatty acid-induced O_2^- formation in the cell-free system (Cross et al. 1984; Miyahara et al. 1987, 1988; see also Sects. 2.1, 5.1.3). In contrast, alkylalcohols do not inhibit O_2^- formation, and the inhibitory effect of alkylamines is antagonized by negatively charged agents (Miyahara et al. 1987, 1988). These results indicate that charge-dependent processes are required for both activation and activity of NADPH oxidase. The mechanism by which alkylamines inhibit NADPH oxidase may involve dissociation of cytosolic components of the enzyme from the plasma membrane (Ohtsuka et al. 1990a).

3.2 Mechanisms Involved in the Activation of NADPH Oxidase by Receptor Agonists

3.2.1 Guanine Nucleotide Binding Proteins (G-Proteins)

Among the large number of intercellular signal molecules which activate the respiratory burst, the chemotactic peptides are those most extensively studied with respect to the characterization of the signal transduction pathways. In many cases, other intercellular signal molecules act through similar mechanisms.

G-proteins are a family of heterotrimeric membrane-attached transducer molecules which functionally couple cell surface receptors to intracellular effector systems such as adenylyl cyclase, cGMP phosphodiesterase of rods and cones, phospholipase C, and ion channels (Stryer and Bourne 1986; Gilman 1987; Casey and Gilman 1988; Neer and Clapham 1988; Milligan 1988; Birnbaumer et al. 1989).

3.2.1.1 Interaction of Plasma Membrane Receptors with G-Proteins in Phagocytes

Many studies showed that formyl peptide receptors and receptors for other chemoattractants such as complement C5a and leukotriene B_4 (LTB_4) mediate cell activation via G-proteins (see also Sect. 3.3.1). Guanine nucleotides modulate the agonist affinity of formyl peptide receptors, i.e., GTP or stable GTP analogues convert a portion of the receptors from a high-affinity to a low-affinity state (Koo et al. 1982, 1983; Snyderman et al. 1984; Snyderman 1984; Sklar et al. 1987; see also Sect. 3.3.1.1). Recently, Gierschik et al. (1989a) have shown that divalent cations and high-affinity agonist binding are not required for a functional interaction of formyl peptide receptors with G-proteins in differentiated HL-60 cells. The formyl peptide N-formyl-L-methionyl-L-leucyl-L-phenylalanine (fMet-Leu-Phe) stimulates binding of stable guanine nucleotides to plasma membranes of human myeloid cells (Matsumoto et al. 1987; Gierschik et al. 1989a). fMet-Leu-Phe, LTB_4, and C5a stimulate high-affinity GTPase activity in neutrophil membranes which is desensitized and resensitized (Hyslop et al. 1984; Okajima et al. 1985; Matsumoto et al. 1986, 1987; Feltner et al. 1986; Wilde et al. 1989; McLeish et al. 1989a,b; see also Sect. 3.3.1.1.3). These chemoattractants also stimulate the photolabeling of 40-kDa proteins in differentiated HL-60 cells using the photoreactive analogue of GTP, [α-^{32}P]GTP azidoanilide (Offermans et al. 1990). The G-proteins which couple to formyl peptide receptors are ADP-ribosylated by pertussis toxin both in intact cells and in isolated plasma membranes, and this covalent modification uncouples receptors from G-proteins (Bokoch and Gilman 1984; Ohta et al. 1985; Verghese et al. 1986c; Gierschik and Jakobs 1987). In plasma membranes, these G-proteins are also substrates for cholera toxin-induced ADP riboylation under certain experimental conditions (Verghese et al. 1986c; Gierschik and Jakobs 1987; Iiri et al. 1989).

Treatment of phagocytes with pertussis toxin or cholera toxin is associated with a reduction of chemoattractant-stimulated GTPase activity in comparison to control membranes (Okajima et al. 1985; Feltner et al. 1986;

McLeish et al. 1989a). A mixture of purified isoforms of the inhibitory G-protein for adenylyl cyclase G_i has been reported to reconstitute chemotactic peptide-stimulated GTPase activity in neutrophil membranes (Okajima et al. 1985). Purified formyl peptide receptors and G-proteins incorporated into phospholipid vesicles functionally interact as well (Williamson et al. 1988).

Stable GTP analogues stimulate phosphoinositide degradation in permeabilized neutrophils and in neutrophil membranes. This process is enhanced by fMet-Leu-Phe and is inhibited by pertussis toxin (C.D. Smith et al. 1985, 1986; Bradford and Rubin 1986). Chemoattractant-induced phosphoinositide degradation in pertussis toxin-treated membranes has been reported to be reconstituted with a purified mixture of G_i and with G_o, the major G-protein of the brain, and both preparations of G-proteins are similarly effective (Kikuchi et al. 1986). This nonselectivity of the effects of G-proteins indicates that further reconstitution studies will have to be performed with clearly defined subtypes of G-proteins.

The G-proteins coupling to formyl peptide receptors have been purified from neutrophils and HL-60 cells (Dickey et al. 1987; Oinuma et al. 1987; Uhing et al. 1987; Gierschik et al. 1986, 1987; Polakis et al. 1988). Myeloid cells contain two major G-proteins of the G_i family, i.e., G_{i2} and G_{i3}, whose α-subunits have molecular masses of 41 and 40 kDa, with $G_{i\alpha2}$ being the major pertussis toxin substrate in phagocytes (Suki et al. 1987; Didsbury and Snyderman 1987; Murphy et al. 1987; Goldsmith et al. 1987; Milligan 1988; Rudolph et al. 1989a). The results of a recent study by Gierschik et al. (1989b) suggest that formyl peptide receptors interact functionally with both G_{i2} and G_{i3}. Specific granules of neutrophils contain large amounts of a 40- to 41-kDa pertussis toxin substrate, suggesting that this intracellular pool of G-proteins is translocated to the plasma membrane (Rotrosen et al. 1988; see also Sects. 2.5, 3.3.1.1, 5.1.2). In addition, human neutrophils contain substantial amounts of cytosolic $G_{i\alpha2}$, but its functional meaning is not known (Rosenthal et al. 1987; Rudolph et al. 1989a,b; see also Sect. 5.1.4). Finally, myeloid cells contain G_s, the G-protein which activates adenylyl cyclase (see Sect. 4.1). Recently, Strathmann et al. (1989) identified cDNAs of yet unknown α-subunits of G-proteins. These results raise the question of whether these or other G-proteins, in addition to the known ones, are involved in signalling processes in phagocytes (see also below).

Table 3 summarizes some data concerning the involvement of G-proteins in the activation of phagocytes by various classes of receptor agonists (see also Sect. 3.3.1). Until recently, almost all receptors in phagocytes have been assumed to couple functionally to pertussis toxin-sensitive G-proteins, but an increasing number of recent studies clearly point to the importance of pertussis toxin-insensitive signal transduction

Table 3. Pertussis toxin (PT)-sensitive and -insensitive activation of phagocytes by various agonists

Agonists	Receptor involved	Cellular effects	PT-sensitivity	Selected references
Pertussis toxin-sensitive				
Complement C5a	C5a receptor	Ca^{2+} mobilization, O_2^{-} formation	+	Gennaro et al. (1984), Shirato et al. (1988)
NAP-1 (IL-8)	IL-8 receptor	Ca^{2+} mobilization, O_2^{-} formation	+	Thelen et al. (1988b), Besemer et al. (1989)
GM-CSF	GM-CSF receptor	Induction of *c-fos* mRNA, potentiation of PAF-induced Ca^{2+} mobilization, stimulation of GTPase and protein tyrosine phosphorylation, secretion of myeloperoxidase and lactoferrin; potentiation of arachidonic acid-induced O_2^{-} formation	+	Corey and Rosoff (1989), Gomez-Cambronero et al. (1989a,b), McColl et al. (1989), Richter et al. (1989), Sha'afi et al. (1989)
Elastin peptides	Elastin peptide receptor	Phosphoinositide degradation, Ca^{2+} mobilization, O_2^{-} formation	+	Varga et al. (1989)
LTB4	LTB4 receptor	O_2^{-} formation, phosphoinositide degradation, Ca^{2+} mobilization	+	Molski et al. (1984), Mong et al. (1986), Holian (1986), Andersson et al. (1986b)
LTC4/LTD4	LTC4/LTD4 receptors	Ca^{2+} mobilization	+	Koo et al. (1989)
Propionic acid	Unknown	Ca^{2+} mobilization	+	Naccache et al. (1988b)
Cytochalasin B	Unknown	NBT reduction	+	Elferink et al. (1990b)

Table 3. (continued)

Pertussis toxin-sensitive and/or -insensitive

Agonists	Receptor involved	Cellular effects	PT-sensitivity	Selected references
Chemotactic formyl peptides	Formyl peptide receptor	Phosphoinositide degradation, release of arachidonic acid, Ca^{2+} mobilization, O_2^- formation	+	Bokoch and Gilman (1984), Okajima and Ui (1984), Okamura et al. (1985)
		Priming for enhanced O_2^- formation	+	Karnad et al. (1989)
Synthetic lipopeptides	Lipopeptide receptor? Direct activation of G-proteins?	O_2^- formation Lysozyme release	Partial −	Seifert et al. (1990) Seifert et al. (1990)
PAF	PAF receptor	O_2^- formation, phosphoinositide degradation, exocytosis, aggregation	+	Lad et al. (1985c), Naccache et al. (1986), Huang (1988)
		Phosphoinositide degradation, Ca^{2+} mobilization	−	Naccache et al. (1986), Barzaghi et al. (1989), Ng and Wong (1989)
TNF	TNF receptor	Activation of high-affinity GTPase in HL-60 cells	+	Inamura et al. (1988)
		H_2O_2 formation in adherent neutrophils	−	Meurer and MacIntyre (1989)
		Potentiation of O_2^- formation	−	Berkow and Dodson (1988)
		Myeloperoxidase secretion	+	Richter et al. (1989)
		Lactoferrin secretion	Partial	Richter et al. (1989)
PGE$_1$	Prostaglandin receptor	cAMP accumulation in human monocytes	Partial	Griese et al. (1990)
Isoproterenol	β-adrenoceptors	cAMP accumulation in human monocytes	Partial	Griese et al. (1990)
Substance P	Formyl peptide receptor? Substance P receptor? Direct activation of G-proteins?	H_2O_2 formation	Partial	Serra et al. (1988)

Table 3. (continued)

Agonists	Receptor involved	Cellular effects	PT-sensitivity	Selected references
IgG immune complexes	IgG Fc receptor	Ca^{2+} mobilization, $O_2^{\cdot-}$ formation, exocytosis	+, partial, or —	Shirato et al. (1988), Feister et al. (1988)
Surface-bound IgG	IgG Fc receptor	Phosphoinositide degradation, $O_2^{\cdot-}$ formation	—	Blackburn and Heck (1988, 1989)
Soluble IgG aggregates	IgG Fc receptor	$O_2^{\cdot-}$ formation	+	Blackburn and Heck (1988)
Concanavalin A	Plasma membrane glycoproteins	$O_2^{\cdot-}$ formation, $O_2^{\cdot-}$ consumption	—	Verghese et al. (1985a), Rossi et al. (1986), Lad et al. (1986), Lu and Grinstein (1989)
	Unknown	Cap formation	+	Lad et al. (1986)
Asbestos	Unknown	NBT reduction	+	Elferink and Ebbenhout (1988)
Uncoated urate crystals	Unknown	Activation of high affinity GTPase; exocytosis, $O_2^{\cdot-}$ formation; phagocytosis, Ca^{2+} mobilization;	+ partial, or —	Terkeltaub et al. (1990)
Purine nucleotides (ATP, ATP[γS])	Cell type-specific purinoceptor	Phosphoinositide degradation, Ca^{2+} mobilization, $O_2^{\cdot-}$ formation, exocytosis, arachidonic acid release	Partial, or —	Dubyak et al. (1988), Kuhns et al. (1988), Seifert et al. (1989b), Cockcroft and Stutchfield (1989a,b)
		Potentiation of $O_2^{\cdot-}$ formation, exocytosis	—	Kuhns et al. (1988), Wenzel-Seifert and Seifert (1990)
		Aggregation	+	Seifert et al. (1989d)
GTP[γS]	Purinoceptor? Direct activation of G-proteins?	$O_2^{\cdot-}$ formation	Partial or —	Elferink and Deierkauf (1989a)

Table 3. (continued)

Agonists	Receptor involved	Cellular effects	PT-sensitivity	Selected references
Pyrimidine nucleotides (UTP)	Cell type-specific pyrimidinoceptor	O_2^- formation, aggregation, Ca^{2+} mobilization, exocytosis	Partial or +	Seifert et al. (1989d), Seifert and Schultz (1989), Wenzel-Seifert and Seifert (1990)
NaF	None, direct activation of G-proteins	Phosphoinositide degradation, Ca^{2+} mobilization	−	Strnad and Wong (1985b), Strnad et al. (1986),
		O_2^- formation, exocytosis	+	Toper et al. (1987), Gabler et al. (1989)
LPS	LPS receptor?	Potentiation of PMA-induced O_2^- formation	−	Forehand et al. (1989)
		IL-1 production in P388D$_1$ macrophages and U-937 cells	+	Jakway and DeFranco (1986), Daniel-Issakani et al. (1989)
Nonopsonized Candida albicans hyphae	Unknown	Ca^{2+} mobilization, O_2^- formation	Partial	Meshulam et al. (1988)
Influenza A virus	Influenza A virus-binding protein (hemagglutinin)	H_2O_2 formation, Ca^{2+} mobilization	−	Hartshorn et al. (1990a,b)

pathways activated by receptor agonists in these cells. For example, it has been suggested that priming of the respiratory burst by chemotactic peptides, unlike activation of NADPH oxidase by these chemoattractants, involves pertussis toxin-insensitive signal transduction pathways (Karnad et al. 1989; see also Sect. 3.3.1.1.4).

Treatment of neutrophils, differentiated HL-60 cells, and mononuclear phagocytes with pertussis toxin inhibits the functional responses induced by formyl peptides with the exception of priming, e.g., phosphoinositide degradation, increase in cytoplasmic Ca^{2+}, activation of Na^+/H^+ exchange, membrane depolarization, protein phosphorylation, exocytosis, release of arachidonic acid, chemotaxis, shape change, aggregaion, actin polymerization, and O_2^- formation (Molski et al. 1984; Bokoch and Gilman 1984; Okajima and Ui 1984; Lad et al. 1985b,c; Krause et al. 1985; Ohta et al. 1985; Goldman et al. 1985b; Brandt et al. 1985; Spangrude et al. 1985; Okamura et al. 1985; Satoh et al. 1985; Shefcyk et al. 1985; Volpi et al. 1985; Verghese et al. 1985a, 1986a,b; Dillon et al. 1987; Dubyak et al. 1988; Seifert et al. 1989b,d, 1990; McLeish et al. 1989a,b). In contrast to O_2^- formation induced by intercellular signal molecules, e.g., fMet-Leu-Phe, platelet-activating factor (PAF), LTB_4, purine and pyrimidine nucleotides, and C5a, those induced by A 23187, PMA, diacylglycerol, fatty acids, or lectins, which circumvent receptor stimulation, are pertussis toxin insensitive.

Table 3 also lists a number of examples for intercellular signal molecules which activate functions of myeloid cells in a pertussis toxin-sensitive-and/or -insensitive manner, and the table shows that various cell functions activated by one given receptor agonist may show a differential pertussis toxin sensitivity. These data suggest that different classes of G-proteins (pertussis toxin-sensitive and/or pertussis toxin-insensitive) and possibly low molecular mass GTP-binding proteins (see Sect. 3.2.1.2) are involved in the signal transduction pathways. In addition, ADP-ribosylated G-proteins may interact to different degrees with different types of plasma membrane receptors, and specific populations of G-proteins may be involved in the signal transduction pathway activated by a given type of receptor agonist (Dubyak et al. 1988; Ashkenazi et al. 1989). Another attractive hypothesis to explain the (partial) pertussis toxin insensitivity of effects of some stimuli is that certain substances, e.g., mastoparan of wasp venom, compound 48/80, substance P, and positively charged lipopeptides, directly activate G-proteins rather than act through receptors (Higashijima et al. 1988; Seifert et al. 1990; Mousli et al. 1990; see also Sects. 3.3.1.2.1, 3.3.1.2.6).

Studying the effects of various guanine nucleotides on arachidonic acid-induced activation of NADPH oxidase in electropermeabilized human neutrophils, Lu and Grinstein (1990) suggested that two different

GTP-binding proteins are involved in the activation of NADPH oxidase; one may be a pertussis toxin-sensitive G-protein, and the other may be a GTP-binding protein in the vicinity of NADPH oxidase.

3.2.1.2 Low Molecular Mass GTP-Binding Proteins

Human myeloid cells have recently been found to contain various GTP-binding proteins with molecular masses of 20-26 kDa, which are no substrates for pertussis toxin or cholera toxin, i.e., the *ras*-related GTP-binding proteins *rho*, *rac*1, *rac*2, and *rap*1 (Bokoch and Parkos 1988; Didsbury et al. 1989; Polakis et al. 1989; Quinn et al. 1989). Purified *rap*1 from human neutrophils binds guanine nucleotides, and this process is modulated by Mg^{2+} but not by the β/γ-complex of G-proteins or phosphorylation by cAMP-dependent protein kinase (Bokoch and Quilliam 1990). Interestingly, certain small GTP-binding proteins are associated with formyl peptide receptors (Polakis et al. 1989), and cytochrome b_{-245} is associated with *rap*1 (Quinn et al. 1989). Moreover, the 47-kDa protein shows sequence homology with the *ras* p21 GTPase-activating protein (also referred to as *ras*-GAP; Volpp et al. 1989b; Lomax et al. 1989; see also Sects. 5.1.4, 5.1.5).

The GTP-binding proteins *rho*, *rac*1, and *rac*2, but not *rap*1 are substrates for *Clostridium botulinum* ADP ribosyltransferase C3 (Braun et al. 1989; Quilliam et al. 1989; Didsbury et al. 1989). Apparently, *C. botulinum* ADP ribosyltransferase C3 is probably the active component (as a contaminant) in preparations referred to as botulinum D toxin, which by itself does not possess ADP ribosyltransferase activity (Ohashi and Narumiya 1987; Mege et al. 1988a, 1989; Banga et al. 1988; Aktories et al. 1988; Matsuoka et al. 1989; Braun et al. 1989; Quilliam et al. 1989). Treatment of neutrophils with *C. botulinum* ADP ribosyltransferase C3 does not inhibit fMet-Leu-Phe-induced increase in cytoplasmic Ca^{2+}, exocytosis, cytoskeletal changes, or O_2^- formation (Mege et al. 1988a). These data suggest that the substrates for ADP ribosyltransferase C3 are not involved in the regulation of agonist-induced O_2^- formation, at least in the case of chemotactic peptides.

3.2.1.3 NaF

Studies with NaF provided important information on the role of G-proteins in the activation of NADPH oxidase (see also Sect. 5.1.4.2). Rall and Sutherland (1958) had just reported that NaF stimulates cAMP formation in liver homogenates, when Sbarra and Karnovsky (1959) showed that NaF activates the respiratory burst in intact neutrophils. The NaF-induced respiratory burst in human neutrophils is a reversible process and depends on the presence of extracellular Ca^{2+}, and other halides are inactive (Curnutte and Babior 1975; Curnutte et al. 1979; Della Bianca et al. 1988; Gabler et al. 1989). Unlike in human neutrophils, activation of the respiratory burst by NaF in guinea pig phagocytes apparently does not depend on Ca^{2+} (Della Bianca et al. 1988; Toper et al. 1987). During myeloid differention, neutrophils acquire the ability to reduce NBT upon exposure to NaF (Zakhireh and Root 1979). Like PMA, NaF is a very effective activator of O_2^- formation, and respiratory burst shows a lag time and is long lasting (Curnutte et al. 1979; see also Sect. 3.1.1.1). The delayed onset of the NaF-induced respiratory burst may be due to the fact that fluoride must first cross the plasma membrane and bind to its intracellular target prior to inducing cellular activation (Curnutte et al. 1979; Della Bianca et al. 1988). NaF is required at concentrations of about 20–40 mM to maximally activate NADPH oxidase (Curnutte et al. 1979; Della Bianca et al. 1988).

A clue to explain the molecular mechanism by which NaF activates NADPH oxidase was the discovery that NaF stimulates G_s and other G-proteins, and that the effects of NaF are enhanced by aluminium salts (Howlett et al. 1979; Sternweis et al. 1981; Sternweis and Gilman 1982; Gilman 1987). Fluoride in the presence of aluminium, probably as AlF_4^-, has been suggested to activate G-proteins by mimicking the γ-phosphate group of GTP (Bigay et al. 1985; Chabre 1989). However, other yet unknown modes of action of fluoride on G-proteins cannot be ruled out (Stadel and Crooke 1989).

The above results suggest that in neutrophils, NaF may activate G-proteins as in other systems. The data on the pertussis toxin sensitivity of the effects of NaF in neutrophils are not consistent. In human neutrophils, the NaF-induced phosphoinositide degradation, increase in cytoplasmic Ca^{2+}, and respiratory burst are not inhibited by pertussis toxin (Strnad and Wong 1985b; Strnad et al. 1986; Della Bianca et al. 1988). These findings are in agreement with the assumption that pertussis toxin-catalyzed ADP ribosylation of G-proteins of the G_i-family and of G_o inhibits the interaction of G-proteins with activated receptors but does not prevent activation of G-proteins by NaF or stable guanine nucleotides (Jakobs et al. 1984; Gilman 1987; see also Sect. 5.1.4). In contrast, Gabler et al. (1989) reported that NaF-induced O_2^- formation in human neutrophils is pertussis toxin sensitive. The NaF-induced activation of NADPH oxidase and protein kinase C translocation but not the increase in cytoplasmic Ca^{2+} have been reported to be pertussis toxin-sensitive events in guinea pig neutrophils (Toper et al. 1987).

NaF induces phosphoinositide degradation, increase in cytoplasmic Ca^{2+}, and translocation and activation of protein kinase C in phagocytes (Strnad and Wong 1985b; Strnad et al. 1986; English et al. 1986; Hauschildt et al. 1988a; Della Bianca et al. 1988). As the effects of NaF on these parameters occur prior to activation of the respiratory burst, and as the NaF-induced respiratory burst is inhibited by H-7, these data suggest that protein kinase C is involved in the signal transduction pathway (see also Sect. 3.2.2.3). In contrast, recent data indicate that the mechanism of action of NaF on the respiratory burst is more complex. In Ca^{2+}-depleted neutrophils, NaF does not induce phosphoinositide degradation, increase in cytoplasmic Ca^{2+}, or $O_2^{\bullet-}$ formation (Della Bianca et al. 1988). Priming of Ca^{2+}-depleted neutrophils with PMA restores NaF-induced activation of NADPH oxidase but not that of phospholipase C (Della Bianca et al. 1988). These results suggest that activation of NADPH oxidase by NaF does not necessarily depend on phospholipase C activation (see also Sects. 3.2.2.1, 3.2.2.5, 5.1.4). NaF not only activates phospholipase C but also phospholipase D (Olson et al. 1990). There are, however, differences in the sensitivity to inhibition by ethanol of the effects of NaF and fMet-Leu-Phe.

In electropermeabilized neutrophils, NaF induces a respiratory burst which is rapid in onset (Hartfield and Robinson 1990). In this system, Mg^{2+} is required and ATP potentiates the effects of NaF, whereas stable GDP analogues are inhibitory and pertussis toxin is without effect. Protein kinase C-dependent and -independent pathways have both been suggested to be involved in NaF-induced activation of NADPH oxidase in electropermeabilized neutrophils.

With regard to $O_2^{\bullet-}$ formation, NaF and fMet-Leu-Phe may interact in a synergistic or in an antagonistic manner (Wong 1983; Toper et al. 1987). In neutrophils, NaF on one hand stimulates phospholipase C, but on the other hand NaF also activates adenylyl cyclase via G_s (Ham et al. 1983; Wong 1983; Saad et al. 1987; Gilman 1987; Bokoch 1987; see also Sect. 4.1). As an increase in cAMP inhibits $O_2^{\bullet-}$ formation, the competitive activation of G_s and of the G-proteins activating phospholipase C and/or NADPH oxidase may explain the opposite effects of NaF (see also Sect. 4.1).

3.2.2 Protein Kinase C

In phagocytes, chemoattractants induce degradation of phosphatidylinositol 4,5-biphosphate through phospholipase C, resulting in the formation of diacylglycerol and inositol 1,4,5-triphosphate, with subsequent activation of protein kinase C and Ca^{2+} mobilization (Dougherty et al. 1984; Bradford and Rubin 1985) (see also Sect. 3.1.1). Interestingly, human neutrophils also contain phosphatidylinositol 3,4-biphosphate and phosphatidylinositol 3,4,5-triphosphate (Traynor-Kaplan et al. 1989). Upon stimulation with chemoattractants, phosphatidylinositol 3,4,5-triphosphate accumulates, and there is a correlation between accumulation of this phospholipid and O_2^- formation (Traynor-Kaplan et al. 1989). Moreover, the results of recent studies point to the importance of phospholipase D induced degradation of phosphatidylcholine as a source of phosphatidic acid and its dephosphorylation product, diacylglycerol (Exton 1988; Billah et al. 1989; Anthes et al. 1989; see also Sect. 3.2.2.1).

The role of protein kinase C in the activation of NADPH oxidase by receptor agonists is a subject of current debate. On one hand, it has been suggested that chemoattractants, in analogy to phorbol esters and cell-permeant diacylglycerols, activate NADPH oxidase through protein kinase C-mediated phosphorylation of specific proteins (Takenawa et al. 1985; Gavioli et al. 1987; Nath et al. 1989; see also Sect. 3.1.1). There are, indeed, some correlations between receptor-mediated activation of phosphoinositide degradation and protein kinase C and activation of NADPH oxidase. On the other hand, a number of reports have called into question the central role of protein kinase C in receptor-mediated activation of NADPH oxidase.

3.2.2.1 Correlations and Dissociations Between Activation of Phospholipases C and D and Protein Kinase C, on one Hand, and NADPH Oxidase, on the Other

LTB$_4$, ionomycin, concanavalin A (ConA), and fMet-Leu-Phe have been reported to increase cytoplasmic Ca^{2+} and to induce diacylglycerol release, but only the latter two agents activate the respiratory burst (Korchak et al. 1988b). In contrast, other authors reported that LTB$_4$ activates NADPH oxidase in neutrophils, although it is much less effective than fMet-Leu-Phe (see below and Sect. 3.3.1.7.1). The chemotactic peptide-induced accumulation of diacylglycerol is potentiated by CB and is abolished by pertussis toxin (Honeycutt and Niedel 1986). Diacylglycerol accumulation precedes O_2^- formation, and changes in the concentration of diacylglycerol correlate with the kinetics of O_2^- formation (Rider and Niedel 1987). In comparison to fMet-Leu-Phe, LTB$_4$ induces only a small and short-lasting respiratory burst (Truett et al. 1988; Reibman et al. 1988; see also Sect. 3.3.1.7.1). Both agonists induce a rapid increase in the concentration of diacylglycerol, but only fMet-Leu-Phe induces sustained diacylglycerol accumulation which depends on extracellular Ca^{2+} and is enhanced by CB. The diacylglycerol released during this sustained phase may be derived from phosphatidylcholine rather than from phosphoinositides (Truett et al. 1988; Reibman et al. 1988; see also below).

An interesting new approach to study the role of phospholipase C in the regulation of NADPH oxidase was recently presented by Smith et al. (1990). These authors reported that 1-[6-[[17β-3-methoxyestra-1,3,5(10)-trien-17-yl]amino]hexyl]-1H-pyrrole-2,5-dione (U-73122) may inhibit chemoattractant-induced activation of phospholipase C in human neutrophils. This assumption is supported by the finding that U-73122 inhibits agonist-induced production of inositolphosphates and diacylglycerol and the rise in cytoplasmic Ca^{2+}. Apparently, U-73122 does not directly affect protein kinase C (Smith et al. 1990). The inhibition of the above phospholipase C-related events by U-73122 correlates with inhibition of O_2^- formation and exocytosis (Smith et al. 1990).

Human myeloid cells possess a phospholipase D which is activated by chemotactic peptides and catalyzes the degradation of phosphatidylcholine to phosphatidic acid and choline (Exton 1988; Pai et al. 1988a,b, 1989; Truett et al. 1989; Gelas et al. 1989; Billah et al. 1989). Activation of phospholipase D may be a Ca^{2+}- and GTP-dependent process, and phosphatidic acid can be converted to diacylglycerol by a phosphohydrolase (Billah et al. 1989; Anthes et al. 1989). In addition, phosphatidic acid per se may act as a signal molecule (Nayar et al. 1984; Murayama and Ui 1987). Interestingly, phosphatidic acid substituted with short-chain saturated fatty acids has recently been shown to activate the respiratory burst in guinea pig neutrophils in a concentration-dependent manner (Ohtsuka et al. 1989). Activation of NADPH oxidase by phosphatidic acid is independent of Ca^{2+} and may involve phosphorylation of the 47-kDa protein (Ohtsuka et al. 1989; see also Sect. 5.3). A role of phospholipase D in the activation of NADPH oxidase by fMet-Leu-Phe is further supported by the results of a recent report showing that certain aliphatic alcohols inhibit both chemotactic peptide-induced release of phosphatidic acid and O_2^- formation (Bonser et al. 1989; see also Sects. 3.3.1.5, 3.3.2.7). Recently, the inhibitor of phosphatidic acid phosphohydrolase propranolol (Billah et al. 1989) was shown to inhibit the accumulation of diacylglycerol without inhibiting the formation of phosphatidic acid in fMet-Leu-Phe-stimulated human neutrophils (Rossi et al. 1990). In the presence of CB, the chemotactic peptide induces a short-lasting respiratory burst which is reinitiated by propranolol (Rossi et al. 1990). Propranolol also potentiates the fMet-Leu-Phe-induced respiratory burst in the absence of CB; this process is associated with accumulation of phosphatidic acid and inhibition of formation of diacylglycerol (Rossi et al. 1990). These data point to phosphatidic acid having a role as intracellular signal molecule in the activation process of NADPH oxidase by receptor agonists.

Several dissociations between activation of NADPH oxidase by PMA and opsonized zymosan have been documented (Newburger et al. 1980a,b; Andre et al. 1988; Huizinga et al. 1989; Phillips and Hamilton 1989; see also Sects. 3.3.1.5, 3.4.3, 6.2). Recently, Koenderman et al. (1989b,c) reported that there is no correlation between diacylglycerol release and activation of oxygen consumption induced by opsonized zymosan, fMet-Leu-Phe, PAF, or PAF plus fMet-Leu-Phe in human neutrophils (see also Sect. 3.3.1.6). The initial phase of the respiratory burst induced by opsonized zymosan is not accompanied by diacylglycerol release and is not affected by staurosporine, whereas the later phase of zymosan-induced NADPH oxidase activation is associated with diacylglycerol formation and staurosporine sensitivity (Koenderman et al. 1989b,c). Finally, the stimulatory effects of opsonized zymosan on O_2^- formation are less sensitive to inhibition by alcohols than those of fMet-Leu-Phe (Bonser et al. 1989; see also above).

Priming of the respiratory burst by PAF in human eosinophils but not its activation has been shown to be accompanied by diacylglycerol accumulation (Koenderman et al. 1989c). In addition, eosinophilia may be associated with an enhanced respiratory burst in response to zymosan and intermediate accumulation of diacylglycerol (Koenderman et al. 1989c; see also Sect. 3.4.3.1.1). These data suggest that diacylglycerol accumulation in neutrophils and eosinophils may be important in the propagation of the respiratory burst by receptor agonists rather than in its initiation.

Similar to phorbol esters and cell-permeant diacylglycerols, chemotactic peptides induce translocation of protein kinase C from the cytosol to the plasma membrane (Pike et al. 1986; Nishihira et al. 1986; Horn and Karnovsky 1986; Pontremoli et al. 1986b; Christiansen 1988; Christiansen et al. 1988a,b). Translocation of protein kinase C by fMet-Leu-Phe precedes O_2^- formation, but there is no close correlation between translocation of protein kinase C and its activation (Christiansen 1988). In comparison to PMA, the fMet-Leu-Phe-induced association of the kinase with the plasma membrane is less tight and depends on the presence of CB (Pontremoli et al. 1986a; Christiansen et al. 1988a; see also Sect. 3.2.5). Pertussis toxin inhibits fMet-Leu-Phe-induced O_2^- formation in parallel with protein kinase C translocation (Christiansen 1990). Protein kinase C translocated by PMA but not that translocated by fMet-Leu-Phe is active in the absence of Ca^{2+} and added phospholipids (Pontremoli et al. 1986a). It should be emphasized that translocation of protein kinase C to the plasma membrane by fMet-Leu-Phe depends critically on the experimental conditions. For example, Ca^{2+} at physiologically relevant cytoplasmic concentrations in the extraction buffer per se may lead to the association of protein kinase C with the plasma membrane (Phillips et al. 1989). Moreover, the chemotactic peptide-induced translocation of protein kinase C in the absence of CB is observed only when special homogenization techniques are applied (Horn and Karnovsky 1986). Interestingly, fMet-Leu-Phe also induces the translocation of diacylglycerol kinase to the plasma membrane (Ishitoya et al. 1987). This translocation may play a role in the termination of the respiratory burst as this enzyme and protein kinase C compete for diacylglycerol (see also Sects. 3.2.2.4, 3.3.1.1.3).

Chemotactic peptides induce protein phosphorylation in neutrophils (Schneider et al. 1981; Andrews and Babior 1984). The characteristics of protein phosphorylation induced by PMA and fMet-Leu-Phe are similar but not identical (Andrews and Babior 1984). Like PMA, fMet-Leu-Phe induces phosphorylation of the 47-kDa protein, and the kinetics of phosphorylation correlate with $O_2^{\cdot-}$ formation (Schneider et al. 1981; Ohtsuka et al. 1987; Reibman et al. 1988; Badwey et al. 1989a; see also Sect. 3.1.1). Activation of NADPH oxidase by opsonized latex beads, NaF, and A 23187 is also associated with the phosphorylation of the 47-kDa protein (Heyworth and Segal 1986). In contrast, H-7 has been reported to block fMet-Leu-Phe-induced phosphorylation of the 47-kDa protein without blunting $O_2^{\cdot-}$ formation (Sha'afi et al. 1988; see also Sect. 3.2.2.3). Recently, the local anesthetics tetracaine, bupivacaine, cocaine, and lidocaine were shown to inhibit $O_2^{\cdot-}$ formation induced by PMA, A 23187, and fMet-Leu-Phe (Haines et al. 1990). Interestingly, local anesthetics do not affect phosphorylation of the 47-kDa protein, suggesting that phosphorylation of this protein is an insufficient signal for the activation of NADPH oxidase (Haines et al. 1990; see also Sect. 3.1.1).

3.2.2.2 Priming by Activators of Protein Kinase C

PMA at nonstimulatory concentrations primes phagocytes for enhanced $O_2^{\cdot-}$ formation upon exposure to chemoattractants (McPhail et al. 1984a; Bender et al. 1987; Tyagi et al. 1988; Ohsaka et al. 1988; Sha'afi et al. 1988; Smith et al. 1988c; Seifert et al. 1989a). The respiratory burst induced by fMet-Leu-Phe, C5a, PAF, or LTB_4 at low concentrations (1–10 nM) in PMA-primed human neutrophils correlates with oscillations in cell shape (Wymann et al. 1989). In the presence of the chemoattractants at high concentrations (50–100 nM), these oscillations are apparent only in the presence of 17-hydroxywortmannin (see also Sect. 4.3.1). The mechanism of priming by PMA has been suggested to involve membrane depolarization, enhanced generation of diacylglycerol, and activation of protein kinase

C-independent processes (Tyagi et al. 1988; Ohsaka et al. 1988; Sha'afi 1989). In fact, PMA may block agonist-induced activation of phospholipase C in human myeloid cells with parallel potentiation of exocytosis and $O_2^{\cdot-}$ formation, and priming by PMA apparently does not depend on phosphorylation of the 47-kDa protein (Della Bianca et al. 1986a, 1988; Sha'afi et al. 1988; Cockcroft and Stutchfield 1989b; Wenzel-Seifert and Seifert 1990; see also Sect. 3.2.2.5). In contrast, Gay and Stitt (1990) suggested that chemotactic peptides and phorbol esters synergistically activate the respiratory burst through synergistic translocation of protein kinase C to the plasma membrane.

In addition to PMA, OAG potentiates fMet-Leu-Phe-induced $O_2^{\cdot-}$ formation (Dewald et al. 1984; Bass et al. 1987, 1988; Smith et al. 1988a). OAG shortens the lag time, increases the rate of $O_2^{\cdot-}$ formation, and prolongs the respiratory burst (Bass et al. 1988). OAG-induced priming apparently does not depend on extracellular Ca^{2+} and may involve activation of phospholipase A_2 (Dewald et al. 1984; Bauldry et al. 1988). As H-7 inhibits priming by OAG, its effects have been suggested to be mediated by protein kinase C (Smith et al. 1988a). In contrast, Bass et al. (1988) did not observe translocation of protein kinase C to the plasma membrane by OAG, and the protein kinase C inhibitor 1-(5-isoquinolinesulfonyl) piperazine (C-1) did not prevent priming (see also Sect. 3.2.2.3).

In addition to diacylglycerides, human neutrophils contain substantial amounts of alkylacylglycerides (Tyagi et al. 1989a). In neutrophils primed with PMA or CB, but not in unprimed cells, fMet-Leu-Phe induces the release of alkylacylglycerol presumably through activation of phospholipases C and/or D (Anthes et al. 1989; Billah et al. 1989; see also Sect. 3.2.2.1). Alkylacylglycerol may regulate the activity of protein kinase C in an inhibitory or in a stimulatory manner (Ford et al. 1989; Bass et al. 1989). Alkylacylglycerol has been shown to potentiate chemotactic peptide-induced $O_2^{\cdot-}$ formation and to inhibit the stimulatory effects of diacylglycerol (Bauldry et al. 1988; Bass et al. 1989). Priming by alkylacylglycerol may involve activation of phospholipase A_2 and shows properties which are different from those of diacylglycerol-induced priming (Bauldry et al. 1988; Bass et al. 1989).

3.2.2.3 Studies with Protein Kinase C Inhibitors

Many studies with protein kinase C inhibitors have been performed to clarify the role of this kinase in the activation of NADPH oxidase by phorbol esters, diacylglycerols, and especially by intercellular signal molecules. Table 4 summarizes some data on the effects of protein kinase

Table 4. Inhibition of the respiratory burst by protein kinase inhibitors and agents interfering with Ca^{2+}-dependent processes

Inhibitory agent	Respiratory burst stimulated by	Mechanisms discussed	Selected references
U-73122	Chemoattractants	Inhibition of phospholipase C	Smith et al. (1990)
Staurosporine, K-252a	PMA, chemoattractants (controversial, inhibition or potentiation)	Inhibition of protein kinase C, inhibition of other kinases?	Tamaoki et al. (1986), Kase et al. (1987), Smith et al. (1988), Thelen et al. 1988b), Rüegg and Burgess (1989), Combadiere et al. (1990)
Polymyxin B	OAG, PMA, chemotactic peptides (inhibition or none or stimulation)	Inhibition of protein kinase C, inhibition of other kinases? Protein kinase C-independent processes?	Mazzei et al. (1982), Wise et al. (1982), Seifert and Schächtele (1988), Aida et al. (1990)
Sphingosine	Chemotactic peptides, PMA	Inhibition of protein kinase C, unspecific cell damage?	Wilson et al. (1986), Pittet et al. (1987), Bazzi and Nelsestuen (1987), Lambeth et al. (1988), Merrill and Stevens (1989)
Ebselen	PMA	Inhibition of protein kinase C, interaction with SH groups?	Cotreave et al. (1989)
C-1, H-7	Chemotactic peptides, PMA (controversial, inhibition or none)	Inhibition of protein kinase C, inhibition of other kinases?	Hidaka et al. (1984), Gerard et al. (1986), Fujita et al. (1986), Sha'afi et al. (1986), Wright and Hoffman (1986, 1987), Berkow et al. (1987b), Shibanuma et al. (1987), Nath and Powledge (1988), Holian et al. (1988), Seifert and Schächtele (1988), Love et al. (1989), Nath et al. (1989)
Trifluoperazine, chlorpromazine, W-7	Chemotactic peptides, PMA	Inhibition of protein kinase C, calmodulin-dependent processes or calpain I, interaction with hydrophobic domains of NADPH oxidase?	H.J. Cohen et al. (1980b), Alobaidi et al. (1981), Tanaka et al. (1982), Wise et al. (1982), Tomlinson et al. (1984), Wright and Hoffman (1986, 1987), Sakata et al. (1987a), Holian et al. (1988), Seifert and Schächtele (1988), Brumley and Wallace (1989)

Table 4. (continued)

Inhibitory agent	Respiratory burst stimulated by	Mechanisms discussed	Selected references
AMG-C$_{16}$	Chemotactic peptides, PMA	Inhibition of protein kinase C.	Kramer et al. (1989)
Vitamin E	Controversial (inhibition or stimulation)	Inhibition of protein kinase C, other mechanisms?	Harris et al. (1980), Baehner et al. (1982), Butterick et al. (1983), Mahoney and Azzi (1988), Cadenas (1989)
Flavonoids (e.g., quercetin)	Various stimuli	Inhibition of protein kinase C, interaction with hydrophobic domains of NADPH oxidase?	Tauber et al. (1984), Pagonis et al. (1986), Blackburn et al. (1987), Ferriola et al. (1989)
Suramin	PMA	Inhibition of protein kinase C	Mahoney et al. (1990)
Ca^{2+} channel blockers (diltiazem, verapamil, dihydropyridines)	Various stimuli	Inhibition of protein kinase C and other kinases, direct inhibition of NADPH oxidase, local-anesthetic effects	DiPerri et al. (1984), Elferink and Deierkauf (1984, 1988), Della Bianca et al. (1985), Schächtele et al. (1989), Zimmerman et al. (1989)
TMB-8	Chemotactic peptides, PMA	Putative inhibitor of intracellular Ca^{2+} release, other mechanisms?	Smith and Iden (1981), Smolen et al. (1981), Korchak et al. (1984a), Elferink and Deierkauf (1985)
CI-922	Chemotactic peptides, opsonized zymosan, ConA, A23187; PMA (no effect)	Inhibition of calmodulin-dependent processes.	Wright et al. (1987a,b, 1988)
ST 638	Chemotactic peptides, opsonized zymosan, NaF; PMA and A23187 (no effect)	Inhibition of protein tyrosine kinases, inhibition of other kinases?	Berkow et al. (1989), Gomez-Cambronero et al. (1989b)
Biscoclaurine alkaloids	Various stimuli	Inhibition of protein kinase C, stabilization of plasma membrane	Matsuno et al. (1990)

C inhibitors and of agents which interfere with Ca^{2+}-dependent processes on the respiratory burst.

The polycationic cyclic peptide antibiotic polymyxin B inhibits protein kinase C in vitro by interacting with phospholipids (Mazzei et al. 1982; Wise et al. 1982). Polymyxin B does not inhibit cAMP- or cGMP-dependent protein kinases, but unfortunately the specificity of the antibiotic is hampered by the fact that it also inhibits calmodulin-sensitive myosin light-chain kinase and Ca^{2+}-activated K^+ channels (Mazzei et al. 1982; Wise et al. 1982; Varecka et al. 1987). Paradoxically, polymyxin B mimics certain effects of PMA on protein phosphorylation and phospholipid metabolism in HL-60 cells (Kiss et al. 1987; Kiss and Anderson 1989). The effects of polymyxin B on the respiratory burst have not been studied very extensively. In human neutrophils, polymyxin B does not inhibit $O_2^{\cdot-}$ formation induced by PMA and fMet-Leu-Phe, whereas the antibiotic partially inhibits phorbol ester- and chemotactic peptide-induced $O_2^{\cdot-}$ formation in dibutyryl cAMP-differentiated HL-60 cells (Seifert and Schächtele 1988; see also Sect. 3.4.4.1.3). Aida et al. (1990) reported that polymyxin B inhibits PMA-induced $O_2^{\cdot-}$ formation in human neutrophils, is without inhibitory effect on the fMet-Leu-Phe-induced respiratory burst, and potentiates that induced by OAG. The latter effect of polymyxin B has been suggested to be protein kinase C independent (Aida et al. 1990).

Sphingosine is another commonly employed inhibitor of protein kinase C, but the usefulness of sphingoid long-chain bases as protein kinase C inhibitors in studies dealing with intact cells is a matter of debate (Wilson et al. 1986; Bazzi and Nelsestuen 1987; Pittet et al. 1987; Krishnamurthi et al. 1989; Merrill and Stevens 1989). Sphingosine and sphinganine have been shown to inhibit fMet-Leu-Phe- and PMA-induced $O_2^{\cdot-}$ formation and protein phosphorylation in human neutrophils (Wilson et al. 1986). These observations have been confirmed by Pittet et al. (1987), but these authors attributed the inhibitory effects of sphingosine to cytotoxicity rather than to inhibition of protein kinase C. A subsequent study suggested that the addition of sphingosine and sphinganine to cells from a stock solution containing albumin minimizes cytotoxicity (Lambeth et al. 1988).

Staurosporine and K-252a are the most potent inhibitors of protein kinase C presently available, and they act presumably by interfering with the ATP binding site of the kinase (Tamaoki et al. 1986; Kase et al. 1987). In agreement with their effects on purified protein kinase C, staurosporine and K-252a are very potent and effective inhibitors of protein phosphorylation and $O_2^{\cdot-}$ formation induced by PMA, chemoattractants, and lipopeptides in human neutrophils (Smith et al. 1988b; Thelen et al. 1988b; Dewald et al. 1989; Seifert et al. 1990). Unexpectedly, at concentrations in the nanomolar range staurosporine was found to enhance the chemoattractant-induced respiratory burst in human neutrophils, suggesting that protein kinase C may also play an inhibitory role in receptor agonist-induced O_2^- formation (Combadiere et al. 1990; see also Sect. 3.3.2.5). Staurosporine at nanomolar concentrations effectively inhibits PMA-induced O_2^- formation in neutrophils, but subsequent stimulation with fMet-Leu-Phe results in substantial O_2^- formation. (Robinson et al. 1990). This activation of NADPH oxidase is accompanied by inhibition of phosphorylation of the 47-kDa protein. A kinase other than the one activated by PMA may play a role in this activation of NADPH oxidase, and phosphorylation of the 47-kDa protein may be of minor relevance. Unfortunately, staurosporine and K-252a cannot be regarded as specific as they are potent inhibitors of other kinases as well (Smith et al. 1988; Rüegg and Burgess 1989). In addition, staurosporine does not inhibit platelet aggregation induced by certain activators of protein kinase C (Schächtele et al. 1988). Moreover, staurosporine paradoxically induces exocytosis of specific granules from neutrophils and shows some functional similarities with PMA in this regard (Dewald et al. 1989).

The isoquinolinesulfonamide H-7 is one of the most extensively studied protein kinase C inhibitors with respect to the effects on NADPH oxidase (Hidaka et al. 1984). Closely related to H-7 is the isoquinolinesulfonyl-piperazine analogue C-1 (Gerard et al. 1986). Similar to staurosporine, H-7 and C-1 inhibit protein kinase C and other kinases by interfering with their ATP binding sites, and therefore these compounds are not specific pharmacological tools (Hidaka et al. 1984; Gerard et al. 1986; Schächtele et al. 1988). Moreover, H-7 shows effects in intact cells which are apparently unrelated to inhibition of protein kinases (Love et al. 1989).

The results concerning the effects of H-7 and C-1 on O_2^- formation are controversial. Wright and Hoffman (1986, 1987) reported that H-7 inhibits neither PMA- nor fMet-Leu-Phe-induced O_2^- formation in human neutrophils. Berkow et al. (1987b) and Sha'afi et al. (1988) showed that H-7 inhibits the stimulatory effect of PMA but not that of fMet-Leu-Phe on the respiratory burst. In contrast, other investigators reported that H-7 inhibits, at least in part, the PMA- and fMet-Leu-Phe-induced respiratory burst (Fujita et al. 1986; Sha'afi et al. 1986; Shibanuma et al. 1987; Nath and Powledge 1988; Holian et al. 1988; Seifert and Schächtele 1988). The effects of H-7 are apparently species specific, as H-7 inhibits PMA-induced priming in rabbit but not in human neutrophils (Sha'afi et al. 1988; see also Sect. 3.2.2.2).

C-1 inhibits PMA-induced O_2^- formation but not that induced by fMet-Leu-Phe or C5a (Gerard et al. 1986). In contrast, Nath and Powledge (1988) and Nath et al. (1989) reported that C-1 inhibits fMet-Leu-Phe-induced O_2^- formation in a temperature-dependent manner. Finally, C-1 has been shown to enhance fMet-Leu-Phe-induced exocytosis and to inhibit chemotaxis (Harvath et al. 1987; Salzer et al. 1987).

The antipsychotic drugs chlorpromazine and trifluoperazine and the naphthalenesulfonamide N-(6-aminohexyl)-5-chloro-1-naphthalenesulfonamide (W-7) are used primarily as inhibitors of calmodulin, but they also inhibit protein kinase C and the Ca^{2+}-dependent protease calpain I (Tanaka T. et al. 1982; Wise et al. 1982; Schatzman et al. 1983; Tomlinson et al. 1984; Brumley and Wallace 1989). Surprisingly, the data on the effects of these drugs on the respiratory burst are less controversial than those on other and more potent inhibitors of protein kinase C. A substantial number of reports showed that chlorpromazine, trifluoperazine, and W-7 inhibit both the phorbol ester- and chemoattractant-induced $O_2^{\bullet-}$ formation and phosphorylation of the 47-kDa protein (Elferink 1979; Cohen et al. 1980b; Alobaidi et al. 1981; Smith et al. 1981; Takeshige and Minakami 1981; Elferink and Deierkauf 1985; Heyworth and Segal 1986; Wright and Hoffman 1986, 1987; Shibanuma et al. 1987; Seifert and Schächtele 1988; Holian et al. 1988). Paradoxically, W-7 per se has been reported to induce a short-lasting respiratory burst in alveolar macrophages (Holian et al. 1988). Taking into consideration the inhibitory profile of these substances, it has been suggested that the effects of these compounds on $O_2^{\bullet-}$ formation are due to inhibition of calmodulin-dependent processes (Alobaidi et al. 1981; Smith et al. 1981; Takeshige and Munakami 1981; Smolen et al. 1982; Wright and Hoffman 1986, 1987; Shibanuma et al. 1987) or to inhibition of protein kinase C (Shibanuma et al. 1987; Heyworth and Segal 1986; Holian et al. 1988). In addition to inhibiting proximal parts of the signal transduction process, W-7 and trifluoperazine may directly interfere with components of NADPH oxidase (Cohen et al. 1980b; Alobaidi et al. 1981; Sakata et al. 1987a; Seifert and Schächtele 1988; see also Sect. 5.1.3.3).

The ether lipid, 1-O-hexadecyl-2-O-methylglycerol (AMG-C_{16}) is a recently introduced inhibitor of protein kinase C (Kramer et al. 1989). AMG-C_{16} has been reported not to inhibit cAMP- or Ca^{2+}/calmodulin-dependent protein kinases (Kramer et al. 1989). In addition, AMG-C_{16} is apparently not cytotoxic and does not interfere with fMet-Leu-Phe-induced phosphoinositide degradation and increase in cytoplasmic Ca^{2+} (Kramer et al. 1989). Moreover, the drug apparently does not directly interfere with NADPH oxidase (Kramer et al. 1989). AMG-C_{16} inhibits the phorbol ester- and chemotactic peptide-induced respiratory burst in human neutrophils (Kramer et al. 1989). With respect to phorbol esters, there is a close correlation between inhibition of the respiratory burst and phosphorylation of the 47-kDa protein by AMG-C_{16}. The correlation between inhibition of NADPH oxidase and phosphorylation of the 47-kDa protein is less stringent with fMet-Leu-Phe, suggesting that AMG-C_{16}-sensitive and -insensitive signal transduction pathways are involved in chemotactic peptide-induced O_2^- formation (Kramer et al. 1989).

Both inhibitory and stimulatory effects of vitamin E on the respiratory burst have been reported (Baehner et al. 1982; Butterick et al. 1983; Leb et al. 1985), and neutrophils from vitamin E-deficient rats show increased oxygen consumption and H_2O_2 release in comparison to control cells (Harris et al. 1980). The inhibitory effects of vitamin E on O_2^- formation are possibly due among others to inhibition of protein kinase C and to its radical-scavenging properties (Mahoney and Azzi 1988; Cadenas 1989).

Flavonoids are plant-derived compounds with antiallergic and anti-inflammatory properties (Middleton 1984). Various flavonoids, e.g., kaempferol, morin, fisetin and quercetin, inhibit the respiratory burst in human neutrophils induced by soluble and particulate stimuli (Tauber et al. 1984; Pagonis et al. 1986). The ability of flavonoids to inhibit the respiratory burst correlates with their hydrophobicity. The mechanism by which these agents inhibit NADPH oxidase may involve interference with protein kinase C-mediated protein phosphorylation (Blackburn et al. 1987). This assumption is supported by the finding that flavonoids inhibit purified protein kinase C, but, paradoxically, quercetin has also been reported to show stimulatory effects on this enzyme (Ferriola et al. 1989; Picq et al. 1989).

The selenium-containing heterocyclic compound ebselen inhibits PMA-induced O_2^- formation in human neutrophils (Cotgreave et al. 1989). Ebselen may inhibit protein kinase C or may directly inhibit NADPH oxidase through interaction with SH groups (see also Sect. 4.3.3).

The biscoclaurine alkaloids cepharanthine, tetrandrine, and isotetrandine inhibit O_2^- formation in guinea pig neutrophils induced by various stimuli including PMA and fMet-Leu-Phe (Matsuno et al. 1990). Evidence was presented by these authors that the effects of biscoclaurine alkaloids are mediated through an inhibition of protein kinase C.

The effects of retinoids on the respiratory burst are very complex. On the one hand, retinoids have been suggested to suppress O_2^- formation through inhibition of protein kinase C, but on the other, retinoids have also been reported to be effective activators of the respiratory burst (see Sect. 3.3.2.4).

The hexa-anionic hydrophobic compound, suramin, activates purified protein kinase C in the presence of Ca^{2+} and in the absence of phospholipid and inhibits the enzyme activated by phospholipid, Ca^{2+} and diacylglycerol (Mahoney et al. 1990). The inhibitory effect of suramin on protein kinase C may be due to competition with ATP (Mahoney et al. 1990). In addition, suramin at very high concentrations inhibits the phorbol ester-induced respiratory burst in human neutrophils (Mahoney et al. 1990).

The above data clearly show that studies with protein kinase C inhibitors are difficult to interpret. Neither the failure nor the effectiveness of a compound to inhibit O_2^- formation answers the question conclusively whether protein kinase C plays a role in the process or not. Among the factors which contribute to this unsatisfying situation are the lack of specificity of protein kinase C inhibitors, cell type, and stimulus specificities of NADPH oxidase activation and possibly the involvement of various isoenzymes of protein kinase C, which may possess different physiological roles and differential sensitivities to inhibitory drugs.

3.2.2.4 Studies with Inhibitors of Diacylglycerol Kinase and Diacylglycerol Lipase

Diacylglycerol kinase catalyzes the phosphorylation of diacylglycerol to phosphatidic acid and may play a role in the termination of protein kinase C activation by removing the former lipid (Abdel-Latif 1986). Some years ago, 6-[2{4-[(4-fluorophenyl)phenylmethylene]-1-piperidinyl}ethyl]-7-methyl-5H-thiazolo-[3,2-al]pyrimidin-5-one (R 59 022) was introduced as an inhibitor of diacylglycerol kinase which apparently does not affect the activity of other enzymes of phosphoinositide metabolism (de Chaffoy de Courcelles et al. 1985). In intact platelets, R 59 022 amplifies OAG-induced activation of protein kinase C and thrombin-induced release of diacylglycerol (de Chaffoy de Courcelles et al. 1985). These data suggest that R 59 022 functions in an analogous manner as do inhibitors of phosphodiesterases which potentiate agonist-induced accumulation of cAMP (see Sect. 4.1).

In neutrophils, R 59 022 enhances O_2^- formation induced by OAG and A 23187 and that induced by the receptor agonists fMet-Leu-Phe, PAF, IgG, opsonized zymosan, and lipopeptide (Dale and Penfield 1987; Muid et al. 1987; Gomez-Cambronero et al. 1987; Mege et al. 1988c; Seifert et al. 1990). Unfortunately, not only inhibition of diacylglycerol kinase but also activation of phospholipase D may contribute to the stimulatory effects of R 59 022 on chemoattractant-induced O_2^- formation (Mahadevappa 1988; see also Sect. 3.2.2.1). In guinea pig neutrophils, R59022 potentiates formyl peptide- but not PMA-induced O_2^- formation and phosphorylation of the 47-kDa protein (Ohtsuka et al. 1990b). Additionally, R59022 potentiates agonist-stimulated formation of diacylglycerol and inhibits the formation of phosphatidic acid. These data indicate that, at least in this system, the effects of R59022 are mediated via inhibition of diacylglycerol kinase with subsequent accumulation of diacylglycerol, resulting in enhanced activation of protein kinase C.

Degradation of diacylglycerol to monoacylglycerol by diacylglycerol lipase is another pathway to remove this intracellular signal molecule (Abdel-Latif 1986). 1,6-Di(O-(carbamoyl)cyclohexanone oximine)hexane (RHC 80267) has been reported to be a potent and relatively selective inhibitor of diacylglycerol lipase (Sutherland and Amin 1982). RHC 80267 has no effect on O_2^- formation induced by fMet-Leu-Phe, IgG, or opsonized zymosan, suggesting that removal of diacylglycerol through diacylglycerol lipase does not play a crucial role in the termination of the agonist-induced respiratory burst (Muid et al. 1987). In contrast, RHC 80267 potentiates OAG-induced O_2^- formation (Dale and Penfield 1987). Finally, indomethacin has been reported to potentiate OAG-induced O_2^- formation, presumably by inhibition of diacylglycerol lipase and kinase (Dale and Penfield 1985, 1987; see also Sect. 4.2.2).

3.2.2.5 The Inhibitory Role of Protein Kinase C
in Receptor Agonist-Induced Cell Activation

In many cell types, protein kinase C plays not only a stimulatory role but also an inhibitory role in agonist-induced cell activation (Nishizuka 1984, 1986, 1988, 1989; Lefkowitz and Caron 1986; Sibley et al. 1987). Phorbol esters uncouple receptors, e.g., β_1-, β_2- and α_1-adrenergic receptors, from intracellular effector systems through protein kinase C-mediated phosphorylation of receptor proteins (Lefkowitz and Caron 1986; Sibley et al. 1987).

Pretreatment of human myeloid cells with PMA blunts phosphoinositide degradation and increase in cytoplasmic Ca^{2+} induced by various intercellular signal molecules, i.e., fMet-Leu-Phe, ATP, PAF, and LTB$_4$ (Naccache et al. 1985a; Della Bianca et al. 1986a; C.D. Smith et al. 1987; Kikuchi et al. 1987; Cockcroft and Stutchfield 1989b; Yamzaki et al. 1989). In addition, activators of protein kinase C may modulate binding of PAF and LTB$_4$ to their receptors (Yamzaki et al. 1989; O'Flaherty et al. 1989; McCarthy et al. 1989). Phorbol esters may disrupt the signal transduction cascade by uncoupling formyl peptide receptors from G-proteins and activated G-proteins from phospholipase C (C.D. Smith et al. 1987; Kikuchi et al. 1987). Phosphorylation of G-proteins may be involved in this desensitization process (Katada et al. 1985; Pyne et al. 1989).

With respect to O_2^- formation, inhibitory effects of phorbol esters are apparently stimulus dependent. On one hand, fMet-Leu-Phe enhances the ability of neutrophils to generate O_2^- upon stimulation with PMA, and

PMA does not prevent subsequent stimulation of O_2^- formation by fMet-Leu-Phe (English et al. 1981b; Bender et al. 1987). In addition, PMA increases the binding of formyl peptides to their receptors in human neutrophils in a concentration-dependent manner (Bender et al. 1987). Moreover, PMA and fMet-Leu-Phe interact in an additive or synergistic manner to activate O_2^- formation in neutrophils or HL-60 cells (Bender et al. 1983; Seifert et al. 1989a; Wenzel-Seifert and Seifert 1990). Finally, pretreatment of neutrophils with PMA results in inhibition of phospholipase C but in potentiation of respiratory burst induced by NaF or fMet-Leu-Phe (Della Bianca et al. 1986a; see also Sects. 3.2.1.3, 3.2.2.2, 3.2.3.1).

On the other hand, it has been reported that PMA decreases the binding of C5a to its receptors and blunts the stimulatory effects of this intercellular signal molecule on O_2^- formation in human neutrophils (Bender et al. 1987). In addition, pretreatment with PMA desensitizes human neutrophils and human peritoneal macrophages to undergo a respiratory burst upon stimulation with opsonized and unopsonized bacteria (Gbarah et al. 1989; see also Sects. 3.3.1.5, 3.3.2.12.1). These data suggest that different types of receptors show differential sensitivity to sensitization and/or desensitization by protein kinase C. However, in comparison to adrenergic receptors (Lefkowitz and Caron 1986; Sibley et al. 1987), much less information on the molecular basis of these processes at phagocyte receptors is available (see also Sects. 3.3.1.1.3, 3.3.1.1.4, 4.1.3).

3.2.3 Calcium and Calmodulin

Ca^{2+} plays an important role as intracellular signal molecule, mainly in the regulation of calmodulin-dependent enzymes (Rasmussen and Waisman 1983; Tomlinson et al. 1984). Following receptor-mediated activation of phospholipase C, inositol 1,4,5-triphosphate is released from phosphatidylinositol 4,5-biphosphate and mobilizes intracellular Ca^{2+} from non-mitochondrial stores (Streb et al. 1983; Berridge and Irvine 1984; DiVirgilio et al. 1985; Krause and Lew 1987; Volpe et al. 1988; Jaconi et al. 1988; Perianin and Snyderman 1989). In addition, fMet-Leu-Phe induces mobilization of plasma membrane-bound Ca^{2+} and induces Ca^{2+} influx from the extracellular space (Nacchache et al. 1979; Schell-Frederick 1984; Rossi et al. 1985; Andersson et al. 1986a; von Tscharner et al. 1986a,b; Di Virgilio et al. 1987; Nasmith and Grinstein 1987b). The question whether the fMet-Leu-Phe-induced Ca^{2+} influx via plasma membrane ion channels depends on an increase in cytoplasmic Ca^{2+} is a subject of present debate (von Tscharner et al. 1986b; Nasmith and Grinstein 1987b). Moreover, inositol 1,3,4,5-tetrakisphosphate has recently been suggested to be involved in receptor-mediated Ca^{2+} influx in myeloid cells (Pittet et al. 1989).

3.2.3.1 Correlations and Dissociations Between Activation of Ca^{2+}-Dependent Processes and NADPH Oxidase

Similar to the role of protein kinase C, the role of Ca^{2+} in the receptor-mediated activation of NADPH oxidase is very controversial (see Sect. 3.2.2.1). There are certain correlations between activation of $O_2^{\cdot-}$ formation and an increase in cytoplasmic Ca^{2+}, and Ca^{2+} ionophores, i.e., A 23187 and ionomycin, may activate the respiratory burst in various types of phagocytes (Schell-Frederick 1974; Romeo et al. 1975; McPhail and Snyderman 1983; Dale and Penfield 1985, 1987; Wymann et al. 1987b; Christiansen et al. 1988a; Seifert et al. 1989c; Dahlgren and Follin 1990). In addition, Ca^{2+} ionophores prime phagocytes for an enhanced respiratory burst upon subsequent stimulation with chemotactic peptides, cytokines, cell-permeant diacylglycerols, phorbol esters, and other stimuli (McPhail and Snyderman 1983; McPhail et al. 1984a; Dale and Penfield 1984; Robinson et al. 1984; Strnad and Wong 1985a; Finkel et al. 1987; Dahlgren 1989; Koenderman et al. 1989a; see also Sects. 3.1.1, 3.3.1.3).

With regard to the chemoattractant-induced respiratory burst, an increase in cytoplasmic Ca^{2+} precedes $O_2^{\cdot-}$ formation. Inhibition of the increase in cytoplasmic Ca^{2+} and removal of extracellular Ca^{2+} are associated with suppression of fMet-Leu-Phe-induced $O_2^{\cdot-}$ formation (Serhan et al. 1983; Hallett and Campbell 1984; Nakagawara et al. 1984; Lew et al. 1984b; Sklar and Oades 1985; Lazzari et al. 1986; Dahlgren 1987; Hruska et al. 1988; Seifert et al. 1990). Extracellular Ca^{2+} restores the ability of phagocytes to generate $O_2^{\cdot-}$ upon exposure to chemotactic peptides (Stickle et al. 1984). In contrast, the zymosan-induced respiratory burst in Kupffer's cells does not depend on extracellular Ca^{2+} (Dieter et al. 1988).

The putative inhibitor of intracellular Ca^{2+} release, 3,4,5-trimethoxybenzoic acid 8-(diethylamino)-octyl ester (TMB 8), inhibits PMA- and fMet-Leu-Phe-induced $O_2^{\cdot-}$ formation (Matsumoto et al. 1979; Smith and Iden 1981; Smolen et al. 1981; Smolen 1984; Korchak et al. 1984a,b; Elferink and Deierkauf 1985). In addition, organic Ca^{2+} channel blockers of different chemical classes, e.g., verapamil, diltiazem, and dihydropyridines at very high concentrations, inhibit $O_2^{\cdot-}$ formation induced by a variety of stimuli, including PMA and fMet-Leu-Phe in a stereo-unspecific manner (Di Perri et al. 1984; Elferink and Deierkauf 1984; Della Bianca et al. 1985; Irita et al. 1986; Elferink and Deierkauf 1988; Zimmerman et al. 1989). Apparently, these drugs do not inhibit $O_2^{\cdot-}$ formation via blockade of voltage-dependent Ca^{2+} channels. This view is supported by the fact that neutrophils possess Ca^{2+}-activated cation channels which are not sensitive to dihydropyridines, and high-affinity binding sites for dihydropyridines are also missing in myeloid cells (von Tscharner et al. 1986a; Pennington et al. 1986; Mitsuhashi et al. 1989). Verapamil does not inhibit fMet-Leu-Phe-induced Ca^{2+} influx from the extracellular space, and the concentrations of Ca^{2+} channel blockers required to inhibit $O_2^{\bullet-}$ formation are much higher than those required for the blockade of Ca^{2+} channels in other tissues (Elferink and Deierkauf 1984; Pennington et al. 1986). Organic Ca^{2+} channel blockers inhibit protein kinase C, may directly interfere with components of NADPH oxidase, or may show anesthetic-like membrane effects (Della Bianca et al. 1985; Irita et al. 1986; Elferink and Deierkauf 1988; Schächtele et al. 1989). Paradoxically, the dihydropyridine felodipine has been shown to stimulate phosphorylation of protein kinase C substrates in platelets (Sutherland and Walsh 1989). 3,7-Dimethoxy-4-phenyl-N-1H-tetrazol-5-yl-4Hfuro[3,2-b]indole-2-carboxamide (CI-922) has been reported to inhibit A 23187- and receptor agonist-induced $O_2^{\bullet-}$ formation presumably through interference with calmodulin-dependent processes (Wright et al. 1987a,b). The effects of the calmodulin antagonists W-7, chlorpromazine, and trifluoperazine and of purified calmodulin on $O_2^{\bullet-}$ formation are described in Sects. 3.2.2.3 and in 5.1.3.3.

Neutrophils possess a Na^+/Ca^{2+} exchanger which mediates Ca^{2+} influx in the resting state (Simchowitz et al. 1990). The order of effectiveness of various cations (e.g., La^{3+}, Zn^{2+}, Sr^{2+}, Cd^{2+}) and analogues of amiloride (e.g., benzamil, phenamil) in inhibiting Na^+/Ca^{2+} exchange and fMet-Leu-Phe-induced $O_2^{\bullet-}$ formation is similar. Additionally, the above substances inhibit the fMet-Leu-Phe-induced rise in cytoplasmic Ca^{2+}. These data suggest that Na^+/Ca^{2+} exchange contributes to the chemotactic peptide-mediated increase in cytoplasmic Ca^{2+} and is involved in the activation of NADPH oxidase.

During the past few years a rapidly increasing number of studies have provided evidence for the assumption that cytoplasmic Ca^{2+} does not play a key role or may even be of no relevance in receptor-mediated activation of the respiratory burst. Some of the evidence available in the literature pointing against a crucial role of Ca^{2+} in the regulation of NADPH oxidase by receptor agonists is summarized below. In 1983, Pozzan et al. showed that fMet-Leu-Phe and iononycin induce similar increases in cytoplasmic Ca^{2+}, whereas only the chemotactic peptide is an effective activator of $O_2^{\bullet-}$ formation. Interestingly, fMet-Leu-Phe induces a maximal increase in cytoplasmic Ca^{2+} without activating $O_2^{\bullet-}$ formation (Korchak et al. 1984b). In 1985, Apfeldorf et al. reported that a murine monoclonal antibody induces an increase in cytoplasmic Ca^{2+} but not $O_2^{\bullet-}$ formation in human neutrophils.

In adherent neutrophils and cultured human monocytes, the fMet-Leu-Phe-induced O_2^- formation and increase in cytoplasmic Ca^{2+} do not correlate (Rebut-Bonneton et al. 1988; Bernardo et al. 1988), and fMet-Leu-Phe activates oxygen consumption in electropermeabilized and Ca^{2+}-depleted human neutrophils (Grinstein and Furuya 1988). 1,25-Dihydroxyvitamin D_3-differentiated U-937 cells generate O_2^- upon exposure to opsonized zymosan and PMA but not upon stimulation with fMet-Leu-Phe, although fMet-Leu-Phe induces an increase in cytoplasmic Ca^{2+} (Polla et al. 1989). In neutrophil cytoplasts, fMet-Leu-Phe induces a respiratory burst independently of extracellular Ca^{2+} (Torres and Coates 1984). The fact that fMet-Leu-Phe, C5a, LTB$_4$, and PAF induce a rapid increase in cytoplasmic Ca^{2+}, but that only fMet-Leu-Phe and C5a are effective activators of NADPH oxidase, provides another example for the dissociation of Ca^{2+} mobilization and O_2^- formation (Hartiala et al. 1987). With respect to priming of O_2^- formation by PAF, both Ca^{2+}-dependent and -independent pathways may exist (Koenderman et al. 1989a). In human blood monocytes, an increase in cytoplasmic Ca^{2+} is not sufficient for the activation of O_2^- formation (Kemmerich and Pennington 1988).

The onset of the respiratory burst by fMet-Leu-Phe, PAF, LTB$_4$, or C5a is faster than that by PMA or ionomycin, suggesting that chemoattractants and the latter two agents activate NADPH oxidase by different mechanisms (Wymann et al. 1987b). PMA reduces the lag time of chemotactic peptide-induced H_2O_2 formation, and chemoattractants have been suggested to activate the respiratory burst through Ca^{2+}/protein kinase C-dependent and -independent mechanisms (Wymann et al. 1987b; Dewald et al. 1988; see also Sects. 3.2.2.2, 4.3.1). In human neutrophils primed with PMA, the fMet-Leu-Phe-induced O_2^- formation is potentiated, whereas the release of inositol triphosphate and increase in cytoplasmic Ca^{2+} are blocked due to inhibition of phospholipase C (Della Bianca et al. 1986a; see also Sects. 3.2.2.1, 3.2.2.2, 3.2.2.5). In addition, Ca^{2+}-depleted neutrophils primed with PMA undergo a respiratory burst in the absence of phosphoinositide turnover (Grzeskowiak et al. 1986). Furthermore, ConA plus fMet-Leu-Phe activate NADPH oxidase in Ca^{2+}-depleted human neutrophils without activating phospholipase C or increasing cytoplasmic Ca^{2+} (Rossi et al. 1986). NaF also does not induce phosphoinositide degradation, Ca^{2+} mobilization, or oxygen consumption in Ca^{2+}-depleted neutrophils, but priming with PMA restores the ability of NaF to activate NADPH oxidase without an effect on phospholipase C (Della Bianca et al. 1988; see also Sects. 3.2.1.3, 5.1.4.2). Finally, the results of a recent study indicate that priming of the respiratory burst by fMet-Leu-Phe involves phospholipase C- and Ca^{2+}-independent signal transduction pathways (Karnad et al. 1989) (see also Sects. 3.2.1.2, 3.3.1.1.4).

3.2.4 Phospholipase A_2 and Arachidonic Acid

Intercellular signal molecules, e.g., chemotactic peptides, PAF, and ATP, induce the release of arachidonic acid in a variety of phagocytes, e.g., guinea pig, rabbit and human neutrophils, HL-60 cells, and guinea pig macrophages, presumably through activation of phospholipase A_2 (Bromberg and Pick 1983; Bokoch and Gilman 1984; Okajima and Ui 1984; Ohta et al. 1985; Tao et al. 1989; Nakashima et al. 1989; Cockcroft and Stutchfield 1988b). In contrast, other authors reported that fMet-Leu-Phe is only a poor stimulus for arachidonic acid release unless cells are primed with OAG or A 23187 (Clancy et al. 1983; Billah and Siegel 1987; Bauldry et al. 1988). Interestingly, recent studies indicate that G-proteins are involved in the regulation of phospholipase A_2 in various cell types including phagocytes (Okajima and Ui 1984; Ohta et al. 1985; Burch et al. 1986; Jelsema and Axelrod 1987; Jelsema 1987; Axelrod et al. 1988; Nakashima et al. 1989; Cockcroft and Stutchfield 1989b). fMet-Leu-Phe, C5a, and LTB_4 have been reported to activate phospholipase A_2 in membranes of rabbit neutrophils in a Ca^{2+}-dependent manner (Bormann et al. 1984), but this finding was not confirmed in a subsequent study (Matsumoto et al. 1988).

3.2.4.1 Correlations and Dissociations Between Activation of Arachidonic Acid Release and NADPH Oxidase

As is the case of protein kinase C and cytoplasmic Ca^{2+}, there is evidence for and against the hypothesis that arachidonic acid or one of its lipoxygenase products serves as intracellular signal molecule for the activation of NADPH oxidase. On one hand, various chemically unrelated stimuli induce the release of arachidonic acid and a respiratory burst in guinea pig macrophages (Bromberg and Pick 1983). Most importantly, unsaturated fatty acids activate $O_2^{\cdot-}$ formation both in intact phagocytes and in cell-free systems (Badwey et al. 1981, 1984; Bromberg and Pick 1983; Boukili et al. 1986; see also Sects. 3.1.2, 5.1.3). In addition, the respiratory burst induced by various agents including fMet-Leu-Phe has been reported to be enhanced by exogenous unsaturated fatty acids and/or exogenous phospholipase A_2 (Lackie and Lawrence 1987; Ginsburg et al. 1989). Certain lysophosphatides potentiate the respiratory burst as well (Ginsburg et al. 1989). In contrast, inhibitory effects of fatty acids on receptor-mediated $O_2^{\cdot-}$ formation have also been observed (see Sect. 4.1.2).

A phospholipase A_2-activating protein which possesses antigenic and biochemical similarities with mellitin activates $O_2^{\cdot-}$ formation and release of arachidonic acid in human neutrophils in a concentration-dependent manner without inducing cytotoxicity (Bomalaski et al. 1989). Moreover, inhibitors of phospholipase A_2, e.g., p-bromophenacyl bromide and mepacrine (quinacrine), and inhibitors of lipoxygenases, e.g., ETYA, nordihydroguaiaretic acid, esculetin, and BW755C, have been reported to inhibit the agonist-induced respiratory burst (Bokoch and Reed 1979; Smolen and Weissmann 1980; Rossi et al. 1981b; Kaplan et al. 1984; Maridonneau-Parini and Tauber 1986; Maridonneau-Parini et al. 1986; Sakata et al. 1987b). Interestingly, deficiency of polyunsaturated fatty acids is associated with decreased chemotactic peptide-induced O_2^- formation, possibly due to perturbation of arachidonic acid metabolism (Palmblad et al. 1988b; Gyllenhammar and Palmblad 1989). Recently, arachidonic acid has been suggested to play a role in the mobilization of Ca^{2+} from intracellular stores in human neutrophils (Beaumier et al. 1987). Finally, the antileprosy agent, clofazimine, potentiates O_2^- formation and arachidonic acid release in human neutrophils induced by various stimuli including fMet-Leu-Phe and PMA (Anderson et al. 1988). As the potentiating effect of clofazimine is abolished by p-bromophenacyl bromide, it was suggested that clofazimine potentiates O_2^- formation by a phospholipase A_2-dependent mechanism, and that this priming effect on the respiratory burst contributes to its antimycobacterial activity.

On the other hand, it has been reported that there is no close correlation between the effects of lipoxygenase inhibitors on O_2^- formation and production of LTB_4 (Ozaki et al. 1986b), and blockade of arachidonic acid release in neutrophils by combining inhibitors of diacylglycerol kinase, diacylglycerol lipase, and phospholipase A_2 (R 59 022 plus RHC 80267 plus indomethacin) does not substantially affect O_2^- formation induced by IgG or opsonized zymosan (Muid et al. 1988). Several dissociations between the release of arachidonic acid and activation of the respiratory burst have been observed in murine macrophages (Tsunawaki and Nathan 1986). As is the case for protein kinase C inhibitors, the specificity of some commonly used inhibitors of arachidonic acid metabolism is of concern. For example, p-bromophenacyl bromide, quinacrine, nordihydroguaiaretic acid, and ETYA may inhibit the respiratory burst through nonspecific mechanisms or via suppression of glucose uptake from the extracellular space rather than through inhibition of arachidonic acid metabolism (Schultz et al. 1985; Tsunawaki and Nathan 1986). In addition, several inhibitors of lipoxygenases and phospholipase A_2 at nontoxic concentrations do not inhibit the respiratory burst in murine peritoneal macrophages (Schultz et al. 1985). Moreover, ETYA and esculetin may directly inhibit NADPH oxidase (Ozaki et al. 1986; Seifert and Schultz 1987a; see also Sect. 5.1.3). The results of studies with glucocorticoids, which inhibit the release of arachidonic acid, are also controversial (see Sect. 4.2.1).

3.2.5 Cytoskeleton

Cyclic alterations in morphology, e.g., lamellipod extensions and retractions, which are regulated by the cytoskeleton, have recently been suggested to play a part in chemoattractant-induced $O_2^{\cdot-}$ formation (Wymann et al. 1989). The major component of the cytoskeleton in neutrophils is actin, which exists in a globular monomeric form (G-actin) and in a double helical form (F-actin; Omann et al. 1987a; Sandborg and Smolen 1988). Cytochalasins, especially CB, are widely used experimental tools to study the role of actin filaments in the regulation of NADPH oxidase. Cytochalasins are fungal metabolites which permeate cell membranes and cause morphological and metabolic alterations, e.g., inhibition of glucose transport, in various cell types (Korn 1982). Cytochalasins bind to actin filaments and inhibit their elongation (Flanagan and Lin 1990; Brown and Spudich 1981).

3.2.5.1 Cytochalasins

CB and other cytochalasins potentiate chemoattractant-induced $O_2^{\cdot-}$ formation (Lehmeyer et al. 1979; O'Flaherty et al. 1980; Williams and Cole 1981; Cooke et al. 1985; Al-Mohanna et al. 1987). In addition, CB has been shown to potentiate receptor-mediated release of diacylglycerol and increase in cytoplasmic Ca^{2+}, and the concentrations of CB which half-maximally inhibit actin polymerization and potentiate $O_2^{\cdot-}$ formation, are similar (Honeycutt and Niedel 1986; Treves et al. 1987; Al-Mohanna and Hallett 1987).

The mechanism by which cytochalasins potentiate receptor-mediated $O_2^{\cdot-}$ formation is only incompletely understood. The exposure of phagocytes to the chemoattractants fMet-Leu-Phe or C5a is associated with actin polymerization and an increase in cytoskeleton-bound actin (White et al. 1983a; Sha'afi and Molski 1987; Sklar et al. 1985a; Howard and Wang 1987; Omann et al. 1987a; Banks et al. 1988). This receptor-mediated process is interrupted by CB (White et al. 1983; Omann et al. 1987a). LTB$_4$ and PAF but not fMet-Leu-Phe induce rapid oscillations of actin polymerization (Omann et al. 1989). The mechanism by which chemoattractants induce actin polymerization, has been suggested not to involve activation of phospholipase C or protein kinase C or the increase in ctyoplasmic Ca^{2+} but rather more direct regulation by G-proteins (Banks et al. 1988; Bengtsson et al. 1988; Downey et al. 1989; Rao et al. 1989; Therrien and Naccache 1989; Omann et al. 1989). Not only CB but also dihydro-CB, which apparently does not interfere with glucose transport, potentiates chemotactic peptide-induced O_2^- formation in human neutrophils (Jesaitis et al. 1986). The effect of dihydro-CB on O_2^- formation is maximal when added to phagocytes prior to fMet-Leu-Phe. A competitive antagonist at formyl peptide receptors inhibits the effect of dihydro-CB, suggesting that permanent stimulation of formyl peptide receptors is essential for potentiation of O_2^- formation. Dihydro-CB enhances binding of fMet-Leu-Phe to formyl peptide receptors and inhibits the formation of slowly dissociating complexes of agonist-occupied receptors with the cytoskeleton (Jesaitis et al. 1984, 1985, 1986; Omann et al. 1987a,b). In addition, dihydro-CB inhibits desensitization of formyl peptide receptors, which process is correlated with the association of receptors to the cytoskeleton. Thus, internalization of formyl peptide receptors may play a role in the termination of neutrophil responses to fMet-Leu-Phe, and this process is prevented by cytochalasins (Jesaitis et al. 1986; see also Sect. 3.3.1.1.3).

Neutrophils possess cryptic formyl peptide receptors which are expressed upon storage of the cells at room temperature, and this process results in enhanced O_2^- formation (Dahlgren et al. 1987). In contrast, this phenomenon is not seen in dimethyl sulfoxide-differentiated HL-60 cells (see also Sects. 1, 3.3.1.1.4, 3.4.4.1.3). We observed that CB is a considerably less effective potentiator of fMet-Leu-Phe-induced O_2^- formation in dimethyl sulfoxide-differentiated HL-60 cells than in neutrophils (unpublished results), suggesting that the regulation of expression of formyl peptide receptors is different in the two cell types. This assumption is also supported by the finding that chemotactic peptide-induced O_2^- formation in differentiated HL-60 cells is more sensitive to homologous desensitization than that in human neutrophils (Lee et al. 1989; McLeish et al. 1989b; Seifert et al. 1989b; see also Sect. 3.3.1.1.3).

In electropermeabilized neutrophils, phalloidin inhibits depolymerization of actin following stimulation with formyl peptides, and this process is accompanied by inhibition of NADPH oxidase (Al-Mohanna and Hallett 1990). These data suggest that agonist-induced actin polymerization plays a part in the termination of the respiratory burst (Al-Mohanna and Hallett 1990; see also Sect. 3.3.1.1.3).

Certain cytochalasins including CB have been reported to activate, at least to a limited extent, the respiratory burst in rabbit alveolar macrophages, guinea pig neutrophils and differentiated HL-60 cells (Nakagawara and Minakami 1975; Takeshige et al. 1980; Okamura et al. 1980; Bentley and Reed 1981; Sugimoto et al. 1982; Wenzel-Seifert and Seifert 1990). In human neutrophils, CB is only a weak activator of O_2^- formation, but it substantially activates NBT reduction (Elferink et al. 1990b). The involvement of G-proteins in CB-induced NBT reduction is suggested by the finding that pertussis toxin inhibits this process (Elferink et al. 1990). There are some reports in the literature that cytochalasins also affect $O_2^{\cdot-}$ formation induced by stimuli which circumvent receptor stimulation. For example, CB potentiates OAG-induced $O_2^{\cdot-}$ formation but not that induced by phorbol esters (Lehmeyer et al. 1979; O'Flaherty et al. 1980; Ozaki et al. 1986a). In contrast, cytochalasin E has been reported to alter the kinetics of PMA-induced $O_2^{\cdot-}$ formation (Badwey et al. 1982). Finally, cytochalasins have been reported to inhibit the respiratory burst induced by various agents, e.g., digitonin, latex beads, substance P, and opsonized zymosan (Williams and Cole 1981; Hallett and Campbell 1983; Serra et al. 1988; Elferink and Deierkauf 1989b; see also Sect. 3.3).

3.2.5.2 Botulinum C2 Toxin

The problems associated with the use of cytochalasins to study the role of actin filaments in receptor-mediated activation of $O_2^{\cdot-}$ formation, i.e., the lack of specificity of the substances, are avoided by the use of the binary toxin of certain *Clostridium botulinum* strains, botulinum C2 toxin. Botulinum C2 toxin possesses ADP ribosyltransferase activity and prevents actin polymerization by ADP-ribosylating G-actin of various cell types including platelets and neutrophils (Aktories et al. 1986a,b, 1987). ADP-ribosylated actin acts as a capping protein and inhibits further actin polymerization (Weigt et al. 1989). In neutrophils, botulinum C2 toxin inhibits fMet-Leu-Phe-induced actin polymerization without substantially altering the binding or dissociation dynamics of formyl peptides or Ca^{2+} mobilization (Al-Mohanna et al. 1987; Norgauer et al. 1988, 1989). However, botulinum C2 toxin slows endocytosis of ligand/receptor complexes (Norgauer et al. 1989). In analogy to CB, botulinum C2 toxin potentiates the fMet-Leu-Phe-, ConA-, and PAF-induced respiratory burst in human and rat neutrophils (Al-Mohanna et al. 1987; Norgauer et al. 1988). The effects of botulinum C2 toxin and CB are additive only at submaximally stimulatory concentrations of either agent, and the effect of botulinum C2 toxin is not evident in CB-treated cells (Al-Mohanna et al. 1987; Norgauer et al. 1988). These data support the view that CB and botulinum C2 toxin potentiate receptor agonist-induced $O_2^{\cdot-}$ formation, at least in part, by a mechanism which they have in common. In contrast, botulinum C2 toxin does not potentiate PMA-induced $O_2^{\cdot-}$ formation (Norgauer et al. 1988).

3.2.5.3 Miscellaneous Agents

Substances which disrupt microtubules also enhance chemotactic peptide-induced $O_2^{\cdot-}$ formation. Among these substances are colchicine, which is used in the treatment of acute gouty arthritis, the antifungal agent griseofulvin, and the anti-neoplastic agents, vincristine, vinblastine and podophyl-lotoxin (Kitagawa and Takaku 1982). In contrast, Al-Mohanna and Hallett (1987) did not find a stimulatory effect of colchicine on fMet-Leu-Phe-in-duced O_2^- formation. Minta and Williams (1986) also reported on a lack of inhibitory effect of colchicine on O_2^- formation in human neutrophils. Somewhat unexpectedly, deuterium oxide, which is assumed to stabilize microtubules, has been reported to potentiate fMet-Leu-Phe-induced O_2^- formation as well (Kitagawa and Takaku 1982).

3.2.6 Cyclic Nucleotides

3.2.6.1 The Role of cAMP

The role of chemotactic peptides in the regulation of adenylyl cyclase, which catalyzes the formation of cAMP from ATP, is controversial. Whereas Verghese et al. (1985b) and Bokoch (1987) reported that chemotactic peptides neither inhibit nor stimulate adenylyl cyclase in neutrophil plasma membranes, Saad et al. (1987) reported an inhibition of adenylyl cyclase by fMet-Leu-Phe at concentrations above 1 μM in a pertussis toxin-sensitive manner. It should be emphasized, however, that these concentrations of fMet-Leu-Phe are considerably higher than those required to induce neutrophil activation (Seifert et al. 1989a,b,d).

Neutrophils possess cAMP-dependent protein kinase, and several proteins are phosphorylated by this kinase (Tsung et al. 1975; Helfman et al. 1983; Huang et al. 1983a,b; Kramer et al. 1988a). In addition, neutrophils possess high- and low-affinity cAMP phosphodiesterases (Grady and Thomas 1986).

In intact phagocytes, fMet-Leu-Phe induces a transient increase in cAMP (Jackowski and Sha'afi 1979; Simchowitz et al. 1980a,b; Smolen et al. 1980; Pryzwansky et al. 1981; Verghese et al. 1985b; Elliott et al. 1986; Cronstein et al. 1988; Iannone et al. 1989). The chemotatic peptide-induced rise in cAMP is apparently not due to direct activation of adenylyl cyclase (Verghese et al. 1985b). The results of a recent study indicate that the increase in cAMP subsequent to stimulation with fMet-Leu-Phe is due to endogenous adenosine which may activate adenylyl cyclase through adenosine A_2 receptors (Iannone et al. 1989; see also Sect. 4.1.1.4).

It has been discussed whether the chemotactic peptide-induced in-crease in cAMP represents a stimulatory signal in neutrophil activation (Jackowski and Sha'afi 1979; Simchowitz et al. 1980a,b; Smolen et al. 1980; Pryzwansky et al. 1981). Under certain experimental conditions, the fMet-Leu-Phe-induced increase in cAMP and O_2^- formation are dissociated

(Smolen et al. 1980). In addition, the inhibitor of adenylyl cyclase, 9-(tetrahydro-2-furyl)adenine (SQ 22,536), blunts the fMet-Leu-Phe-induced increase in cAMP, whereas $O_2^{\cdot-}$ formation is not abolished (Harris et al. 1979; Simchowitz et al. 1983). Thus, an increase in cAMP is apparently not a critical signal in the activation of $O_2^{\cdot-}$ formation.

Other authors suggested that the chemoattractant-induced rise in cAMP is an inhibitory signal (Hopkins et al. 1983; Korchak et al. 1984d; Claesson and Feinmark 1984; Verghese et al. 1985b). Formation of prostaglandins of the E series subsequent to chemoattractant-induced release of arachidonic acid may contribute, at least in part, to cAMP-induced inhibition of the respiratory burst (Mallery et al. 1986; Bjornson et al. 1989; see also Sects. 3.2.4, 4.1.1.1).

3.2.6.2 The Role of cGMP

In addition to cAMP, cGMP plays a role in the regulation of certain cell functions (Goldberg and Haddox 1977; Waldman and Murad 1987; Tremblay et al. 1988). The formation of cGMP from GTP is catalyzed by soluble and particulate guanylyl cyclases, and cGMP activates a cGMP-dependent protein kinase. In addition, cGMP may directly modulate the activity of ion channels and phosphodiesterases (Waldman and Murad 1987; Tremblay et al. 1988).

Human neutrophils possess a soluble guanylyl cyclase which requires Mn^{2+} or Mg^{2+} and is stimulated by NO-generating compounds, e.g., sodium nitroprusside (Lad et al. 1985e). In intact neutrophils, fMet-Leu-Phe does not induce an increase in cGMP (Smolen et al. 1980). In comparison to protein kinase C, neutrophils possess a considerably lower activity of cGMP-dependent protein kinase (Helfman et al. 1983, Pryzwansky et al. 1990).

Stimulation of neutrophils with chemotactic peptides is accompanied by the association of cGMP-dependent protein kinase to specific components of the cytoskeleton (Pryzwansky et al. 1990). Human mononuclear phagocytes and neutrophils show both cAMP and cGMP phosphodiesterase activity (Prigent et al. 1990).

In comparison to protein kinase C, Ca^{2+}, and cAMP, the role of cGMP in the regulation of NADPH oxidase has been only very poorly studied and is obscure. Several years ago, exocytosis induced by A 231876 or opsonized zymosan in human neutrophils was suggested to be associated with an increase in cGMP (Ignarro and George 1974; Smith and Ignarro 1975). Human neutrophils and monocytes have been shown to possess high-affinity binding sites for muscarinic agonists (Dulis et al. 1979; Lopker et al. 1980), and muscarinic cholinergic agonists have been reported to enhance exocytosis and cGMP accumulation (Ignarro and George 1974; Smith and Ignarro 1975; Weissmann et al. 1975). In addition, carbachol has been reported to activate O_2^- formation in human neutrophils (Fülöp et al. 1988). We reexamined the latter issue and found that carbachol at concentrations between 10 nM and 10 μM does not activate O_2^- formation in human neutrophils in the presence or absence of CB. In addition, carbachol at these concentrations does not inhibit or stimulate O_2^- formation induced by fMet-Leu-Phe at 10 nM-1 μM (unpublished results; see also Sects. 1, 3.3.1.2.4, 6.2.4). The cell-permeant analogue of cGMP, dibutyryl cGMP, has been reported to inhibit chemoattractant-induced exocytosis and O_2^- formation (Fujita et al. 1984; Schröder et al. 1989), whereas dibutyryl cGMP does not affect zymosan-induced O_2^- formation in rat neutrophils (Smith et al. 1980). A differential sensitivity of various receptors to desensitization has also been observed with PMA (Bender et al. 1987; see also Sect. 3.2.2.5). We found that both dibutyryl cAMP and dibutyryl cGMP inhibit O_2^- formation induced by fMet-Leu-Phe in human neutrophils (Ervens et al. 1991). Dibutyryl cGMP is more effective than dibutyryl cAMP to inhibit O_2^- formation induced by fMet-Leu-Phe at a submaximally effective concentration (50 nM) but does not affect O_2^- formation induced by fMet-Leu-Phe at a maximally effective concentration (1 μM) (Ervens et al. 1991). In contrast, dibutyryl cGMP potentiates O_2^- formation induced by C5a at submaximally and maximally effective concentrations and dibutyryl cGMP antagonizes inhibition of O_2^- formation caused by dibutyryl cAMP. Dibutyryl cGMP inhibits PAF-induced O_2^- formation to a lesser extent than dibutyryl cAMP and has no effect on that induced by LTB$_4$. Dibutyryl cAMP and dibutyryl cAMP have no effect on O_2^- formation induced by NaF, γ-hexachlorocyclohexane, PMA, arachidonic acid, and A 23187 (Ervens et al. 1991). These data suggest that dibutyryl cAMP generally desensitizes chemoattractant-stimulated O_2^- formation (see also Sect. 4.1). Dibutyryl cGMP desensitizes fMet-Leu-Phe- and PAF-stimulated O_2^- formation but sensitizes C5a-induced O_2^- formation (see also Sect. 3.3.1.5.1). The lack of effect of cyclic nucleotides on O_2^- formation induced by agents other than receptor agonists indicates that cAMP and cGMP modulate early steps of signal transduction processes initiated by chemoattractants (Ervens et al. 1991).

Serotonin plays a role in the pathogenesis of inflammatory processes (Owen 1987) and has been reported to increase the concentration of cGMP in human monocytes (Sandler et al. 1975a,b; Williams et al. 1986). Serotonin enhances the PMA- induced respiratory burst in resident mouse peritoneal macrophages and in PU5-1.8-F7 macrophages (Silverman et al. 1985; see also Sect. 3.3.2.3). In human neutrophils, serotinin at concentrations up to 10 μM does not activate O_2^- formation in the presence or absence of CB, but serotonin shows a weak inhibitory effect on chemotactic peptide-induced O_2^- formation (Seifert, unpublished results).

Activation of the respiratory burst by GM-CSF, elastin peptides, methionine enkephalin, and tuftsin has been claimed to be accompanied by an increase in cGMP (Stabinsky et al. 1980; Fülöp et al. 1986; Foris et al. 1986; Coffey et al. 1988; see also Sects. 3.3.1.3.5.2, 3.3.1.4.3, 3.3.2.3). Finally, sodium nitroprusside has been reported to induce a respiratory burst in guinea pig peritoneal macrophages, but this effect has been suggested to be independent of guanylyl cyclase activation (Pick and Keisari 1981).

In addition to the cyclic purine nucleotides cAMP and cGMP, the cyclic pyrimidine nucleotide cCMP, which is an endogenous substance in mammalian cells, was suggested to modulate activation of NADPH oxidase (Ervens and Seifert 1991). This assumption is supported by the finding that a cell-permeant analogue of cCMP differentially modulates O_2^- formation in human neutrophils stimulated by various agents; that stimulated by fMet-Leu-Phe is inhibited, that stimulated by PAF and γ-hexachlorocyclohexane is potentiated, and that stimulated by NaF, A23187, PMA, and arachidonic acid is unaffected. Additionally, evidence was presented by these authors that cAMP, cGMP, and cCMP are functionally nonequivalent.

3.2.7 Protein Tyrosine Phosphorylation and Protein Phosphatases

During the past 2 years, substantial evidence has been accumulated that phosphorylation of tyrosine residues of proteins and dephosphorylation of proteins by (tyrosine) phosphatases play a role in the activation of the respiratory burst. In contrast, protein tyrosine phosphorylation may be less crucial for activation of actin polymerization and exocytosis (Trudel et al. 1990). Human neutrophils and HL-60 cells possess protein tyrosine kinase and phosphotyrosine phosphatase activity, and the latter enzyme is inhibited subsequent to stimulation with fMet-Leu-Phe or PMA (Kraft and Berkow 1987; Boutin et al. 1989). In rabbit neutrophils, chemotactic peptides induce tyrosine phosphorylation of various proteins in a pertussis toxin-sensitive manner, whereas PMA is inactive in this respect (C.K. Huang et al. 1988). In human neutrophils, both chemotactic peptides and phorbol esters have been reported to induce tyrosine phosphorylation of proteins (Gomez-Cambronero et al. 1989b). The inhibitor of protein tyrosine kinase α-cyno-3-ethoxy-4-hydroxy-5-phenylmethyl-cinnamamide (ST 638) inhibits the cytosolic but not the particulate protein tyrosine kinase in human neutrophils (Berkow et al. 1989). ST 638 has been reported to inhibit O_2^- formation induced by fMet-Leu-Phe, opsonized zymosan, and

NaF but not that induced by PMA or A 23187, suggesting that tyrosine phosphorylation of proteins plays a role in the G-protein-mediated activation of NADPH oxidase (Berkow et al. 1989; Gomez-Cambronero et al. 1989b). Formyl peptide-induced protein tyrosine phosphorylation may involve an increase in cytoplasmic Ca^{2+} and, at least in part, H-7-sensitive protein kinases (Huang et al. 1990). Erbstatin is an inhibitor of protein tyrosine kinases isolated from culture fluid of *Streptomyces viridosporus* (Naccache et al. 1990). This compound inhibits fMet-Leu-Phe-induced protein tyrosine phosphorylation, cytosolic acidification, and O_2^- formation, whereas actin polymerization and the increase in cytoplasmic Ca^{2+} and exocytosis are not inhibited by erbstatin. Additionally, erbstatin does not inhibit O_2^- formation stimulated by PMA and A 23187. In electropermeabilized human neutrophils, the stable GTP analogue and activator of G-proteins, guanosine 5'-0-[3-thio]triphosphate (GTP[γS]), activates oxygen consumption in the presence of Mg^{2+} and ATP (Nasmith et al. 1989; see also Sect. 5.1.4). In addition, GTP[γS] but not a cell-permeant diacylglycerol induces tyrosine phosphorylation of various proteins, suggesting that G-proteins are involved in the regulation of protein tyrosine kinases (Nasmith et al. 1989).

A role of protein (tyrosine) phosphatases in the regulation of NADPH oxidase is supported by the finding that ATP or adenosine 5'-0-[3-thio]triphosphate (ATP[γS]) is required for activation of the respiratory burst by fMet-Leu-Phe in electropermeabilized human neutrophils, and that activation of oxygen consumption in the presence of ATP but not in the presence of ATP[γS] is blocked upon addition of formyl peptide antagonists (Nasmith et al. 1989; Grinstein et al. 1989). In addition, ATP[γS] but not ATP or the nonphosphorylating analogue of ATP, adenosine 5'[β, γ-imido]triphosphate ([β, γNH]ATP), per se induces a respiratory burst in electropermeabilized human neutrophils. These data suggest that ATP[γS] induces thiophosphorylation and activation of regulatory proteins, and that these thiophosphoproteins are resistant to dephosphorylation (Eckstein 1985; Grinstein et al. 1989). Apparently, specific protein kinases are active in neutrophils in the absence of stimuli, and the accumulation of phosphoproteins but not that of thiophosphoproteins can be prevented by active protein phosphatases (Grinstein et al. 1989; see also Sects. 3.3.1.8; 5.1.4.3). This view is supported by the recent finding that the inhibitor of phosphatases, vanadate, induces a respiratory burst in electropermeabilized human neutrophils which process is associated with protein tyrosine phosphorylation (Grinstein et al. 1990).

Platelet-derived growth factor (PDGF) is an important intercellular signal molecule for the activation of various cell types of mesenchymal origin and may play a role in malignant cell transformation (Hunter and

Cooper 1985). The plasma membrane receptor for PDGF possesses protein tyrosine kinase activity (Ek and Heldin 1982; Hunter and Cooper 1985). The role of PDGF in the activation of the respiratory burst is controversial. Tzeng et al. (1984) reported that PDGF at physiologically relevant concentrations activates $O_2^{\bullet-}$ formation in human neutrophils to a similar extent as does fMet-Leu-Phe or C5a. In addition, PDGF has been shown to increase cytoplasmic Ca^{2+} and to induce exocytosis, adherence, and aggregation in neutrophils (Tzeng et al. 1984). In contrast, Nathan (1987) did not find a stimulatory effect of PDGF on the respiratory burst in adherent human neutrophils. We also did not observe stimulatory effects of PDGF on $O_2^{\bullet-}$ formation in suspended human neutrophils (unpublished results). Inhibitory effects of PDGF on the respiratory burst were also reported. Wilson et al. (1987) found that PDGF at picomolar concentrations inhibits $O_2^{\bullet-}$ formation in human neutrophils induced by chemoattractants, whereas the respiratory burst stimulated by PMA and arachidonic acid is not affected.

3.2.8 Proteases

Proteolytic processes are discussed to play a role in the regulation of the respiratory burst. Some experiments with exogenous proteases support a role of proteolytic processes in the activation of NADPH oxidase, and exogenous proteases may be of pathophysiological relevance as potentiators and/or activators of the respiratory burst in inflammatory processes. Neutrophil membranes possess chymotrypsinlike protease and neutral endopeptidase activity (Tsung et al. 1978; Duque et al. 1983; Painter et al. 1988). Exogenous cathepsin G, chymotrypsin, and elastase have been reported to potentiate fMet-Leu-Phe-induced $O_2^{\bullet-}$ formation in neutrophils (Kusner and King 1989). $O_2^{\bullet-}$ formation stimulated by PMA or arachidonic acid is potentiated by certain proteases as well (Kusner and King 1989). A monoclonal antibody which inhibits chymotrypsinlike proteases has been shown to inhibit fMet-Leu-Phe-induced $O_2^{\bullet-}$ formation (King et al. 1987). In addition, various exogenous proteases such as trypsin, chymotrypsin, pronase or papain, or the neutrophil proteases elastase and cathepsin G have been reported to enhance the respiratory burst in macrophages (Johnston et al. 1981; Speer et al. 1984). Finally, chymotrypsin and trypsin induce a respiratory burst in isolated rat glomeruli (Basci and Shah 1987; see Sect. 3.4.4.2.2).

The interpretation of studies dealing with the effects of protease inhibitors is complicated by the fact that some of these substances may show other effects than inhibition of proteases. Various inhibitors of proteases, e.g., phenylmethylsulfonyl fluoride (PMSF), tosyl-L-phenylalanyl chloromethyl ketone (TPCK), and aprotinin, have been reported to inhibit O_2^- formation induced by various stimuli including fMet-Leu-Phe and proteases in various types of phagocytes (Kitagawa et al. 1979, 1980a; Goldstein et al. 1979; Hoffman and Autor 1982; Duque et al. 1983; Basci and Shah 1987). However, the effects of chloromethyl ketones on O_2^- formation may be attributable to inhibition of SH groups rather than to inhibition of serine proteases (Tsan 1983). Recently, Conseiller and Lederer (1989) have suggested that the inhibitory effects of TPCK on O_2^- formation are not due to reduction of the cellular content of SH groups or to inhibition of proteases or protein kinase C but due to interference with protein components which are involved the maintenance of O_2^- formation. In fact, TPCK binds to a protein with an apparent molecular mass of 15 kDa in human neutrophils (Conseiller et al. 1990). However, the identity of this protein remains to be clarified. PMSF has been suggested to interfere with fMet-Leu-Phe-induced actin polymerization and O_2^- formation through protease-independent mechanisms as well (Rao and Castranova 1988). In addition soy bean trypsin inhibitor has been reported to inhibit O_2^- forma- tion (Abramovitz et al. 1983b; Basci and Shah 1987). However, it has been shown that superoxide dismutase present in soy bean trypsin inhibitor scavenges O_2^- and explains the "inhibition' of O_2^- formation (Abramovitz et al. 1983a). Finally, the protease-binding glycoprotein, α_2-macroglobulin, has been reported to inhibit O_2^- formation in murine peritoneal macro- phages (Hoffman et al. 1983; Sottrup-Jensen 1989).

3.3 Activation of NADPH Oxidase by Various Classes of Stimuli

The effects of activators of protein kinase C on the respiratory burst are described in Sect. 3.1.1 (see Table 2). Table 3 summarizes those activators of phagocytes which presumably act through pertussis toxin-sensitive and/or -insensitive G-proteins (see also Sects. 3.2.1, 3.3.1). Among the activators of the respiratory burst are peptides, proteins, lipid mediators, microbial agents, particulate agents, and drugs. Tables 5–9 summarize data concerning activation and/or potentiation of the respiratory burst by these agents.

Table 5. Activation of the respiratory burst by various peptides and proteins (I)

Agent	Effect (cell type)	Mechanisms discussed	Selected references
Chemoattractants and agents acting through related mechanisms			
Chemotactic formyl peptides	Activation of O_2^- formation (neutrophils, mononuclear cells)	Phosphoinositide degradation, activation of protein kinase C, Ca^{2+}-dependent processes, release of arachidonic acid, protein kinase C- and Ca^{2+}-independent pathways, activation of phospholipase D, rearrangement of the cytoskeleton, protein tyrosine phosphorylation, pH changes, membrane depolarization, direct activation by G-proteins, and/or low molecular mass GTP-binding proteins	Korchak and Weissmann (1978), Bromberg and Pick (1983), Pozzan et al. (1983), Simchowitz (1985a), Della Bianca et al. (1986a), Honeycutt and Niedel (1986), Jesaitis et al. (1986), Grzeskowiak et al. (1986), Badwey et al. (1989a), Bonser et al. (1989), Karnad et al. (1989), Nasmith et al. (1989)
	Priming for enhanced O_2^- formation (neutrophils)	Phospholipase C- and Ca^{2+}-independent mechanisms; role of low molecular mass GTP-binding proteins?	Bokoch and Parkos (1988), Didsbury et al. (1989), Quinn et al. (1989), Karnad et al. (1989)
Substance P	Activation and potentiation of O_2^- and H_2O_2 formation (neutrophils, macrophages)	Partial similarities to chemotactic peptides and cytokines; direct activation of G-proteins?	Serra et al. (1988), Perianin et al. (1989), Wozniak et al. (1989)
Gramicidin	O_2^- formation (neutrophils)	Similar to chemotactic peptides	Jacob (1988)
Elastin peptides	Activation of H_2O_2 formation (monocytes)	Phosphoinositide degradation, Ca^{2+} mobilization, similar to chemotactic peptides	Fülöp et al. (1986), Varga et al. (1989)
Synthetic lipopeptides [Pam3Cys-Ser(Lys)4]	Activation and potentiation of O_2^- formation (neutrophils)	Similarities and dissimilarities to chemotactic peptides; direct activation of G-proteins? Role of phospholipase C, protein kinase C and Ca^{2+} mobilization? Modulation of formyl peptide receptor expression?	Hauschildt et al. (1988b), Steffens et al. (1989), Seifert et al. (1990)

Table 5. (continued)

Agent	Effect (cell type)	Mechanisms discussed	Selected references
Wasp venom chemotactic peptides	Activation of O_2^- formation (neutrophils)	Similar to chemotactic peptides; direct activation of G-proteins?	Nagashima et al. (1990)
IgG immune complexes	Activation of H_2O_2 formation (neutrophils, macrophages)	Partial similarities to chemotactic peptides	Johnston et al. (1984), Young et al. (1984), Sato et al. (1987), Willis et al. (1988), Tosi and Berger (1988), Blackburn and Heck (1988), Shirato et al. (1988)
IgA immune complexes	H_2O_2 and O_2^- formation (neutrophils; monocytes)	Specific IgA receptors, synergism with complement, mechanism similar to IgG?	Gorter et al. (1987, 1989), Shen and Collins (1989)
Complement C5a	Activation of O_2^- formation (neutrophils, mononuclear cells)	Similar to chemotactic peptides	Gennaro et al. (1984), Deli et al. (1987), Wymann et al. (1987b), Banks et al. (1988), Shirato et al. (1988)
Complement C3b and C3bi	Controversial (neutrophils and mononuclear cells)	No effect or activation of the respiratory burst. Involvement of CR1 receptor which binds C3b and CR3 receptor which binds C3bi. Important role in adherence-mediated activation of the respiratory burst	Roos et al. (1981), Gordon et al. (1985), Hoogerwerf et al. (1990), Shappell et al. (1990), Entman et al. (1990)
NAP-1 (IL-8)	Activation of O_2^- formation (neutrophils)	Similar to chemotactic peptides	Thelen et al. (1988b), Baggiolini et al. (1989)
Cytokines			
TNF	Activation and potentiation of O_2^- formation (neutrophils)	Increase in cytoplasmic Ca^{2+}, alteration of formyl peptide receptor expression, activation of protein kinase C? Phosphorylation of a 64-kDa protein, actin polymerization, membrane depolarization; adherence of importance	Tsujimoto et al. (1986), Nathan (1987), Berkow and Dodson (1988), Atkinson et al. (1988), Kownatzki et al. (1988b), Berkow et al. (1989), Yuo et al. (1989a)

Table 5. (continued)

Agent	Effect (cell type)	Mechanisms discussed	Selected references
GM-CSF	Activation and potentiation of O_2^- formation (neutrophils)	Potentiation of fMet-Leu-Phe-induced membrane depolarization and release of arachidonic acid and diacylglycerol. Protein tyrosine phosphorylation, increase in cGMP, modulation of formyl peptide receptor expression, priming for Ca^{2+}-dependent activation of NADPH oxidase; adherence of importance, de novo protein synthesis	Weisbart et al. (1986), Sullivan et al. (1987, 1988, 1989a,b), English et al. (1988), Naccache et al. (1988a), Coffey et al. (1988), Gomez-Cambronero et al. (1989a,b), Edwards et al. (1989), Nathan (1989), Sha'afi et al. (1989), Tyagi et al. (1989b)
G-CSF	Activation and potentiation of O_2^- formation (neutrophils)	Increase in membrane depolarization, adherence of importance. Mechanism largely unknown	Kitagawa et al. (1987), Yuo et al. (1987, 1989b), Nathan (1989)
IL-6	Potentiation of O_2^- formation (neutrophils)	Not known	Borish et al. (1989)
IL-1α, IL-1β	Controversial (neutrophils)	No effect, activation or potentiation of the respiratory burst, mechanism not known	Georgilis et al. (1987), Ozaki et al. (1987), Sullivan et al. (1989)
IFN-α, IFN-β	Controversial (mononuclear cells, eosinophils)	No effect, stimulatory or inhibitory effects on the respiratory burst, mechanism unknown	Nathan et al. (1984), Garotta et al. (1986), Ding et al. (1988), Yoshie et al. (1989)
IFN-γ	Activation (macrophages) and potentiation (macrophages, neutrophils) of O_2^- and H_2O_2 formation	De novo protein synthesis, Ca^{2+} mobilization, activation of protein kinase C, increased expression of cytochrome b_{-245}, alterations of kinetic properties of NADPH oxidase, modulation of expression of binding sites for PMA and formyl peptide receptors	Cassatella et al. (1985, 1988), Hamilton et al. (1985), Hamilton and Adams (1987), Thelen et al. (1988a), Humphreys et al. (1989), Yoshie et al. (1989)

Table 6. Activation of the respiratory burst by various peptides and proteins (II)

Agent	Effect (cell type)	Mechanisms discussed	Selected references
Proteins with enzymatic activities			
Phospholipase C (exogenous, e.g., from *Clostridum perfringens* or *Bacillus cereus*)	Activation of O_2 consumption and H_2O_2 formation (macrophages, neutrophils)	Degradation of membrane phospholipids, activation of protein kinase C	Pick and Keisari (1981), Grzeskowiak et al. (1985), Styrt et al. (1989)
Phospholipase A_2 (exogenous)	Potentiation of $O_2^{\bullet-}$ formation in neutrophils	Arachidonic acid release	Lackie and Lawrence (1987)
Phospholipase A_2-activating protein	Activation of $O_2^{\bullet-}$ formation (neutrophils)	Arachidonic acid release	Bomalaski et al. (1989)
Neuraminidase (exogenous)	Potentiation of $O_2^{\bullet-}$ formation (neutrophils)	Removal of sialic acid, facilitation of Fc receptor-mediated activation	Henricks et al. (1982), Suzuki et al. (1982)
Proteases (exogenous, e.g., trypsin, chemotrypsin, pronase, cathepsin)	Controversial, potentiation or inhibition of $O_2^{\bullet-}$ formation (macrophages, neutrophils, mesangial cells)	Unknown	Johnston et al. (1981), Speer et al. (1984), Basci and Shah (1987), Kusner and King (1989)
Botulinum C2 toxin	Potentiation of $O_2^{\bullet-}$ formation and chemiluminescence (neutrophils)	ADP-ribosylation of actin, inhibition of actin polymerization, some similarities to CB	Al-Mohanna et al. (1987), Norgauer et al. (1988, 1989)
Adenosine desaminase	Potentiation of $O_2^{\bullet-}$ formation (neutrophils, eosinophils)	Removal of endogenous adenosine as an inhibitor of the respiratory burst	Schmeichel and Thomas (1987), Yukawa et al. (1989)

Table 6. (continued)

Agent	Effect (cell type)	Mechanisms discussed	Selected references
Miscellaneous peptides and proteins			
Laminin	Potentiation of $O_2^{\bullet-}$ formation (neutrophils)	Increase in formyl peptide receptor expression	Yoon et al. (1987), Pike et al. (1989)
Collagen	$O_2^{\bullet-}$ formation (neutrophils)	Unknown	Monboisse et al. (1987)
Tamm-Horsfall glycoprotein	Activation of chemiluminescence (neutrophils)	Similarities to IgG and C3 components, release of leukotrienes	Horton et al. (1990)
Opioid peptides, morphine, naloxone	Controversial (neutrophils, macrophages)	No effect, stimulatory or inhibitory effects on the respiratory burst, mechanism unknown	Simpkins et al. (1985, 1986), Sharp et al. (1985, 1987), Foris et al. (1986), Nagy et al. (1988), Seifert et al. (1989a)
Somatotropin	Potentiation of $O_2^{\bullet-}$ formation (macrophages)	Unknown	Edwards et al. (1988)
PDGF	Controversial (activation or inhibition of $O_2^{\bullet-}$ formation or no effect) (neutrophils)	Ca^{2+} mobilization, protein tyrosine phosphorylation?	Tzeng et al. (1984), Nathan (1987), Wilson et al. (1987)
Muramyl dipeptide, serotonin	Potentiation of $O_2^{\bullet-}$ formation (mononuclear cells)	Serotonin/muramyl dipeptide receptor activation?	Pabst and Johnston (1980), Pabst et al. (1982), Silverman et al. (1985)
Tuftsin	$O_2^{\bullet-}$ formation (macrophages), NBT reduction (neutrophils)	Ca^{2+} mobilization, changes in cyclic nucleotide concentrations?	Spirer et al. (1975), Fridkin et al. (1977), Stabinsky et al. (1980), Tritsch and Niswander (1982)
Fetal bovine serum	Potentiation of chemiluminescence (alveolar macrophages)	Cytokines and/or growth factors present in serum?	Hayakawa et al. (1989)

Table 7. Activation of the respiratory burst by receptor agonists and related agents

Agent	Effect (cell type)	Mechanisms discussed	Selected references
Lipid mediators			
PAF	Activation and potentiation of $O_2^{\bullet-}$ and H_2O_2 formation, (neutrophils, mononuclear cells)	Phosphoinositide degradation, Ca^{2+} mobilization, release of arachidonic acid, similar to chemotactic peptides; short-lasting activation	Naccache et al. (1986), Huang S.J. et al. (1988), Storch et al. (1988), Barzaghi et al. (1989), Tao et al. (1989)
LTB$_4$	Activation and potentiation of $O_2^{\bullet-}$ and H_2O_2 formation (neutrophils, mononuclear cells)	Similar to PAF, very short-lasting activation	Molski et al. (1984), Dewald and Baggiolini (1985, 1986), Holian (1986), Andersson et al. (1986b)
LTA$_4$	Potentiation of $O_2^{\bullet-}$ formation (neutrophils)	Conversion to LTB$_4$	Beckham et al. (1985)
LTC$_4$	Activation of $O_2^{\bullet-}$ and H_2O_2 formation (macrophages)	Similar to LTB$_4$, direct activation of protein kinase C?	Hartung (1983), Hansson et al. (1986), Shearman et al. (1989), Koo et al. (1989)
Lipoxin A	Activation of $O_2^{\bullet-}$ formation (neutrophils)	Receptor-mediated process? direct activation of protein kinase C?	Serhan et al. (1984), Hansson et al. (1986)
5-HETE	Potentiation of $O_2^{\bullet-}$ formation (neutrophils)	Receptor-mediated process? Ca^{2+} mobilization, modulation of protein kinase C activity?	O'Flaherty et al. (1985a), O'Flaherty and Nishihira (1987), Badwey et al. (1988)
Microbial agents			
LPS	Activation and potentiation of $O_2^{\bullet-}$ and H_2O_2 formation, (neutrophils, macrophages)	Modulation of formyl peptide receptor expression and membrane potential; synthesis of PAF, Ca^{2+} mobilization, phosphoinositide degradation, release of arachidonic acid, de novo protein synthesis, protein myristoylation	Wightman and Raetz (1984), Cooper et al. (1984), Kitagawa and Johnston (1985), Goldman et al. (1986), Aderem et al. (1986a), Hamilton and Adams (1987), Prpic et al. (1987), Worthen et al. (1988), Forehand et al. (1989)

Table 7. (continued)

Agent	Effect (cell type)	Mechanisms discussed	Selected references
Lipoteichoic acid (from *Streptococcus faecalis*)	Activation of O_2^- formation (monocytes)	Activation of phospholipase A_2, increase in cytoplasmic Ca^{2+}	Tarsi-Tsuk and Levy (1990)
Nonopsonized *Candida albicans* hyphae	Activation of O_2^- formation (neutrophils)	Phosphoinositide degradation, Ca^{2+} mobilization	Meshulam et al. (1988)
Opsonized zymosan	Activation of O_2^- and H_2O_2 formation and hexose-monophosphate shunt (macrophages and neutrophils)	Activation of complement and Fc receptors, phosphoinositide degradation, some similarities and dissimilarities to chemotactic peptides; role of protein kinase C controversial	Goldstein et al. (1975, 1976), Roos et al. (1981), Smith et al. (1984b), Ezekowitz et al. (1985), Deli et al. (1987), Andre et al. (1988), Bonser et al. (1989), Koenderman et al. (1989b,c)
Histamine (zymosan-bound)	Activation of O_2^- formation (macrophages)	Stimulation of H_1 receptors?	Diaz et al. (1979)
Nonopsonized zymosan	Activation of O_2^- and H_2O_2 formation and hexose-monophosphate shunt (murine macrophages and neutrophils)	Mannose/fucose-specific plasma membrane receptors?	Danley et al. (1981), Berton and Gordon (1983b), Sugar and Field (1988)
	Poor activation of respiratory burst (human neutrophils and macrophages)		Goldstein et al. (1975), Roos et al. (1981), Ezekowitz et al. (1985), Meshulam et al. (1988)

Table 7. (continued)

Agent	Effect (cell type)	Mechanisms discussed	Selected references
Miscellaneous agents			
Lectins (e.g., ConA)	Activation of $O_2^{\bullet-}$ and H_2O_2 formation and O_2 consumption (neutrophils, macrophages)	Similar to chemotactic peptides but differences with respect to Ca^{2+} mobilization, degradation of membrane phospholipids, and protein kinase C translocation; activation of pertussis toxin-insensitive G-proteins without Ca^{2+} mobilization. Signal transduction components specific for ConA, phosphorylation reactions	Cohen et al. (1980), Cohen et al. (1984), Rossi et al. (1986), Costa-Casnellie et al. (1986), Korchak et al. (1988a), Balsinde and Mollinedo (1988), Lu and Grinstein (1989)
Adenosine	Activation (macrophages) and potentiation (neutrophils) of $O_2^{\bullet-}$ formation	Stimulation of adenosine receptors? (A_1 receptors)?	Tritsch and Niswander (1983), Ward et al. (1988c), Salmon and Cronstein (1990)
Purine and pyrimidine nucleotides	Activation and potentiation of $O_2^{\bullet-}$ formation (neutrophils, HL-60 cells)	Phosphoinositide degradation, Ca^{2+} mobilization, translocation of protein kinase C, role of ectoprotein kinases? modification of functional state of cytosolic components of NADPH oxidase	Kuhns et al. (1988), Ward et al. (1988c), Dubyak et al. (1988), Seifert et al. (1989a,b), Dusenbery et al. (1989), Balazovich and Boxer (1990), Axtell et al. (1990)
NaF	Activation of $O_2^{\bullet-}$ formation (neutrophils)	Activation of G-proteins with subsequent phosphoinositide degradation and Ca^{2+} mobilization, direct activation of G-proteins without Ca^{2+} mobilization	Strnad and Wong (1985b), Strnad et al. (1986), English et al. (1986, 1989), Seifert et al. (1986), Toper et al. (1987), Della Bianca et al. (1988)

Table 8. Activation of the respiratory burst by ionophores, particles and agents interfering with the cytoskeleton

Agent	Effect (cell type)	Mechanisms discussed	Selected references
Ca²⁺ ionophores			
Ionomycin, A 23187	Activation and potentiation of O₂⁻ formation (neutrophils, mononuclear cells)	Ca²⁺ influx, activation of various Ca²⁺-dependent processes	Schell-Frederick (1974), Matsumoto et al. (1979), McPhail et al. (1981), Smolen (1984), Di Perri et al. (1984), Finkel et al. (1987), Koendermann et al. (1989a)
Particulate stimuli			
Latex particles	Activation of O₂ consumption, H₂O₂ formation (neutrophils)	Activation of protein kinase C? different from that of chemotactic peptides	Segal and Coade (1978), Hallett and Campbell (1983)
Asbestos	Activation and potentiation of O₂⁻ formation, (neutrophils, macrophages)	Ca²⁺ dependency, involvement of G-proteins? poorly understood processes	Donaldson and Cullen (1984), Cantin et al. (1988), Elferink and Ebbenhout (1988)
Quartz	Potentiation of O₂⁻ formation (macrophages)	Unknown	Cantin et al. (1988)
Urate crystals	Activation of O₂⁻ formation (neutrophils)	Ca²⁺ mobilization, synthesis of leukotrienes	Abramson et al. (1982), Simchowitz et al. (1982), Poubelle et al. (1987)
Agents interacting with the cytoskeleton			
Deuterium oxide	Potentiation of O₂⁻ formation (neutrophils)	Stabilization of microtubules?	Kitagawa and Takaku (1982)
Cytochalasins (e.g., CB)	Activation and potentiation of O₂⁻ formation (neutrophils and macrophages)	Inhibition of actin polymerization, modulation of formyl peptide receptor expression	Sugimoto et al. (1982), White et al. (1983a), Jesaitis et al. (1986), Omann et al. (1987a), Al-Mohanna et al. (1987)

Table 9. Activation of the respiratory burst by miscellaneous agents

Agent	Effect (cell type)	Mechanisms discussed	Selected references
Fatty acids	Activation and potentiation of $O_2^{\bullet-}$ formation (neutrophils, macrophages)	Hydrophobic interaction with plasma membrane, detergentlike effect, mimic effects of fatty acids released by activation of receptors, activation of phospholipase C, Ca^{2+} mobilization or protein kinase C activation, modulation of membrane fluidity	Kakinuma and Minakami (1978), Badwey et al. (1981, 1984), Curnutte et al. (1984), McPhail et al. (1984b), Murakami and Routtenberg (1985), K. Murakami et al. (1986), Seifert et al. (1988c), Lackie and Lawrence (1987)
Bile and bile salts (e.g. lithocholate)	Potentiation of $O_2^{\bullet-}$ formation (neutrophils)	Unknown interaction with plasma membrane components	Dahm and Roth (1990)
Digitonin	Activation of $O_2^{\bullet-}$ formation (neutrophils)	Binding to cholesterol, Ca^{2+} influx	Cohen and Chovaniec (1978a,b)
Lysophosphatides	Potentiation of $O_2^{\bullet-}$ formation (neutrophils)	Unknown	Ginsburg et al. (1989)
α, γ and δ-Hexachlorocyclohexane	Activation and potentiation of $O_2^{\bullet-}$ formation (neutrophils, macrophages)	Phosphoinositide degradation, Ca^{2+} mobilization	Holian et al. (1984), English et al. (1986), Kuhns et al. (1986, 1988)
Retinoids	Controversial (neutrophils, HL-60 cells)	No effect, stimulation, potentiation or inhibition of $O_2^{\bullet-}$ formation; modulation of membrane fluidity, activation or inhibition of protein kinase C, activation of phosphoinositide degradation	Camisa et al. (1982), Taffet et al. (1983), Ohkubo et al. (1984), Cooke and Hallett (1985), Badwey et al. (1986, 1989b), Lochner et al. (1986)
Mammalian lignan (2,3-dibenzylbutane-1,4-diol)	Potentiation of $O_2^{\bullet-}$ formation (neutrophils)	Activation of Ca^{2+}- and calmodulin-dependent processes	Morikawa et al. (1990)

Agent	Effect (cell type)	Mechanisms discussed	Selected references
1,25-Dihydroxy vitamin D_3	Priming for enhanced $O_2^{\bullet-}$ and H_2O_2 formation (monocytes, macrophages)	Induction of de novo protein synthesis?	M.S. Cohen et al. (1986), Gluck and Weinberg (1987), Gavison and Bar-Shavit (1989)
Room temperature versus 4 °C	Potentiation of $O_2^{\bullet-}$ formation (neutrophils)	Increased expression of formyl peptide receptors	Dahlgren et al. (1987), Tennenberg et al. (1988), English et al. (1988)
Exsudation from circulation	Potentiation of $O_2^{\bullet-}$ formation (neutrophils)	Increased expression of formyl peptide receptors	Zimmerli et al. (1986)
Clofazimine	Potentiation of $O_2^{\bullet-}$ formation (neutrophils)	Stimulation of arachidonic acid release	Anderson et al. (1988)
W-7	Activation of $O_2^{\bullet-}$ formation (macrophages)	Unknown	Holian et al. (1988)
Bleomycin	Potentiation of $O_2^{\bullet-}$ formation (macrophages)	Unknown	Conley et al. (1986)
Glycerol	Activation of $O_2^{\bullet-}$ formation (neutrophils, macrophages)	Unknown	Kaneda and Kakinuma (1986)
Thymol	Activation of $O_2^{\bullet-}$ formation (neutrophils)	Different from that of phorbol esters?	Suzuki et al. (1987), Suzuki and Furuta (1988)
Extracellular K^+	Controversial (neutrophils)	Activation or inhibition of the respiratory burst, membrane depolarization, Ca^{2+} mobilization?	Rossi et al. (1981a), Martin et al. (1988), Sullivan et al. (1989a)
Propionic acid	Activation of $O_2^{\bullet-}$ formation (primed neutrophils)	Ca^{2+} mobilization, acidification of cytosol	Naccache et al. (1988b), Sullivan et al. (1989a)
Vanadate	Oxygen consumption	Inhibition of protein tyrosine phosphatases, enhancement of protein tyrosine phosphorylation	Grinstein et al. (1990)

3.3.1 Agents Presumably Acting Through Plasma Membrane Receptors and G-Proteins

Among this group of activators of NADPH oxidase are bacterial formyl peptides and structurally related peptides, cytokines, extracellular matrix proteins, complement components and immunoglubulins, PAF, leukotrienes, and extracellular nucleotides (see Tables 3, 5–7).

3.3.1.1 Formyl Peptides

Bacteria initiate proteins synthesis with N-fMet, and formylated peptides have been purified from culture fluid of *Escherichia coli* and *Staphylococcus aureus* (Carp 1982; Marasco et al. 1984; Rot et al. 1987). Bacteria-derived N-formyl peptides are chemoattractants for neutrophils and mononuclear phagocytes, and fMet-Leu-Phe is probably the most extensively studied formyl peptide (Schiffmann et al. 1975; Bennett et al. 1980b). N-Formylated peptides are also derived from mitochondria of eukaryotic cells as they use N-fMet for initiation of protein synthesis as well (Carp 1982). Thus, formyl peptides may be of relevance as activators of the respiratory burst in vivo not only in bacterial infections but also in other processes which are associated with the destruction of endogenous cell structures (Carp 1982; see also Sect. 1). Various derivatives of formyl peptides, which are useful pharmacological tools to study the properties of formyl peptide receptors, have been synthesized (Showell et al. 1976, 1981; Kraus et al. 1984; Allen et al. 1986a).

3.3.1.1.1 Formyl Peptide Receptors

Formyl peptides bind to specific formyl peptide receptors in phagocytes (Williams et al. 1977; Sha'afi et al. 1978; Schiffmann et al. 1980; Zigmond and Tranquillo 1986; Walter and Marasco 1987). The formyl peptide receptor is a glycosylated 50- to 70-kDa molecule and has been solubilized, purified, and reconstituted into phospholipid vesicles (Niedel et al. 1980; Dolmatch and Niedel 1983; Baldwin et al. 1983; Hoyle and Freer 1984; Malech et al. 1985; Marasco et al. 1985; Allen et al. 1986b, 1989; Huang 1987). Recently, the cDNA sequence for the formyl peptide receptor which encodes a protein of 350 amino acids was isolated (Boulay et al. 1990). In analogy to β-adrenoreceptors and retinal rhodopsin (Sibley et al. 1987), the formyl peptide receptor apparently possesses seven hydrophobic membrane-spanning regions (Boulay et al. 1990).

Formyl peptide receptors undergo dynamic alterations upon occupation with agonist,and they have been identified not only in the plasma membrane but also in intracellular compartments, especially in specific granules (Fletcher and Gallin 1983; Gardner et al. 1986). This intracellular pool serves as a reserve of receptors, which may be translocated to the plasma membrane upon stimulation (Fletcher and Gallin 1983; Gardner et al. 1986). Following occupation with agonists, formyl peptide receptors become associated with the cytoskeleton and are internalized (Niedel et al. 1979; Sklar et al. 1984; Anderson and Niedel 1984; Painter et al. 1984). Internalized receptor-ligand complexes may then be transported to intracellular compartments and may be degraded (Niedel et al. 1979; Anderson and Niedel 1984; Painter et al. 1984; see also Sect. 3.2.5.1, 3.3.1.1.3).

3.3.1.1.2 Activation of $O_2^{\cdot-}$ Formation

Upon exposure to chemotactic peptides, neutrophils and mononuclear phagocytes of various species undergo a reversible respiratory burst (Holian and Daniele 1979, 1981; Yasaka et al. 1982; Dewald and Baggiolini 1985; Jesaitis et al. 1986; Seifert et al. 1989b). The extent of activation of the respiratory burst by chemotactic peptides is both species and cell type specific, and some types of phagocytes, e.g., bovine neutrophils, do not possess functional formyl peptide receptors (Gary et al. 1978; Styrt 1989). There is considerable interindividual variability in the effectiveness of formyl peptides to induce $O_2^{\cdot-}$ formation in human neutrophils (Seifert et al. 1991). At 37°C, fMet-Leu-Phe initiates $O_2^{\cdot-}$ formation in human neutrophils with a lag time of about 10 s, which time increases with decreasing temperature (Sklar et al. 1981a). Activation of $O_2^{\cdot-}$ formation requires the presence of formyl peptides at higher concentrations and higher percentages of receptor occupancy with agonist than those necessary for increase in cytoplasmic Ca^{2+}, membrane depolarization, and chemotaxis (Yuli et al. 1982; Sklar et al. 1984). Membrane fluidization by aliphatic alcohols such as pentanol and butanol increases the apparent affinity of formyl peptide receptors and enhances chemotaxis, whereas $O_2^{\cdot-}$ formation is depressed (Yuli et al. 1982; see also Sects. 3.2.2.1, 3.3.2.7). In contrast, the polyene antibiotic, amphotericin B, decreases the affinity of formyl peptide receptors and inhibits chemotaxis, whereas $O_2^{\cdot-}$ formation is unaffected (Lohr and Snyderman 1982). The high-affinity state of the formyl peptide receptor has been suggested to transduce chemotaxis, whereas the low-affinity state of the receptor has been suggested preferentially to mediate activation of NADPH oxidase (Yuli et al. 1982; Gallin and Seligmann 1984; see also Sect. 3.2.1.1). Interestingly, fMet-Leu-Phe per se induces changes in membrane fluidity, which may play a role in the regulation of receptor/cytoskeleton interactions, Ca^{2+} fluxes, and $O_2^{\cdot-}$ formation (Cherenkevich et al. 1982a,b; Valentino et al. 1986). Maintenance of $O_2^{\cdot-}$

formation requires continuous de novo formation of agonist-receptor complexes (Sklar et al. 1981b, 1984; Rossi et al. 1983). Finally, NH_3, which is metabolite of certain bacteria, has been reported to decrease the affinity of formyl peptide receptors for agonists (Coppi and Niederman 1989; see also Sect. 3.4.2.2).

3.3.1.1.3 Termination and Desensitization of O_2^- Formation

The mechanisms involved in termination of the respiratory burst are apparently complex (see also Sects. 3.1.1.4, 3.2.2.4, 3.2.2.5, 3.2.5, 6.2.1). O_2^- formation ceases while a substantial number of formyl peptide receptors remain occupied with agonist, and the addition of formyl peptide antagonists to phagocytes after initiation of the respiratory burst by agonists rapidly terminates this response (Rossi et al. 1983; Sklar et al. 1981b, 1984b; Seifert et al. 1989b; Grinstein et al. 1989). The respiratory burst is associated with the oxidation of methionine residues of neutrophil proteins, and oxidation of the methionine residue of fMet-Leu-Phe may play a role in the cessation of O_2^- formation, as the inhibition of degradation of chemotactic peptides enhances the respiratory burst (Clark 1982; Fliss et al. 1983; Rossi et al. 1983). fMet-Leu-Phe sulfoxide and fMet-Leu-Phe sulfone bind less avidly to formyl peptide receptors and are much less potent activators of O_2^- formation than fMet-Leu-Phe itself (Harvath and Aksamit 1984). In addition, proteolytic degradation of agonist may be involved in the termination of biological responses to chemotactic peptides (Yuli and Snyderman 1986; Painter et al. 1988).

The repeated exposure of neutrophils and differentiated HL-60 cells to chemotactic peptides shifts the concentration response curve for various cellular functions including O_2^- formation to the right and decreases the maximum extent of the response, a process referred to as homologous desensitization (English et al. 1981a; Seligmann et al. 1982; McPhail et al. 1984a; Lefkowitz and Caron 1986; Sibley et al. 1987; Seifert et al. 1989b; Lee et al. 1989; McLeish et al. 1989b). The maximum rate of O_2^- formation is considerably reduced when the agonist is presented over a period of several minutes, and desensitization to fMet-Leu-Phe is accompanied by a decrease in phosphoinositide degradation and Ca^{2+} mobilization (De Togni et al. 1985a,b). In human neutrophils, O_2^- formation is less sensitive to homologous desensitization than aggregation (Lee et al. 1989). Homologous desensitization of fMet-Leu-Phe-induced O_2^- formation in HL-60 cells is associated with a substantial decrease in the number of formyl peptide receptors without alteration in affinity (McLeish et al. 1989b). In addition, homologous desensitization is accompanied by a functional alteration in the interaction of formyl peptide receptors with G-proteins (Mc-

Leish et al. 1989b). Possibly, phosphorylation of formyl peptide receptors is involved in their uncoupling from G-proteins, as has been shown for other systems (Lefkowitz and Caron 1986; Sibley et al. 1987; Mueller and Sklar 1989; see also Sects. 3.2.2.5, 4.1.3). Desensitization is a reversible process as removal of the agonist restores responsiveness to formyl peptides in a time-dependent manner, indicating that the signal transduction components including NADPH oxidase are not irreversibly altered (English et al. 1981a).

In PMA- or fMet-Leu-Phe-stimulated rat neutrophils, the amount of O_2^- generated per cell is inversely related to the cell number in the assay cuvette, suggesting that NADPH oxidase activation is a self-limiting process, possibly due to the formation of H_2O_2 (Mege et al. 1986). Inactivation of protein kinase C by H_2O_2 may play a role in this process (Gopalakrishna and Anderson 1989). In contrast, Rajkovic and Williams (1985b) did not find an inhibitory effect of H_2O_2 on O_2^- formation.

Eklund and Gabig (1990) isolated a lipid thiobis ester from cytosol of unstimulated human neutrophils which deactivates NADHPH oxidase obtained from PMA- or opsonized zymosan-stimulated phagocytes. In addition, this ester deactivates NADPH oxidase in the cell-free system in a reversible manner (Eklund and Gabig 1990). The authors suggested that this compound may play a role as endogenous inhibitor of the respiratory burst (see also Sect. 3.3.1.1.3).

3.3.1.1.4 Sensitization of O_2^- Formation

In addition to desensitization, homologous sensitization of formyl peptide-induced O_2^- formation has been described. fMet-Leu-Phe at submaximally effective concentrations can prime for itself, leading to enhanced NADPH oxidase activation upon reexposure to the agonist (McPhail and Snyderman 1984; Pontremoli et al. 1989; Karnad et al. 1989; see also Sect. 3.3.1.1.3). Recent results indicate that pertussis toxin-insensitive signal transduction pathways may play a role in this homologous sensitization (Karnad et al. 1989).

Physical conditions may affect the extent of the respiratory burst induced by fMet-Leu-Phe, and these factors may be of considerable importance for the correct interpretation of experimental results (see also Sects. 1, 6.2). Neutrophils isolated at room temperature show increased expression of plasma membrane formyl peptide receptors in comparison to cells isolated at 4°C, and these differences correlate with the maximal rates of O_2^- generation (Tennenberg et al. 1988; Dahlgren et al. 1987). We have repeatedly observed that neutrophils isolated from buffy coat stored overnight at 4°C show a greater respiratory burst upon stimulation with fMet-

Leu-Phe than cells isolated from fresh buffy coat (unpublished results). Interestingly, the number of formyl peptide receptors in guinea pig and human exsudate neutrophils is severalfold higher than in the corresponding blood neutrophils, and this process is associated with priming for enhanced O_2^- formation upon exposure to the chemotactic peptides (Zimmerli et al. 1986).

3.3.1.2 Miscellaneous Peptides

3.3.1.2.1 Substance P

The tachykinin substance P is an undecapeptide, functions as neurotransmitter, and is present in certain peripheral endings of sensory neurones. Substance P has been suggested to play a role in the pathogenesis of neurogenic inflammation (Pernow 1983; Foreman and Jordan 1984; Foreman 1987; Wozniak et al. 1989). Substance P is structurally related to formyl peptides in its C-terminal portion. Substance P binds to specific plasma membrane receptors which can be divided into subtypes (Watson 1984, 1987; Regoli et al. 1987). High-affinity binding sites for substance P have been identified on guinea pig macrophages (Marasco et al. 1981). Substance P has been reported to bind to formyl peptide receptors in rabbit neutrophils, and antagonists at formyl peptide receptors compete with substance P (Marasco et al. 1981; Bonora et al. 1986; Watson 1987). In addition, a C-terminal formyl tetrapeptide analogue of substance P is a partial agonist at formyl peptide receptors with respect to activation of exocytosis (Bonora et al. 1986). Substance P has been reported to activate the respiratory burst in guinea pig macrophages and in human neutrophils (Hartung and Toyka 1983; Serra et al. 1988). The C-terminal octapeptide is a more effective activator of the respiratory burst than substance P itself, and the N-terminal fragment is inactive (Serra et al. 1988). The concentrations of substance P required to activate neutrophils are considerably higher than those occurring in vivo (Serra et al. 1988). However, substance P at concentrations lower than those required to activate NADPH oxidase potentiates the respiratory burst induced by fMet-Leu-Phe or C5a, suggesting that the tachykinin may act as priming agent rather than as activator of NADPH oxidase (Perianin et al. 1989; Wozniak et al. 1989; see also Sects. 3.3.1.6, 3.3.1.7, 3.3.1.8).

Activation of NADPH oxidase by substance P is accompanied by phospholipase C activation and Ca^{2+} mobilization. CB enhances substance P-induced phosphoinositide turnover and Ca^{2+} mobilization, but, somewhat surprisingly, CB inhibits the respiratory burst (Serra et al. 1988; see also

Sect. 3.2.5.1). Unlike the respiratory burst induced by fMet-Leu-Phe, that induced by substance P is only partially pertussis toxin-sensitive (Serra et al. 1988). These data indicate that substance P and chemotactic peptides activate NADPH oxidase by similar but not identical mechanisms. Interestingly, substance P at similarly high concentrations as those required for the activation of the respiratory burst has very recently been shown directly to activate G-proteins (Serra et al. 1988; Mousli et al. 1990). Thus, it is attractive to study in more detail the possibility that substance P activates neutrophils by other mechanisms than by "substance P receptors" (see also Sect. 3.3.1.2.6).

Priming of the respiratory burst by substance P is rapid in onset and reaches a maximum after 15–60 min (Wozniak et al. 1989). Priming by substance P is temperature dependent, is not abolished by removal of the agonist, and is accompanied by an increase in fMet-Leu-Phe-induced formation of lipoxygenase products of arachidonic acid (Wozniak et al. 1989; see also Sect. 3.2.4.1). Thus, priming of the respiratory burst by substance P shows some similarities to the effects of certain cytokines (see also Sect. 3.3.1.3.5).

3.3.1.2.2 Gramicidin

Gramicidins are linear pentadecapeptide ethanolamide antibiotics with a formyl group at the N-terminus (Bamberg et al. 1976). Gramicidin forms transmembrane ion channels, leading to depolarization and cell activation (Hladky and Haydon 1972; Bamberg et al. 1976; Jacob 1988). Recently, gramicidin from *Bacillus brevis* was reported to induce increase in cytoplasmic Ca^{2+}, exocytosis, and O_2^- formation in rabbit peritoneal neutrophils (Jacob 1988). Gramicidin is similarly potent but less effective than fMet-Leu-Phe. A competitive antagonist at formyl peptide receptors prevents activation by gramicidin, suggesting that this peptide is partial agonist at formyl peptide receptors (Jacob 1988).

3.3.1.2.3 Tuftsin

The tetrapeptide tuftsin (Thr-Lys-Pro-Arg) is part of a leukophilic γ-globulin and is released from the carrier molecules by proteases (Nishioka et al. 1972, 1973a; Najjar 1983; Goldman and Bar-Shavit 1983). Tuftsin has been purified and has been chemically synthesized (Nishioka et al. 1972, 1973a,b). Tuftsin regulates growth, phagocytosis, immunogenic responses, and motility of myeloid cells (Najjar and Nishioka 1970; Tzehoval et al. 1978; Goldman and Bar-Shavit 1983; Najjar 1983; Bump and Najjar 1988). The N-terminus of substance P is structurally related to tuftsin (Bar-Shavit et al.

1980; Serra et al. 1988). Neutrophils and monocytes possess high-affinity binding sites for tuftsin which cross-react with substance P, suggesting that tuftsin receptors may be considered a subtype of substance P receptors (Stabinsky et al. 1978; Bar-Shavit et al. 1980; Fridkin and Gottlieb 1981; Watson 1984). Tuftsin has been shown to activate O_2^- formation in murine macrophages with a biphasic concentration-response function (Tritsch and Niswander 1982; several years ago it was reported to stimulate NBT dye reduction in human neutrophils (Spirer et al. 1975; Fridkin et al. 1977), whereas the N-terminal fragment of substance P does not induce a respiratory burst in human neutrophils (Serra et al. 1988). Upon reexamination of this topic, we did not find a stimulatory effect of tuftsin up to $100 \, \mu M$ on O_2^- formation in human neutrophils, regardless of whether fMet-Leu-Phe or CB were present or not (unpublished results).

3.3.1.2.4 Opioid Peptides and Morphine

Opioid peptides have been suggested to be involved in the regulation of cell functions of the immune systeme (Foster and Moore 1987; Sibinga and Goldstein 1988). Human neutrophils have been reported to possess high-affinity binding sites for dihydromorphine (Lopker et al. 1980). Various opioid peptides, e.g., β-endorphin, dynorphin, and methionine enkephalin and morphine have been claimed to activate the respiratory burst in neutrophils and macrophages (Sharp et al. 1985, 1987; Foris et al. 1986; Nagy et al. 1988). The respiratory burst induced by opioids has been reported to be long lasting and to follow biphasic concentration-response functions. In contrast, morphine, methionine enkephalin, β-endorphin, and the opioid antagonist naloxone have also been reported to inhibit the fMet-Leu-Phe- or PMA-induced respiratory burst (Simpkins et al. 1985, 1986; Moon et al. 1986; Diamant et al. 1989). Other authors, however, reported that various opioids including morphine, β-endorphin, and methionine enkephalin show no stimulatory effect on the respiratory burst in human neutrophils and HL-60 cells (Diamant et al. 1989; Seifert et al. 1989a). In addition, we did not obtain any positive evidence for an inhibitory role of opioids in the regulation O_2^- formation in human neutrophils and HL-60 cells (Seifert et al. 1989a; see also Sects. 1, 6.2.4).

3.3.1.2.5 Somatotropin

The adenohypophyseal hormone somatotropin regulates growth processes. Recently, native and recombinant forms of somatotropin were shown to potentiate the opsonized zymosan-induced respiratory burst in porcine blood-derived mononuclear phagocytes (Edwards et al. 1988). Somatotropin is similarly effective as IFN-γ (see also Sect. 3.3.1.3.2), and the effects of somatotropin are abolished by treatment with an antibody specific for somatotropin (Edwards et al. 1988). In addition, administration of somatotropin to hypophysectomized rats at concentrations that significantly stimulate growth primes peritoneal macrophages for an enhanced respiratory burst (Edwards et al. 1988).

3.3.1.2.6 Lipopeptides

In addition to LPS, the outer cell wall of gram-negative bacteria contains lipoprotein (Braun 1975; see also Sect. 3.3.2.2). Native lipoprotein and synthetic lipopeptides are effective activators of lymphocytes and macrophages (Melchers et al. 1975; Bessler and Ottenbreit 1977; Hauschildt et al. 1988b; Reitermann et al. 1989; Deres et al. 1989; Steffens et al. 1989). Stimulation of B-lymphocytes by lipopeptides is apparently independent of phospholipase C and protein kinase C activation, whereas lipopeptide-induced activation of macrophages may involve both phospholipase C-dependent and -independent pathways (Hauschildt et al. 1988b; Steffens et al. 1989).

The synthetic lipoamino acid N-palmitoyl-S-[2,3-bis(palmitoyloxy)-($2RS$)-propyl]-(R)-cysteine (Pam$_3$Cys), which is derived from the N-terminus of bacterial lipoprotein, attached to (S)-seryl-(S)-lysyl-(S)-lysyl-(S)-lysyl-(S)-lysine [Pam$_3$Cys-Ser-(Lys)$_4$] (Reitermann et al. 1989), activates $O_2^{\cdot-}$ formation and lysozyme release in human neutrophils with an effectiveness amounting to about 15% of that of fMet-Leu-Phe (Seifert et al. 1990). In contrast, the lipopeptides Pam$_3$Cys-Ala-Gly, Pam$_3$Cys-Ser-Gly, Pam$_3$Cys-Ser, Pam$_3$Cys-OMe, and Pam$_3$Cys-OH do not activate $O_2^{\cdot-}$ formation, suggesting that positively charged amino acids are important for stimulatory effects of lipopeptides on NADPH oxidase (Seifert et al. 1990; see also below). Pertussis toxin inhibits Pam$_3$Cys-Ser-(Lys)$_4$-induced $O_2^{\cdot-}$ formation by 85%, whereas lipopeptide-induced exocytosis is pertussis toxin-insensitive (Seifert et al. 1990; see also below). $O_2^{\cdot-}$ formation induced by Pam$_3$Cys-Ser-(Lys)$_4$ and fMet-Leu-Phe is enhanced by CB, PMA, and R 59 022 (Seifert et al. 1990; see also Sects. 3.2.2, 3.2.5.1). Various activators of adenylyl cyclase and removal of extracellular Ca^{2+} differently inhibit $O_2^{\cdot-}$ formation by fMet-Leu-Phe and Pam$_3$Cys-Ser-(Lys)$_4$ (Seifert et al. 1990; see also Sects. 3.2.3, 4.1). Unlike $O_2^{\cdot-}$ formation induced by fMet-Leu-Phe, that induced by Pam$_3$Cys-Ser-(Lys)$_4$ is not augmented by PAF, UTP or TNF-α (Seifert et al. 1990). Pam$_3$Cys-Ser-(Lys)$_4$ synergistically enhances fMet-Leu-Phe-induced $O_2^{\cdot-}$ formation and primes neutrophils to respond to the chemotactic peptide at nonstimulatory concentrations

(Seifert et al. 1990). Evidence has been presented for the assumption that PAF and extracellular purine and pyrimidine nucleotides, on one hand, and lipopeptides, on the other, are functionally nonequivalent potentiators of the respiratory burst (Seifert et al. 1990; see also Sects. 3.3.1.6, 3.3.1.8).

The above data suggest that $Pam_3Cys-Ser-(Lys)_4$ activates neutrophils through G-proteins, involving pertussis toxin-sensitive and -insensitive processes (see also Sect. 3.2.1). The signal transduction pathways activated fMet-Leu-Phe and $Pam_3Cys-Ser-(Lys)_4$ are similar but apparently not identical.

The physiological role of lipoproteins in the regulation of neutrophil functions in vivo is not yet known, but in inflammatory processes bacterial lipoproteins and chemotactic peptides may interact synergistically to activate $O_2^{\bullet-}$ formation, leading to enhanced bactericidal activity.

There are certain structural and functional similarities between $Pam_3Cys-Ser-(Lys)_4$, on one hand, and substance P and mastoparan, on the other. These substances carry positive charges, and they mimic certain but not all aspects of receptor agonist-induced cell activation (Serra et al. 1988; Higashijima et al. 1988; Seifert et al. 1990; Mousli et al. 1990). In addition, both substance P and $Pam_3Cys-Ser-(Lys)_4$-induced activation of human neutrophils is partially pertussis toxin-insensitive (Serra et al. 1988; Seifert et al. 1990). These data raise the question whether lipopeptides interact directly with G-proteins as do substance P and mastoparan (Seifert et al. 1990b). Interestingly, the effects of mastoparan may also be partially pertussis toxin-insensitive (Higashijima et al. 1988), but to our knowledge there is no report in the literature dealing with the effects of mastoparan on the respiratory burst in intact phagocytes (see also Sect. 5.1.4). Surprisingly, we found that in cell-free systems derived from HL-60 cells mastoparan inhibits arachidonic acid-induced $O_2^{\bullet-}$ formation in the absence and in the presence of guanine nucleotides (unpublished results; see also Sect. 5.1.4).

Recently, certain wasp venom chemotactic peptides were reported to stimulate the respiratory burst and exocytosis in human neutrophils (Nagashima et al. 1990). Substitution of lysine for proline at the 7th position of these peptides results in a substantial loss of stimulatory activity, suggesting that cationic functions in these peptides impair the activation of NADPH oxidase (Nagashima et al. 1990). Whether or not activation of NAPDH oxidase by these peptides is mediated through plasma membrane receptors or through direct interaction with G-proteins remains to be clarified (see also Sect. 3.3.1.2.6).

Other cationic peptides play a role in the regulation of NADPH oxidase as well. Eosinophil granule major basic protein is a cationic polypeptide with an apparent molecular mass of 13.8 kDa which is localized in eosinophil granules (Moy et al. 1990). This protein activates the respiratory burst in neutrophils as assessed by chemiluminescence and $O_2^{\cdot-}$ formation. In addition, eosinophil granule major basic protein interacts synergistically with fMet-Leu-Phe or PAF to activate NADPH oxidase. Moy et al. (1990) suggested that the interaction between this protein and neutrophils contributes to the pathogenesis of some reactions in allergy.

3.3.1.2.7 Endothelin-1

Endothelin-1 is a very potent vasoconstrictor and enhances fMet-Leu-Phe-induced $O_2^{\cdot-}$ formation about twofold (Ishida et al. 1990). The priming effect of endothelin-1 requires an incubation time of 10 min to become evident and is apparently independent of an increase in cytoplasmic Ca^{2+}.

3.3.1.3 Cytokines

Cytokines are a heterogeneous group of peptide intercellular signal molecules which regulate functions of various cells of the immune system, including those of phagocytes, and are produced by a variety of cell types such as lymphocytes, mononuclear phagocytes, endothelial cells, and fibroblasts (Murray and Cohn 1980; Billingham 1987; Dinarello and Mier 1987; Martin and Resch 1988; Mizel 1989; Groopman et al. 1989).

In the past few years, much progress has been achieved concerning the role of cytokines in the regulation of NADPH oxidase. These studies have been greatly facilitated by the availability of human recombinant cytokines. The number of cytokines which is assumed to be involved in the regulation of the NADPH oxidase is increasing continuously. Interestingly, there is a substantial heterogeneity in the signal transduction pathways activated by various cytokines, and in many instances, the molecular mechanisms underlying the effects of cytokines on NADPH oxidase are still incompletely understood. Finally, the results of some studies suggest that stimulatory effects of cytokines on NADPH oxidase are of therapeutic relevance.

3.3.1.3.1 Interleukin-1

IL-1 is produced by mononuclear phagocytes and by neutrophils, activates T-lymphocytes, stimulates the production of other cytokines, and induces fever (Martin and Resch 1988; Canning and Neill 1989; Groopman et al. 1989). Human neutrophils possess high-affinity binding sites for IL-1 and internalize this cytokine (Parker et al. 1989). The role of IL-1 in the regulation of the respiratory burst is controversial. Some authors reported that IL-1 has no stimulatory effect on the respiratory burst in macrophages and neutrophils (Georgilis et al. 1987; Ding et al. 1988, Dularay et al. 1990), whereas others reported stimulatory effects of IL-1 on neutrophil activation, including potentiation of fMet-Leu-Phe-induced O_2^- formation (R.J. Smith et al. 1985, 1987; Sullivan et al. 1989). Recombinant IL-1α has been reported to induce H_2O_2 formation in human neutrophils and to potentiate the respiratory burst induced by opsonized zymosan, which latter effect is evident after a treatment for 10 min (Ozaki et al. 1987). In contrast, Sullivan et al. (1989) reported that recombinant IL-1β but not IL-1α primes neutrophils for enhanced O_2^- formation upon exposure to chemotactic peptides. Interestingly, IL-1α has been reported to induce a long-lasting respiratory burst in adherent human skin fibroblasts (Meier et al. 1989; see also Sect. 3.4.4.2.4).

3.3.1.3.2 Interferon-γ

IFN-γ is produced by stimulated T-lymphocytes and plays an important role in the activation of phagocytes (Lengyel 1982; Hamilton and Adams 1987). High-affinity binding sites for IFN-γ have been identified on mononuclear phagocytes and neutrophils (Celada et al. 1984; Hamilton and Adams 1987; Hansen and Finbloom 1990). Purified and recombinant IFN-γ prime mononuclear phagocytes of various species including man for an enhanced respiratory burst and formation of reactive nitrogen oxide intermediates (R-NO; Nathan et al. 1983, 1984; Murray et al. 1985a,b; Iyengar et al. 1987; Thelen et al. 1988a; Ding et al. 1988; see also Sect. 3.4.1). The enhanced respiratory burst in IFN-γ primed macrophages may be of importance for their antiprotozoal activity (Murray et al. 1985b). In human neutrophils, IFN-γ enhances the respiratory burst induced by ConA, fMet-Leu-Phe, immune complexes, and PMA (Berton et al. 1986c; Cassatella et al. 1988). In human monocytes, IFN-γ has been reported to potentiate O_2^- formation induced by PMA but not that induced by fMet-Leu-Phe (Thelen et al. 1988a).

The effects of IFN-γ are slow in onset and require incubation periods longer than 1 h (Berton et al. 1986c; Hamilton and Adams 1987; Perussia et al. 1987; Cassatella et al. 1988; Thelen et al. 1988a; Ding et al. 1988). On one hand, activation of phagocytes by IFN-γ has been suggested to involve Ca^{2+}- and protein kinase C-dependent mechanisms (Hamilton et al. 1985; Hamilton and Adams 1987). In U-937 cells, IFN-γ induces a rapid increase in cytoplasmic Ca^{2+} and the formation of inositol phosphates (Klein et al. 1990). However, the authors pointed out that additional mechanisms are likely to be involved in the activation of U-937 cells by IFN-γ. On the other hand, Cassatella et al. (1988) did not find alterations in the formation of inositol phosphates or changes in Ca^{2+} transients in IFN-γ-treated human neutrophils. In addition, Thelen et al. (1988a) did not find stimulatory effects of IFN-γ on cytoplasmic Ca^{2+} or on the cellular content of protein kinase C in human monocytes.

Priming of neutrophils by IFN-γ depends on the presence of serum in the incubation medium and is ablished by actinomycin and cycloheximide, suggesting that de novo protein synthesis is required for potentiation of the respiratory burst (Berton et al. 1986c; Cassatella et al. 1988). In fact, IFN-γ has been shown to induce the biosynthesis of specific proteins in human neutrophils (Humphreys et al. 1989; Rubin et al. 1989). Exposure of neutrophils, monocyte-derived macrophages, U-937 cells, and HL-60 cells to IFN-γ results in an increase in transcription of the β-subunit of cytochrome b_{-245}, and the effect of IFN-γ on β-chain expression is synergistically enhanced by TNF (Newburger et al. 1988; Cassatella et al. 1989b; see also Sects. 3.3.1.3.8, 6.1).

Cassatella et al. (1985) reported that treatment of human macrophages with IFN-γ is associated with a decrease in the K_m for NADPH of NADPH oxidase, whereas V_{max} is unaffected. In contrast, Thelen et al. (1988a) did not find changes in the affinity of NADPH oxidase for NADPH in IFN-γ-treated human monocytes. In addition, these authors did not observe changes in the cellular content of cytochrome b_{-245}. In human neutrophils, the cellular content of cytochrome b_{-245} and the kinetic properties of NADPH oxidase do not change upon treatment with IFN-γ (Cassatella et al. 1988). Treatment of the human eosinophilic cell line EoL-1 with IFN-γ results in an increase in the number of binding sites for phorbol esters and in enhanced expression of formyl peptide receptors (Yoshie et al. 1989; see also Sect. 3.4.4.1.1). These data indicate that the effects of IFN-γ on the respiratory burst are cell type specific.

3.3.1.3.3 Interferon-α and Interferon-β

On one hand, IFN-α and IFN-β have been reported to antagonize the stimulatory effects of IFN-γ on the respiratory burst in mononuclear phagocytes (Nathan et al. 1984; Garotta et al. 1986; Ding et al. 1988). In addition, IFN-α but not IFN-β has been reported to inhibit the transcription of the β-subunit of cytochrome b_{-245} (Newburger et al. 1988). On the other hand, IFN-α shows stimulatory effects on the respiratory burst in EoL-1 cells (Yoshie et al. 1989). Moreover, IFN-α and IFN-β have been reported to enhance LPS- and bacteria-induced priming of the respiratory burst in J774 murine macrophages (Tosk et al. 1989).

3.3.1.3.4 Tumor Necrosis Factor

Stimulated macrophages produce TNF-α, which has a molecular mass of 17 kDa, and which induces tumor cell killing, cachexia, and lethal shock. Activated B-lymphocytes secrete a structurally and functionally related cytokine, referred to as TNF-β or lymphotoxin (Aggarwal et al. 1985; Beutler and Cerami 1986, 1989; Urban et al. 1986; Klebanoff et al. 1986; Berkow and Dodson 1988; Ferrante et al. 1988; Imamura et al. 1988). Human neutrophils and U-937 cells possess high-affinity binding sites for TNF, and the TNF receptor may have a molecular mass of 100-120 kDa (Shalaby et al. 1987; Stauber and Aggarwal 1989). TNF is rapidly internalized following binding to plasma membrane receptors (Shalaby et al. 1987).

TNF enhances monocyte cytotoxicity and neutrophil phagocytosis, inhibits chemotaxis and promotes neutrophil adherence (Shalaby et al. 1985, 1987; Philip and Epstein 1986; Kharazmi et al. 1988; Kownatzki et al. 1988b, 1989). TNF has been reported to activate O_2^- formation in human neutrophils in a concentration-dependent manner (Tsujimoto et al. 1986; Yuo et al. 1989a). TNF-induced O_2^- formation is not accompanied by membrane potential changes, and Ca^{2+} mobilization and is inhibited by an increase in cAMP (Yuo et al. 1989; see also Sect. 4.1). In contrast, other authors reported that TNF does not substantially activate the respiratory burst in neutrophils (Klebanoff et al. 1986; Berkow et al. 1987c; Berkow and Dodson 1988; Ferrante et al. 1988). The ability of TNF to induce a respiratory burst in human neutrophils apparently depends on the state of adherence of the cells. Kownatzki et al. (1988b, 1989) and Neumann and Kownatzki et al. (1989) showed that TNF is a poor activator of O_2^- formation in suspended neutrophils, whereas the cytokine is a very potent and effective stimulus in adherent cells (see also Sect. 3.4.2). In human neutrophils and in EoL-1 cells, TNF potentiates O_2^- formation induced by various stimuli including opsonized zymosan, fMet-Leu-Phe, and PMA (Lebanoff et al. 1986; Berkow et al. 1987c; Larrick et al. 1987; Shalaby et al. 1987; Kharazmi et al. 1988; Berkow and Dodson 1988; Ferrante et al. 1988; Yoshie et al. 1989). In contrast, Yuo et al. (1989a) did not find a stimulatory effect of TNF on PMA-induced O_2^- formation in human neutrophils. The potentiating effect of TNF on O_2^- formation requires a preincubation period of approximately 15 min to become maximal, and a monoclonal antibody against TNF inhibits TNF-induced priming (Berkow et al. 1987c; Atkinson et al. 1988; Yuo et al. 1989a). In addition to neutrophils, TNF induces a respiratory burst in macrophages and primes these cells for enhanced O_2^- formation (Hoffman and Weinberg 1987; Ding et al. 1988). In EoL-1 cells, the effects of TNF on chemiluminescence are potentiated by IFN-γ and are additively augmented by IFN-α (Yoshie et al. 1989).

The mechanism by which TNF activates phagocytes is under current investigation and is not yet conclusively established. In HL-60 cells, TNF stimulates GTP[γS] binding to plasma membranes and stimulates a high-affinity GTPase in a pertussis toxin-sensitive manner (Imamura et al. 1988). We did not find a stimulatory effect of TNF-α on high affinity GTPase activity in membranes obtained from undifferentiated and dibutyryl cAMP-differentiated HL-60 cells (unpublished results). In addition, pertussis toxin blocks TNF-induced release of myeloperoxidase in human neutrophils, whereas TNF-induced lactoferrin release is only slightly inhibited by the toxin (Richter et al. 1989). Moreover, pertussis toxin does not inhibit TNF-α-induced H_2O_2 formation in adherent human neutrophils, suggesting that activation of the respiratory burst by TNF does not involve pertussis toxin sensitive G-proteins (Meurer and MacIntyre 1989; Berkow and Dodson 1988; see also Sect. 3.2.1).

Activation of O_2^- formation by TNF does not depend on extracellular Ca^{2+} but may require cytoplasmic Ca^{2+} (Tsujimoto et al. 1986; Richter et al. 1989). In contrast, Meurer and McIntyre (1989) reported that the effect of TNF-α depends on the presence of extracellular Ca^{2+}. Removal of TNF after priming does not abolish its potentiating effect (Ferrante et al. 1988). Activation of the respiratory burst by TNF in human neutrophils is not accompanied by the formation of inositol phosphates and release of arachidonic acid (Laudanna et al. 1990; see also Sects. 3.2.2.1, 3.2.3.1). TNF has been reported not to induce changes in formyl peptide receptor expression, in the kinetics of NADPH oxidase, or in protein kinase C activity (Berkow and Dodson 1988). Interestingly, a 64-kDa protein with unknown function has been reported to be phosphorylated upon stimulation with TNF (Berkow and Dodson 1988), and TNF may cause conversion of low- and high-affinity formyl peptide receptors to a single class of binding sites with intermediate affinity without a change in the number of plasma membrane formyl peptide receptors (Atkinson et al. 1988). Increases in F-actin and potentiation of the formyl peptide-induced membrane depolarization may be additional mechanisms by which TNF primes phagocytes for enhanced O_2^- formation (Berkow et al. 197c; Yuo et al. 1989a; see also Sect. 3.4.2.2).

3.3.1.3.5 Colony-Stimulating Factors

The colony-stimulating factors are a family of cytokines secreted by various cell types including lymphocytes, mononuclear phagocytes, endothelial cells and fibroblasts (Morstyn and Burgess 1988; Groopman et al. 1989). Among these cytokines are GM-CSF, granuloycte colony-stimulating factor (G-CSF), macrophage colony-stimulating factor (M-

CSF), and the multicolony-stimulating factor (IL-3). These cytokines are glycoproteins which regulate the production, differentiation, and functional maturation of precursor cells of the myeloid and monocytic lineage (Lopez et al. 1983; Gasson et al. 1984; Welte et al. 1985; Wong et al. 1985; Metcalf 1985, 1986; Souza et al. 1986; Groopman et al. 1989). In contrast to G-CSF and M-CSF, GM-CSF has dual effects on both cell lineages (Lopez et al. 1983; Gasson et al. 1984; Welte et al. 1985; Wong et al. 1985; Metcalf 1985, 1986; Souza et al. 1986; Groopman et al. 1989). Plasma membrane receptors for these intercellular signal molecules have been characterized in a variety of myeloid cell lines including human neutrophils and HL-60 cells (Walker and Burgess 1985; Nicola and Peterson 1986; Gasson et al. 1986; DiPersio et al. 1988). Most attention concerning the role of colony-stimulating factors in the regulation of NADPH oxidase has focused on the effects of GM-CSF in human neutrophils, whereas considerably less information is available on the corresponding effects of G-CSF, M-CSF, and IL-3.

M-CSF is a potentiator of the respiratory burst in macrophages (Wing et al. 1985). M-CSF and IL-3 have been reported not to activate the respiratory burst in adherent human neutrophils (Nathan 1989). Recently, Phillips and Hamilton (1990) showed that M-CSF inhibits priming of the respiratory burst in murine macrophages by various agents such as GM-CSF, TNF-α, IFN-γ, and LPS.

Granulocyte Colony-Stimulating Factor. G-CSF is a potent activator of the respiratory burst in adherent human neutrophils but not in adherent monocytes or in suspended neutrophils (Nathan 1989) (see also Sect. 3.4.3). In human neutrophils, G-CSF potentiates fMet-Leu-Phe- and lectin-induced O_2^- formation but not that induced by PMA or Ca^{2+} ionophores (Kitagawa et al. 1987; Yuo et al. 1987, 1989b; Ohsaka et al. 1989). The effect of G-CSF on O_2^- formation requires a preincubation time of only 5–10 min (Kitagawa et al. 1987; Yuo et al. 1987). G-CSF has been reported to stimulate agonist-induced membrane depolarization, but the cytokine apparently does not affect the cytoplasmic Ca^{2+} concentration, number, or affinity state of formyl peptide receptors (Yuo et al. 1989b). The effect of G-CSF is temperature dependent, is apparently independent of de novo protein synthesis, and is desensitized in a homologous manner (Yuo et al. 1989b; Ohsaka et al. 1989; see also Sect. 3.3.1.1.3).

Granulocyte/Macrophage Colony-Stimulating Factor. In neutrophils, GM-CSF inhibits migration and promotes adhesion, phagocytosis, and release of arachidonic acid metabolites (Arnaout et al. 1986; Fleischmann et al. 1986; Dahinden et al. 1988). GM-CSF potentiates O_2^- formation induced by fMet-Leu-Phe, C5a, PAF, and LTB$_4$ but not that induced by PMA or opsonized zymosan in these phagocytes (Weisbart et al. 1985, 1986, 1987; Lopez et al. 1986; Nathan 1989; Sha'afi et al. 1989; Mege et al. 1989). GM-CSF-induced potentiation of O_2^- formation is time dependent and requires 1–2 h to become maximal (Weisbart et al. 1985, 1986; English et al. 1988). Similar to G-CSF and TNF, GM-CSF is a very potent and effective activator of the respiratory burst in human neutrophils, which adhere to serum- or plasma-derived proteins or to the basement membrane protein laminin, whereas in suspended neutrophils the cytokine per se does not activate NADPH oxidase (Nathan 1989; see also Sect. 3.4.3).

Pertussis toxin-sensitive G-proteins may be involved in GM-CSF-induced activation of neutrophils as the toxin inhibits GM-CSF-stimulated expression of c-*fos* mRNA and GM-CSF-induced potentiation of Ca^{2+} mobilization in human neutrophils (McColl et al. 1989; Mege et al. 1989). Moreover, pertussis toxin inhibits GM-CSF-induced protein tyrosine phosphorylation and potentiation of arachidonic acid-induced O_2^- formation (Corey and Rosoff 1989; Gomez-Cambronero et al. 1989b; see also Sect. 3.2.7). Furthermore, plasma membranes from GM-CSF-treated neutrophils show higher basal and chemoattractant-stimulated GTPase activities than control membranes, and this effect is pertussis toxin-sensitive as well (Sha'afi et al. 1989; Gomez-Cambronero et al. 1989a). Finally, pertussis toxin has been reported to inhibit GM-CSF-induced release of lactoferrin and myeloperoxidase in human neutrophils (Richter et al. 1989).

The mechanism by which GM-CSF activates the respiratory burst is not associated with actin polymerization and does not take place in the presence of CB (Mege et al. 1989). GM-CSF is without effect in neutrophil cytoplasts, suggesting that granule and/or nucleus components are essential (Mege et al. 1989). As cycloheximide has been reported not to inhibit priming by GM-CSF, this process has been suggested not to depend on the de novo synthesis of proteins (Mege et al. 1989). In contrast, Edwards et al. (1989) showed that GM-CSF induces de novo synthesis of proteins in human neutrophils.

Priming by GM-CSF has been suggested to be independent of membrane potential changes, Ca^{2+} mobilization, phosphoinositide degradation, and translocation or activation of protein kinase C, but GM-CSF may prime neutrophils for enhanced diacylglycerol release and increase in cytoplasmic Ca^{2+} upon exposure to chemoattractants (Sullivan et al. 1987, 1989b; English et al. 1988; Naccache et al. 1988a; Mege et al. 1989; Corey and Rosoff 1989; Richter et al. 1989; Tyagi et al. 1989b). In addition, GM-CSF enhances fMet-Leu-Phe-induced release of arachidonic acid, suggesting that phospholipases play a role in the priming process (Corey and Rosoff 1989; see also Sect. 3.2.4). Enhanced activation of phospholipase D may be involved in priming of neutrophils by GM-CSF as well (Bourgoin et al. 1990). GM-CSF also enhances fMet-Leu-Phe-induced membrane depolarization and cell acidification (Sullivan et al. 1987, 1988; Naccache et al. 1988a). Naccache et al. (1988a) did not find an effect of GM-CSF on the intracellular pH. In contrast, Gomez-Cambronero et al. (1989a) reported that GM-CSF increases the intracellular pH in human neutrophils. Moreover, potentiation of $O_2^{\cdot-}$ formation by GM-CSF has been suggested to involve an increase in cGMP (Coffey et al. 1988; see also Sect. 3.2.6.2). Finally, GM-CSF has been reported to potentiate chemoattractant-induced release of PAF, which in turn, may potentiate $O_2^{\cdot-}$ formation as an autocrine signal molecule (Wirthmueller et al. 1989; Yamazaki et al. 1989; see also Sect. 3.3.1.6).

An increase in cytoplasmic Ca^{2+} per se has been suggested to be not sufficient substantially to activate the respiratory burst by chemoattractants (see also Sect. 3.2.3). Priming with GM-CSF alters the regulatory processes for NADPH oxidase activation in such a way that stimulation with Ca^{2+}-elevating agents results in a greatly enhanced respiratory burst (Naccache et al. 1988a,b; Sullivan et al. 1989a). Priming of phagocytes by GM-CSF does not require extracellular Ca^{2+} and is not reversible upon removal of the cytokine (English et al. 1988). Incubation of neutrophils with GM-CSF is associated with an increase in the number of low-affinity formyl peptide receptors (Weisbart et al. 1986). With respect to $O_2^{\cdot-}$ formation, circulating neutrophils are hyporesponsive to fMet-Leu-Phe, and the responsiveness increases with the expression of formyl peptide receptors after leaving the circulation (English et al. 1988; see also Sect. 3.3.1.1.4). This up-regulation of formyl peptide receptors is markedly enhanced by GM-CSF, suggesting that GM-CSF primes neutrophils for an enhanced respiratory burst, at least in part, via alteration in the expression of formyl peptide receptors. Additionally, priming by GM-CSF may involve functional alterations at the level of G-proteins, as this cytokine potentiates the NaF-induced respiratory burst in intact neutrophils and the one induced by GTP analogues in electropermeabilized cells (McColl et al. 1990). Results similar to those for GM-CSF were obtained with TNF-α (McColl et al. 1990; see also Sect. 3.3.1.3.4).

GM-CSF-induced cell activation is associated with the tyrosine phosphorylation of various proteins with molecular masses of 40–118 kDa (Sha'afi et al. 1989; Gomez-Cambronero et al. 1989b). The protein tyrosine phosphorylation patterns induced by fMet-Leu-Phe and GM-CSF are not identical, and the stimulatory effect of GM-CSF on $O_2^{\cdot-}$ formation is suppressed by the protein tyrosine kinase inhibitor ST 638 (Gomez-Cambronero et al. 1989b). A 40-kDa protein which is a substrate for tyrosine phosphorylation has been suggested to be $G_i\alpha_2$, and the 78- or

92-kDa substrate may be the GM-CSF receptor (Gomez-Cambronero et al. 1989b; see also Sect. 3.2.7).

In macrophages, GM-CSF also potentiates the respiratory burst (Reed et al. 1987). The signal transduction pathways activated by GM-CSF in neutrophils and in macrophages may be different. Unlike neutrophil activation, macrophage activation by GM-CSF is obviously unrelated to the stimulation of phospholipases (Corey and Rosoff 1989; Coleman et al. 1989). In addition, macrophage activation by GM-CSF is accompanied by an activation of adenylyl cyclase, whereas in neutrophils, GM-CSF apparently leads to an inhibition of adenylyl cyclase (Coffey et al. 1988; Coleman et al. 1989; see also Sects. 3.2.6.1, 4.1).

3.3.1.3.6 Interleukin-8
Neutrophil-activating peptide 1 (NAP-1, also referred to as IL-8) is a cytokine produced by various cell types including human mononuclear phagocytes (Peveri et al. 1988; Thelen et al. 1988b; Baggiolini et al. 1989). IL-8 consists of 72 amino acids and shows little homology to other cytokines (Walz et al. 1987; Lindley et al. 1988; Furuta et al. 1989). Human neutrophils possess low- and high-affinity binding sites for IL-8 which are different from the receptors for formyl peptides and GM-CSF (Besemer et al. 1989). IL-8 is an effective activator of human neutrophils, and it increases cytoplasmic Ca^{2+} and induces shape change, exocytosis, and O_2^- formation with similar kinetics as do chemotactic peptides (Thelen et al. 1988b). Similar to fMet-Leu-Phe, the respiratory burst induced by IL-8 is ihibited by pertussis toxin, staurosporine, and 17-hydroxy wortmannin (Thelen et al. 1988b; see also Sect. 4.3.1). These data suggest that fMet-Leu-Phe and IL-8 act via G-proteins and by similar signal transduction pathways. IL-8 is a more potent activator of neutrophils than fMet-Leu-Phe and may be of relevance in the pathogenesis of inflammatory processes (Baggiolini et al. 1989). In human monocytes, IL-8 is not stimulatory (Baggiolini et al. 1989).

3.3.1.3.7 Leukocyte Inhibitory Factor and Interleukin-6
Human leukocyte-inhibitory factor (LIF) is produced by activated lymphocytes and binds to neutrophil plasma membranes (Rocklin et al. 1981; Klempner and Rocklin 1982; Meshulam et al. 1982; Masucci et al. 1984). LIF stimulates adherence, phagocytosis, and fMet-Leu-Phe-induced chemotaxis and inhibits random migration (Borish and Rocklin 1985, 1987; Schainberg et al. 1988). In addition, LIF potentiates O_2^- formation induced fMet-Leu-Phe or a Ca^{2+} ionophore (Borish and Rocklin 1985; Borish et al.

1986). The mechanism by which LIF potentiates the respiratory burst may be explained, at least in part, by increased expression of formyl peptide receptors (Borish et al. 1986). Possibly, LIF is identical with one of the above-described cytokines.

Finally, recombinant IL-6 has recently been reported to stimulate exocytosis and to potentiate fMet-Leu-Phe-induced $O_2^{\bullet-}$ formation in neutrophils, but at present little information is available on the signal transduction processes involved (Borish et al. 1989).

3.3.1.3.8 Cytokines and the Respiratory Burst: Therapeutic Implications

The ability of monocytes of cancer patients to generate H_2O_2 upon exposure to PMA is unimpaired in comparison to thealthy subjects, but administration of human recombinant IFN-γ to cancer patients substantially enhances their monocytes' ability to undergo a respiratory burst (Nathan et al. 1985). Alveolar macrophages from patients with acquired immune-deficiency syndrom (AIDS) show no impaired ability to generate H_2O_2 in comparison to healthy subjects, and IFN-γ primes the AIDS patients' macrophages for an enhanced respiratory burst (Murray et al. 1985c). It has been suggested that IFN-γ is useful as a macrophage-activating agent in AIDS patients suffering from opportunistic infections and in certain cancer patients.

In some patients with "variant CGD" (see also Sect. 6.1.3), treatment with IFN-γ results in an increased transcription of the β-subunit of cytochrome b_{-245} and in an increase in V_{max} of NADPH oxidase, whereas the abnormal K_m for NADPH is not altered (Ezekowitz et al. 1987). Subcutaneous injection of IFN-γ to patients with X-chromosomal CGD has been reported to result in an increase in the cellular content of cytochrome b_{-245} and in a substantial and long-lasting enhancement of the phagocytes' ability to generate $O_2^{\bullet-}$ (Ezekowitz et al. 1988). IFN-γ may render myeloid progenitor cells capable of expressing at least in part, a corrected phenotype which is also present in the daughter cells (Ezekowitz et al. 1990). In addition, cultured monocytes of certain patients with X-chromosomal or autosomal recessive CGD acquire the ability to generate $O_2^{\bullet-}$ upon stimulation with PMA subsequent to treatment with IFN-γ (Sechler et al. 1988; see also Sect. 6.1.1).

The ability of colony-stimulating factors to enhance cytotoxic functions of neutrophils may be of therapeutic value in the treatment of life-threatening infections and for the augmentation of host defense in immunodepressed patients (Morstyn and Burgess 1988; Morstyn et al. 1989; Groopman et al. 1989). For example, in AIDS patients, GM-CSF has been reported to enhance phagocytosis and intracellular killing of bacteria (Baldwin et al. 1988). In addition, colony-stimulating factors may induce maturation of myeloid cells including their respiratory burst in patients with myeloid leukemia or myelodysplastic syndrom (Morstyn et al. 1989; Groopman et al. 1989; Geissler et al. 1989). Neutrophils of certain patients with myelodysplastic syndrome show a decreased respiratory burst activity upon exposure to fMet-Leu-Phe, which is significantly increased upon administration of G-CSF (Yuo et al. 1987). Moreover, administration of G-CSF to patients with malignant lymphoma results in a rapid and long-lasting enhancement of the neutrophils' capacity to generate $O_2^{\cdot-}$ upon stimulation with chemotactic peptides (Ohsaka et al. 1989).

3.3.1.4 Matrix Proteins

Extracellular matrix proteins may be of importance for the regulation of the respiratory burst when phagocytes leave the blood stream or when blood vessels are injured, resulting in the exposure of phagocytes to these extracelluar proteins. The effects of laminin, collagen, and elastin on the respiratory burst in phagocytes have been studied separately, but the extent to which these proteins interact is not yet known.

3.3.1.4.1 Laminin

Laminin is a 800- to 1000-kDa glycoprotein which is ubiquitously present in basement membranes (von der Mark and Kühl 1985). Laminin stimulates chemotaxis and adherence of neutrophils (Terranova et al. 1986). These processes are apparently mediated via specific plasma membrane receptors (Yoon et al. 1987). fMet-Leu-Phe and PMA stimulate the expression of laminin receptors on the plasma membrane by mobilization of intracellular receptor pools (Yoon et al. 1987). Conversely, laminin increases the number of formyl peptide receptors at the plasma membrane without changing their affinity state, which process is associated with enhanced fMet-Leu-Phe-stimulated $O_2^{\cdot-}$ formation (Pike et al. 1989; see also Sects. 3.3.1.3..5, 3.4.3).

3.3.1.4.2 Collagen

Collagens are a group of complex and structurally heterogenous matrix proteins (Bornstein and Sage 1980). Certain collagen degradation products are chemotactic for neutrophils (Laskin et al. 1986). In addition, the C-terminal peptide of the $\alpha 1(I)$ chain of collagen has been reported to activate $O_2^{\cdot-}$ formation, chemiluminescence, and exocytosis in human neutrophils (Monboisse et al. 1987). Activation of neutrophils by collagen depends on extracellular Ca^{2+}, but the signal transduction processes underlying the effects this matrix protein, e.g., the role of receptors, G-proteins, and protein kinases, has not yet been reported (Monboisse et al. 1987). Recent data show that two different peptide sequences in the C-terminal portion of the α1(I) chain of collagen are required to mediate neutrophil activation (Monboisse et al. 1990).

3.3.1.4.3 Elastin

Elastin is another extracellular matrix protein, and degradation of elastin may play a role in the pathogenesis of atherosclerosis and emphysema (Fülöp et al. 1986; Varga et al. 1989). Human phagocytes possess high-affinity binding sites for elastin peptides (Varga et al. 1989). Soluble elastin peptides, i.e., k-elastin, have been reported to stimulate phosphoinositide degradation, increase in cytoplasmic Ca^{2+}, and O_2^- formation in human neutrophils and mononuclear phagocytes through pertussis toxin-sensitive G-proteins (Fülöp et al. 1986; Varga et al. 1989).

3.3.1.5 Complement Components and Immunoglobulins

3.3.1.5.1 Complement C5a

C5a is glycoprotein fragment released from component 5 upon activation of the complement cascade and modulates numerous neutrophil functions (Gennaro et al. 1985; Johnson and Chenoweth 1985; Wymann et al. 1987b; Damerau 1987; Jose 1987; Banks et al. 1988; Shirato et al. 1988). Human neutrophils possess about $5–10 \times 10^5$ high-affinity C5a binding sites per cell, i.e., a number comparable to those of formyl peptide and LTB_4 receptors (Huey and Hugli 1985). The C5a receptor of human neutrophils may be a 48- to 52-kDa protein (Chenoweth and Hugli 1978; Johnson and Chenoweth 1985; Rollins and Springer 1985; Huey and Hugli 1985). Human eosinophils also possess C5a receptors, but they show properties different from those of human neutrophils (Gerard et al. 1989).

C5a is a very potent activator of the respiratory burst in human neutrophils, and its effectiveness is comparable to that of chemotactic peptides (Goldstein et al. 1975; Gennaro et al. 1984; Wymann et al. 1987b). Chemotactic peptides and C5a activate NADPH oxidase through similar mechanisms, but these aspects have been studied in greater detail with the former agents (see also Sects. 3.2.2, 3.2.3, 3.2.4, 3.2.5). The respiratory bursts induced by C5a, fMet-Leu-Phe, PAF, and LTB_4 show similarities with respect to kinetics (Wymann et al. 1987b). Similar to activation of O_2^- formation by chemotactic peptides, that induced by C5a depends on extracelullar Ca^{2+}, is associated with an increase in cytoplasmic Ca^{2+} and actin polymerization, and is pertussis toxin sensitive (Gennaro et al. 1984; Shirato et al. 1988; Banks et al. 1988). In contrast, concerning the regulation of

NADPH oxidase by cGMP, there are marked differences between fMet-Leu-Phe and C5a (Ervens et al. 1991; see also Sect. 3.2.6.2). There are additional differences in the regulation of formyl peptide and C5a receptors by activators of protein kinase C (Bender et al. 1987; see also Sect. 3.3.2.5). Moreover, there is a close correlation between the expression of formyl peptide receptors and receptors for C3b, C3bi, and IgG but not between expression of the latter receptors and C5a receptors (van Epps et al. 1990).

3.3.1.5.2 Complement C3b and C3bi

In addition to C5a, C3b and C3bi play a role in the regulation of phagocyte functions. C3b and C3bi bind to CR1 and CR3 receptors, respectively, and their expression is increased by warming of the cells and by stimulation with formyl peptides (Fearon 1980; Fearon and Collins 1983; Berger et al. 1984; Arnaout 1990; Hoogerwerf et al. 1990; van Epps et al. 1990). C3b components act as opsonins of particles and microorganisms and promote adherence and phagocytosis of these particles (Goldstein et al. 1976; Wright and Silverstein 1983; Berger et al. 1984, Andersson et al. 1988; Arnaout 1990; Hoogerwerf et al. 1990; van Epps et al. 1990). Particle-bound C3b and C3bi activate the respiratory burst in human neutrophils (Goldstein et al. 1975; Roos et al. 1981; Gordon et al.; Hoogerwerf et al. 1990), and inhibition by a neutrophil-specific monoclonal antibody of the respiratory burst induced by serum-opsonized zymosan may be due to altered expression of C3b receptors (Nauseef et al. 1983a). Macrophages have been reported to secrete C3b components, resulting in local opsonization of zymosan particles (Ezekowitz et al. 1985). Zymosan coated with C3b components by incubation with human macrophages has been shown to induce a respiratory burst in human neutrophils, whereas unopsonized zymosan is only a very poor activator of NADPH oxidase in these cells (Ezekowitz et al. 1985; see also Sect. 3.3.2.12.1). These data suggest that synthesis and secretion of complement components by macrophages play a role in the opsonization of pathogens and in the interaction of macrophages with neutrophils (Ezekowitz et al. 1985). In contrast to the above results, some authors also reported on a lack of stimulatory effect of C3b and C3bi on the respiratory burst in various types of phagocytes (Wright and Silverstein 1983; Gordon et al. 1985).

3.3.1.5.3 IgG

Immune complexes or particles opsonized with immune complexes induce phagocyte activation, e.g., phagocytosis, exocytosis, and O_2^- formation, with concomitant phosphoinositide degradation, release of arachidonic acid, and increase in cytoplasmic Ca^{2+} and protein phosphorylation (Goldstein et al.

1975; Johnston and Lehmeyer 1976; Johnston et al. 1976, 1984; Yamamoto and Johnston 1984; Green et al. 1984; Young et al. 1984; Sato et al. 1987; Willis et al. 1988; Tosi and Berger 1988; Blackburn and Heck 1988; Shirato et al. 1988). The protein phosphorylation patterns induced by PMA and immune complexes in murine peritoneal macrophages are similar (Johnston et al. 1984). Neutrophil activation by IgG complexes is less dependent on extracellular Ca^{2+} than that induced by C5a (Shirato et al. 1988).

Immunoglobulin concentrates for intravenous injection enhance activation of NADPH oxidase in neutrophils (Marodi et al. 1990). Monomeric human IgG potentiates fMet-Leu-Phe-induced $O_2^{\cdot-}$ formation in human neutrophils through a mechanism which is similar, at least in part, to the one of CB (Aaku et al. 1990; see also Sect. 3.2.5.1). Monomeric human IgG per se does not activate the respiratory burst and does not potentiate $O_2^{\cdot-}$ formation induced by PMA or ConA (Aaku et al. 1990). Moreover, IgG covalently coupled to polyacrylic acid activates $O_2^{\cdot-}$ formation in human neutrophils with an effectiveness comparable to that of PMA and opsonized zymosan (Klauser et al. 1990).

With respect to the role of G-proteins in the IgG-induced activation of the respiratory burst, the results are not consistent. The results of some studies suggest that the physical state of the immune complexes determines which type of signal transduction pathways is activated, as stimulation of the respiratory burst by soluble IgG aggregates, but not by surface-bound IgG, is pertussis toxin sensitive (Blackburn and Heck 1988, 1989; Shirato et al. 1988). In addition, surface-bound IgG stimulates binding of guanine nucleotides and high-affinity GTPase in neutrophil membranes in a pertussis toxin insensitive manner (Blackburn and Heck 1989; see also Sect. 3.2.1). These data suggest that activation of human neutrophils by surface-bound IgG involves pertussis toxin-insensitive G-proteins (Blackburn and Heck 1989). In contrast, Feister et al. (1988) found that IgG-induced $O_2^{\cdot-}$ formation but not exocytosis is inhibited by pertussis toxin.

Phagocytes possess various types of functionally nonequivalent plasma membrane receptors for the Fc region of IgG, and the nomenclature of these receptors is a subject of present discussion (Messner and Jelinek 1970; Silverstein et al. 1977; Fleit et al. 1982; Jones et al. 1985; Willis et al. 1988; Sato et al. 1987; Huizinga et al. 1988; Anderson and Looney 1986; Looney et al. 1986; Tosi and Berger 1988; Shirato et al. 1988; Blackburn and Heck 1988; Unkeless et al. 1988). For example, neutrophils have been reported to express $1\text{--}2 \times 10^4$ "FcII" (40 kDa) and $1\text{--}2 \times 10^5$ "FcIII" (50–80 kDa) receptors per cell (Huizinga et al. 1989). Apparently, the FcII receptor is involved in the activation of the respiratory burst by IgG, as neutrophils of patients with paroxysmal nocturnal hemoglobinuria show strongly reduced

FcIII receptor expression but normal FcII receptor expression and a normal respiratory burst upon stimulation with IgG immune complexes (Huizinga et al. 1989; see also Sect. 6.2.1). Cross-linking of Fc receptors is required for the activation of O_2^- formation, and its maintenance depends on the continuous de novo formation of cross-linked agonist/receptor complexes (Willis et al. 1988; Pfefferkorn and Fanger 1989a). In addition, there is a correlation between the number of cross-linked Fc receptors and NADPH oxidase activity (Pfefferkorn and Fanger 1989a). Subsequently, cross-linked Fc receptors become associated to the cytoskeleton, and deactivation of NADPH oxidase precedes internalization of Fc receptors (Pfefferkorn and Fanger 1989a,b; see also Sect. 3.2.5). Recent data from Crockett-Torabi and Fantone (1990) show that activation of the respiratory burst in human neutrophils by soluble immune complexes involves FcII and FcIII receptors, whereas activation by insoluble immune complexes involves only FcIII receptors. CB potentiates the effects of soluble immune complexes on O_2^- formation, and cAMP-increasing agents and pertussis toxin are inhibitory. By contrast, these substances show no substantial effect on O_2^- formation stimulated by insoluble immune complexes. In this context it should be noted that the stimulatory effects of the phosphatidyl inositol-linked, 55-kDa glycoprotein CD 14 on cytoplasmic Ca^{2+} and the respiratory burst in human neutrophils and monocytes apparently do not involve occupation of Fc receptors (Lund-Johansen et al. 1990). Finally, IgG fragments generated by the action of neutrophil elastase have been shown to inhibit O_2^- formation in these cells induced by PMA and fMet-Leu-Phe but not that triggered by opsonized zymosan (Eckle et al. 1990).

3.3.1.5.4 IgA

In addition to IgG, IgA has been suggested to play a role in the activation of the respiratory burst in human neutrophils. The presence of IgA receptors has been demonstrated on phagocytes of various species, and these receptors may be glycosylated 60-kDa proteins in human neutrophils (Gorter et al. 1988a,b; Albrechtsen et al. 1988). Heat-killed bacteria opsonized with IgA have been reported to induce H_2O_2 formation in human neutrophils, and IgA and complement components may synergistically activate the respiratory burst (Gorter et al. 1987, 1989). Recently, Shen and Collins (1989) have shown that IgA induces O_2^- formation in human monocytes, and that the effects of IgA are mediated through specific receptors.

3.3.1.5.5 Opsonized Particles

Opsonized particles, e.g., bacteria, fungi, zymosan, latex beads, and crystals, are effective activators of the respiratory burst and of phagocytosis in neutrophils and macrophages; the latter process is also referred to as opsonophagocytosis (Allen et al. 1972; Root et al. 1975; Goldstein et al. 1975; Roos et al. 1981; Gudewicz et al. 1982; Abramson et al. 1982; Hendricks et al. 1982; Ezekowitz et al. 1985; Green et al. 1987; Meshulam et al. 1988; Gbarah et al. 1989). Most studies concerning the mechanism of respiratory burst activation by particulate stimuli were carried out with opsonized zymosan (see also Sects. 3.2, 4.1, 4.2). Zymosan is a cell wall component of the yeast *Saccharomyces cerevisiae* and consists of the carbohydrate polymers β-glucan and α-mannan. Among the serum components which adhere to zymosan are complement components and immunoglobulins (Goldstein et al. 1975; Roos et al. 1981; Ezekowitz et al. 1985; Lambeth 1988).

Opsonized zymosan induces a sustained respiratory burst which is delayed in onset (Allen et al. 1972; Root et al. 1975; Goldstein et al. 1975; H.J. Cohen et al. 1981; Gudewicz et al. 1982; Smith et al. 1984b; Gennaro et al. 1984; Wymann et al. 1987b; Andre et al. 1988; Meshulam et al. 1988; Shirato et al. 1988; Banks et al. 1988; Lambeth 1988). Activation of the respiratory burst by opsonized latex particles depends on the concentration and size of the particles, and human neutrophils do not generate O_2^- until a critical ratio of particles to neutrophils is reached (Green et al. 1987). Above this critical value, O_2^- formation varies in a linear manner with the ratio of particles to cells (Green et al. 1987). In addition, the rate of O_2^- formation is a function of the square of the radius of the particles (Green et al. 1987).

Activation of O_2^- formation by opsonized zymosan is preceded by membrane depolarization, is inhibited by N-Ethylmaleimide (NEM) and TMB-8, has been reported to be potentiated by CB, depends on extracellular Ca^{2+}, and is desensitized in a homologous manner (H.J. Cohen et al. 1981; Smith et al. 1984; see also Sect. 3.3.1.1.3). In contrast, Elferink and Deierkauf (1989b) reported on inhibitory effects of CB on O_2^- formation induced by opsonized zymosan (see also Sect. 3.2.5.1). Opsonized zymosan increases cytoplasmic Ca^{2+} predominantly through influx from the extracellular space (Sawyer et al. 1985; Meshulam et al. 1988). Activation of macrophages and neutrophils by opsonized zymosan is associated with activation of phosphoinositide degradation and release of arachidonic acid (Waite et al. 1979; Garcia Gil et al. 1982; Emilsson and Sundler 1984; Leslie and Detty 1986; Meshulam et al. 1988). In addition, opsonized zymosan induces protein kinase C translocation in neutrophils (Deli et al. 1987; see also Sect. 3.2.2). Furthermore, sphingosine, H-7, and K252a inhibit the respiratory burst induced by opsonized bacteria in human neutrophils and macrophages (Gbarah et al. 1989).

Similar to the situation for chemotactic peptides, there are some doubts on the importance of protein kinase C in the activation of NADPH oxidase by opsonized zymosan. For example, rat and murine bone marrow-derived macrophages generate O_2^- only upon stimulation with opsonized zymosan and not upon stimulation with PMA (Andre et al. 1988; Phillips and Hamilton 1989). In addition, the zymosan-induced respiratory burst can be dissociated, at least in part, from phosphoinositide degradation (Koenderman et al. 1989b,c). Furthermore, the respiratory burst in human neutrophils induced by yeast opsonized with IgG or C3b components does not obligatorily depend on the activation of phospholipases A_2, C, or D or on an increase in cytoplasmic Ca^{2+} (Della Bianca et al. 1990; see also Sects. 3.2.2, 3.2.3, and 6.2). Both protein kinase C-dependent and -independent pathways may be involved in the activation of the respiratory burst by IgG-opsonized particles (Gresham et al. 1990). O_2^- formation induced by IgG-opsonized particles alone is apparently protein kinase C-independent, whereas the one induced by the combination of IgG-opsonized particles and cytokines may involve activation of protein kinase C (Gresham et al. 1990).

3.3.1.6 Platelet-Activating Factor

PAF is a mediator of inflammatory and hypersensitivity reactions and is synthesized in a various cell types including endothelium, platelets, mast cells, basophils, mononuclear phagocytes, and neutrophils (Hanahan 1986; Vargaftig and Braquet 1987; Braquet et al. 1987). Interestingly, PAF may induce PAF synthesis in human neutrophils (Yamazaki et al. 1989). Neutrophils and monocytes possess high-affinity binding sites for PAF, and GTP in a concentration-dependent manner decreases PAF binding (O'-Flaherty et al. 1986; Ng and Wong 1986, 1988).

PAF induces phosphoinositide degradation, release of arachidonic acid, Ca^{2+} mobilization, actin polymerization, and O_2^- formation in human neutrophils, eosinophils and mononuclear phagocytes (Shaw et al. 1981; Yasaka et al. 1982; Chilton et al. 1982; Naccache et al. 1986; S.J. Huang et al. 1988; Storch et al. 1988; Barzaghi et al. 1989; Kroegel et al. 1989; Randriamampita and Trautmann 1989; Uhing et al. 1989; Tao et al. 1989; Yamazaki et al. 1989; Parnham et al. 1989; Omann et al. 1989). PAF antagonists inhibit the stimulatory effects of PAF on phagocytes (Rouis et al. 1988; Dent et al. 1989; Barzaghi et al. 1989; Kroegel et al. 1989). Pertussis toxin inhibits the PAF-induced respiratory burst, whereas the PAF-induced activation of phospholipase C in differentiated U-937 cells and the increase in cytoplasmic Ca^{2+} in human monocytes are pertussis toxin-insensitive events (Lad et al. 1985c; Naccache et al. 1986; S.J. Huang et al. 1988; Ng and Wong 1989; Barzaghi et al. 1989). PMA has been reported to suppress or to enhance PAF-induced O_2^{-} formation (Naccache et al. 1985b; Gay et al. 1986; S.J. Huang et al. 1988; see also Sects. 3.2.2,2, 3.2.2.5).

PAF is a considerably less potent and effective activator of O_2^{-} formation in human neutrophils than fMet-Leu-Phe (Dewald and Baggiolini 1986; Gay et al. 1986; Seifert et al. 1989b). This finding is in agreement with the fact that PAF induces more transient activation of phospholipase C and increase in cytoplasmic Ca^{2+} than do chemotactic peptides (Naccache et al. 1986). PAF and fMet-Leu-Phe synergistically activate NADPH oxidase, suggesting that one important physiological function of PAF is to potentiate the effects of chemotactic peptides (Ingraham et al. 1982; Dewald and Baggiolini 1985; Gay et al. 1986; Seifert et al. 1991). This assumption is supported by the recent finding that PAF, generated by thrombin-stimulated endothelial cells, potentiates chemotactic peptide-induced O_2^{-} formation (Vercellotti et al. 1989). This interaction of endothelial cells and neutrophils may play a role in the pathogenesis of tissue injury during sepsis and other thrombin-generating disorders (Vercelotti et al. 1989). We found that uracil or adenine nucleotides may further increase the extent of NADPH oxidase activation induced by fMet-Leu-Phe plus PAF in dibutyryl cAMP-differentated HL-60 cells (unpublished results; see also Sect. 3.3.1.8). With respect to the interaction of PAF with zymosan, the results are controversial. Poitevin et al. (1984) reported that PAF enhances chemiluminescence induced by zymosan, whereas Gay et al. (1986) did not find a synergistic interaction between these stimuli. Priming for enhanced O_2^{-} formation by PAF is not inhibited by removal of the agonist and does not depend on the presence of extracellular divalent cations (Gay et al. 1984, 1986). The mechanism by which PAF primes neutrophils for enhanced O_2^{-} formation may involve increased expression of chemoattractant receptors, increased formation of diacylglycerol, and increase in cytoplasmic Ca^{2+} and Ca^{2+}-independent processes (Shalit et al. 1988; Koenderman et al. 1989a,b; see also Sects. 3.2.2, 3.2.3). Finally, PAF was suggested to play a role as intracellular signal molecule in the activation of NADPH oxidase by formyl peptides as is supported by the finding that antagonists at PAF receptors blunt the fMet-Leu-Phe-induced O_2^{-} formation in rabbit neutrophils (Stewart et al. 1990).

3.3.1.7 Products of the Lipoxygenase Pathway

Hydroperoxyeicosatetraenoic acids, hydroxyeicosatetraenoic acids (HETEs), leukotrienes, and lipoxines are probably involved in the pathogenesis of various diseases such as bronchial asthma, rheumatoid arthritis, and dermatitis (B. Henderson et al. 1987; Salmon and Higgs 1987; Samuelsson et al. 1987; Piper and Samhoun 1987; Barnes and Costello 1987; Barnes et al. 1988). These lipid mediators are synthesized by a variety of cell types including eosinophils, macrophages, monocytes, and neutrophils (Rouzer et al. 1980; Ford-Hutchinson et al. 1980; Sun and McGuire 1984; Verhagen et al. 1984; Goldyne et al. 1984; McIntyre et al. 1987; Haines et al. 1987).

3.3.1.7.1 Leukotrienes

Neutrophils, HL-60 cells, and mononuclear phagocytes possess specific binding sites for LTB_4 (Goldman and Goetzl 1982; Kreisle and Parker 1983; Lin et al. 1984, 1985; Goldman et al. 1985a; Sherman et al. 1988; Cristol et al. 1988). In addition, LTB_4 stimulates high-affinity GTPase activity in plasma membranes of myeloid cells in a pertussis and cholera toxin-sensitive manner (McLeish et al. 1989a; see also Sect. 3.2.1). Similar to fMet-Leu-Phe, LTB_4 stimulates GTP[γS] binding to plasma membranes of differentiated HL-60 cells, but the interaction of LTB_4 receptors with G-proteins is apparently different from that of formyl peptide receptors with G-proteins (McLeish et al. 1989a).

Volpi et al. (1984) suggested that LTB_4 does not stimulate phospholipase C. In contrast, a number of other studies showed that LTB_4 induces phosphoinositide degradation and increases cytoplasmic Ca^{2+} in neutrophils (White et al. 1983b; Lew et al. 1984a, 1987; Holian 1986; Andersson et al. 1986; Mong et al. 1986). Similar to formyl peptide receptors, LTB_4 receptors are associated to the cytoskeleton subsequent to occupancy with agonists (Naccache et al. 1984; see Sect. 3.2.5). In comparison to fMet-Leu-Phe, LTB_4 induces a short-lasting activation of phosphoinositide degradation and increase in cytoplasmic Ca^{2+} and actin polymerization, and it apparently does not induce membrane depolarization (Fletcher 1986; Lew et al. 1987; Omann et al. 1987b, 1989). Activation of phospholipase C and increase in cytoplasmic Ca^{2+} by LTB_4 are pertussis toxin-sensitive events (Molski et al. 1984; Mong et al. 1986; Holian et al. 1986; Andersson et al. 1986b).

LTB_4 induces a respiratory burst in neutrophils and in mononuclear phagocytes (Gagnon et al. 1989; Dewald and Baggiolini 1985, 1986; Seifert et al. 1989d). LTB_4 is a less effective activator of O_2^- formation than PAF in human neutrophils (Palmblad et al. 1984; Sumimoto et al. 1984; Gay et al. 1984; Prescott et al. 1984; Dewald and Baggiolini 1985, 1986; Fletcher 1986; Omann et al. 1987b; Seifert et al. 1989d, 1991). In analogy to fMet-Leu-Phe, activation of $O_2^{\cdot-}$ formation by LTB_4 depends on extracellular Ca^{2+}, is potentiated by CB, and is subject to homologous desensitization (Claesson and Feinmark 1984; Sumimoto et al. 1984). LTB_4 potentiates chemotactic peptide-induced $O_2^{\cdot-}$ formation, but LTB_4 is apparently less effective than PAF in this respect (Gay et al. 1984; Dewald and Baggiolini 1985; Fletcher 1986; Seifert et al. 1989d). We observed that $O_2^{\cdot-}$ formation synergistically induced by fMet-Leu-Phe plus LTB_4 is further stimulated by purine and pyrimidine nucleotides in dibutyryl cAMP-differentiated HL-60 cells (unpublished results; see also Sect. 3.3.1.8). Seifert et al. (1989b) reported on synergistic activation of NADPH oxidase by PAF and LTB_4, but Dewald and Baggiolini (1985) did not find a synergism between these stimuli. Moreover, LTB_4 has been reported not to affect $O_2^{\cdot-}$ formation induced by PMA or opsonized zymosan (Gay et al. 1984). The mechanism by which LTB_4 potentiates $O_2^{\cdot-}$ formation, apparently does not involve alterations in the number or affinity of formyl peptide receptors (Gay et al. 1984; Fletcher 1986).

In comparison to LTB_4, only very limited information is available on the effects of other leukotrienes on the respiratory burst. The unstable epoxide leukotriene A_4 (LTA_4) per se does not stimulate $O_2^{\cdot-}$ formation in human neutrophils but potentiates the fMet-Leu-Phe-induced respiratory burst (Beckham et al. 1985). LTA_4 is a less effective stimulus than LTB_4, supporting the view that LTA_4 rather serves as a precursor for LTB_4 (and LTC_4) than as intercellular signal molecule (Beckham et al. 1985).

The sulphidopeptide leukotrienes LTC_4 and LTD_4 activate phospholipase C and induce Ca^{2+} mobilization in phagocytes (Lew et al. 1987; Koo et al. 1989). Activation of neutrophils by LTC_4 and LTD_4 may involve plasma membrane receptors distinct from LTB_4 receptors (Koo et al. 1989; Thomsen and Ahnfelt-Ronne 1989). In addition, LTC_4 and LTD_4 receptors may couple functionally to pertussis toxin-sensitive G-proteins in these cells (Koo et al. 1989). LTC_4 at concentrations of $1-10$ μM has been reported to stimulate directly protein kinase C (Hansson et al. 1986), whereas Sherman et al. (1989) did not find substantial stimulatory effects of LTC_4 on the γ-isoenzyme of protein kinase C from bovine cerebellum. With respect to the respiratory burst, Hartung (1983) reported that LTC_4 activates NADPH oxidase in guinea pig macrophages, but this mode of activation of NADPH oxidase was not further characterized in detail.

3.3.1.7.2 Lipoxin A and 5-Hydroxyeicosatetraenoic Acid

Lipoxins are formed by the action of 5- and 15-lipoxygenase on arachidonic acid (Samuelsson et al. 1987). Lipoxin A at submicromolar concentrations has been reported to activate O_2^- formation in human neutrophils and to stimulate migration (Serhan et al. 1984; Palmblad et al. 1987). Activation of the respiratory burst by lipoxin A may be directly mediated by protein kinase C, as lipoxin A at concentrations of 1–3 μM has been reported to stimulate this kinase (Hansson et al. 1986; Sherman et al. 1989). Inhibitory effects of lipoxin A on fMet-Leu-Phe-induced phosphoinositide turnover in human neutrophils were also observed (Grandordy et al. 1990). A detailed characterization of the signal transduction mechanisms involved in lipoxin A-induced O_2^- formation, however, is still missing.

In addition to leukotrienes and lipoxins, 5-HETE may play a role in the regulation of O_2^- formation. Various HETEs themselves have little or no effect on O_2^- formation (Goetzl et al. 1980; Shak et al. 1983; O'Flaherty et al. 1985a; O'Flaherty and Nishihara 1987; Badwey et al. 1988). Among various HETEs, 5-HETE has been found to potentiate diacylglycerol- or phorbol ester-induced O_2^- formation (O'Flaherty et al. 1985a; O'Flaherty and Nishihara 1987; Badwey et al. 1988). Neutrophil activation by 5-HETE may be associated with Ca^{2+} mobilization and translocation of protein kinase C from the cytosol to the plasma membrane (O'Flaherty and Nishihara 1987). Badwey et al. (1988) reported that the effect of 5-HETE does not depend on extracellular Ca^{2+} and that synergistic activation of O_2^- formation by 5-HETE and phorbol esters is not associated with a redistribution of protein kinase C. However, 5-HETE stimulates binding of phorbol esters to intact neutrophils, and sphingosine and H-7 inhibit synergistic activation of O_2^- formation (Badwey et al. 1988). These data suggest that 5-HETE potentiates O_2^- formation by modulation of the activity of protein kinase C (see also Sect. 3.2.2.3).

3.3.1.8 Purine and Pyrimidine Nucleotides

Purine and pyrimidine nucleotides are released from various cell types such as neurones, chromaffin cells, platelets, and endothelium (Shirasawa et al. 1983; Butcher et al. 1986; Forsberg et al. 1987; Hardebo et al. 1987). In addition, nucleotides are released into the extracellular space under pathological conditions such as trauma, hypoxia, and cell death (Gordon 1986). Extracellular purine nucleotides interact with purinoceptors, which are classified according to the effectiveness of nucleotides to induce cell activation (Burnstock and Kennedy 1985; Gordon 1986; Williams 1987). The existence of purinoceptors in human myeloid cells is suggested by the finding that ATP binds to human neutrophils in a specific, reversible, and saturable manner (Balazovich and Boxer 1990). Moreover, extracellular nucleotides may also mediate their effects by other mechanisms than through plasma membrane receptors, e.g., by ectoprotein kinase-mediated phosphorylation of membrane proteins (Dusenbery et al. 1988).

In 1982, Ford-Hutchinson showed that ATP effectively induces aggregation of rat neutrophils, but the role of extracellular purine and pyrimidine nucleotides in the regulation of the respiratory burst remained unexplored until the past 2 years. Two recent studies showed that purine and pyrimidine nucleotides induce aggregation of human neutrophils as well (Freyer et al. 1988; Seifert et al. 1989d). In addition, ATP and UTP induce phosphoinositide degradation, release of arachidonic acid, Ca^{2+} mobilization from intracellular stores, and Ca^{2+} influx from the extracellular space in mononuclear phagocytes, human neutrophils, and HL-60 cells (Sung et al. 1985; Steinberg and Silverstein 1987; Greenberg et al. 1988; Kuhns et al. 1988; Dubyak et al. 1988; Cohen et al. 1989; Cockcroft and Stutchfield 1989a,b; Wenzel-Seifert and Seifert 1990). Furthermore, adenine nucleotides induce protein kinase C translocation and phosphorylation of endogenous proteins in human neutrophils (Balazovich and Boxer 1990). ATP also supports proliferation of hemopoietic stem cells in vitro (Whetton et al. 1988).

Various naturally occurring purine and pyrimidine nucleotides, especially ATP, ITP, GTP, and UTP, were recently found to potentiate fMet-Leu-Phe-induced O_2^- formation in human neutrophils and dimethyl sulfoxide-differentiated HL-60 cells (Kuhns et al. 1988; Ward et al. 1988c; Seifert et al. 1989a,b,d). Kuhns et al. (1988) and Seifert et al. (1989b,d) reported that extracellular nucleotides per se do not activate O_2^- formation in human neutrophils, whereas Kuroki and Minakami (1989) reported on a direct stimulatory effect of ATP on O_2^- formation in these cells (see also Sects. 1, 3.3.1.1.4). In dibutyryl cAMP-differentiated HL-60 cells, purine and pyrimidine nucleotides per se induce O_2^- formation (Seifert et al. 1989b). Activation of the respiratory burst by extracellular nucleotides in dibutyryl cAMP-differentiated HL-60 cells is reversible, depends on extracellular Ca^{2+}, is potentiated by CB and chemotactic peptides, and is inhibited by pertussis toxin, the latter finding suggesting that the effects of extracellular nucleotides are mediated via G-proteins (Seifert et al. 1989b). The stimulatory effects of purine and pyrimidine nucleotides on O_2^- formation are desensitized in a homologous manner (Seifert et al. 1989b,d; see also Sect. 3.3.1.1.3). A recent study showed that the effects of adenine nucleotides on O_2^- formation in human neutrophils do not depend on the presence of intracellular granules (Walker et al. 1989). Potentiation of fMet-Leu-Phe-induced O_2^- formation in human neutrophils by extracellular purines shows a specificity for nucleotides which is different from that of other known purinoceptors, i.e., P_{2X} and P_{2Y} purinoceptors (Burnstock and Kennedy 1985; Gordon 1986). Axtell et al. (1990) have put forward the interesting hypothesis that potentiation of fMet-Leu-Phe-induced O_2^- formation by adenine nucleotides is due to modification of cytosolic components of NADPH oxidase (see also Sect. 5.1.5). In rat alveolar macrophages, ATP induces an increase in cytoplasmic Ca^{2+} but does not prime the cells for enhanced O_2^- formation upon exposure to immune complexes (Hagenlocker et al. 1990). These results suggest that an increase in cytoplasmic Ca^{2+} is not sufficient to prime these macrophages for an augmented respiratory burst (Hagenlocker et al. 1990).

ATP-induced O_2^- formation in HL-60 cells is less sensitive to inhibition by pertussis toxin and cAMP-increasing agents than that induced by UTP (Seifert et al. 1989b; see also Sect. 4.1.3). In addition, adenine nucleotide-induced phospholipase C activation, release of arachidonic acid, increase in cytoplasmic Ca^{2+}, exocytosis, and potentiation of O_2^- formation show partial or complete pertussis toxin insensitivity in human neutrophils and HL-60 cells (Kuhns et al. 1988; Dubyak et al. 1988; Cockcroft and Stutchfield 1989a,b; Wenzel-Seifert and Seifert 1990; see also Sect. 3.2.1). However, with regard to the increase in cytoplasmic Ca^{2+} and exocytosis in dibutyryl cAMP-differentiated HL-60 cells, the effects of UTP are also only partially inhibited by pertussis toxin (Wenzel-Seifert and Seifert 1990). The specificity of adenine nucleotides and the corresponding uracil nucleotides to potentiate O_2^- formation in human neutrophils is also quite different (Seifert et al. 1989d). These data suggest that pyrimidine nucleotides do not activate myeloid cells through purinoceptors but through distinct pyrimidinoceptors (Seifert and Schultz 1989).

With regard to the physiological relevance of nucleotide-induced activation of phagocytes, studies reporting on the interaction of neutrophils with platelets are of particular interest. Platelets have been shown to potentiate fMet-Leu-Phe-induced O_2^- formation in human neutrophils, and ATP and ADP have been identified as the stimulatory factors released by platelets (Ward et al. 1988a,b). In contrast, McGarrity et al. (1988a,b, 1989) reported that platelet-derived adenine nucleotides inhibit fMet-Leu-Phe-induced O_2^- formation, and that the conversion of ATP and ADP to adenosine may be responsible, at least in part, for this effect (McGarrity et al. 1989; see also Sect. 4.1.1.4). Finally, Dallegri and collaborators (1989) did not obtain positive evidence for an inhibitory effect of platelets on opsonized zymosan-induced H_2O_2 formation in human neutrophils.

Apparently, the effects of platelets and of ATP on the respiratory burst in neutrophils depend critically on the experimental conditions employed. For example, platelets at low concentrations enhance the respiratory burst, whereas at high concentrations they are inhibitory (Naum et al. 1990). Additionally, the time of contact between neutrophils and platelets is an important determinant.

3.3.1.8.1 The Effects of Guanine Nucleotides

Guanine nucleotides potentiate O_2^- formation not only in intact neutrophils but also in cell-free systems (see also Sect. 5.1.4). As guanine nucleotides modulate the functional state of G-proteins (see Sect. 3.2.1), the question arises whether these nucleotides potentiate O_2^- formation in intact neutrophils *directly* through activation of G-proteins.

The nucleotide specificity for potentiation of O_2^- formation by guanine nucleotides in intact human neutrophils and in cell-free systems is quite different. In intact human neutrophils, GTP[γS], GTP, and guanosine 5'-O-[2-thio]diphosphate (GDP[βS]) are effective potentiators of fMet-Leu-Phe-induced O_2^- formation (Seifert et al. 1989d). Unexpectedly, GDP[βS] was found to enhance the stimulatory effect of GTP[γS] (Seifert et al. 1989d). In contrast, guanosine 5'-[β,γ-imido]triphosphate ([β,γ-NH]GTP) and guanosine 5'-[β,γ-methylene]triphosphate ([β,γ-CH$_2$]GTP) do not potentiate O_2^- formation in intact phagocytes (Seifert et al. 1989d).

In cell-free systems, GTP[γS] and [β,γ-NH]GTP are similarly effective potentiators of O_2^- formation, whereas [β,γ-CH$_2$]GTP and GTP are much less effective (Seifert et al. 1986, 1988b). In addition, GDP[βS] does not enhance O_2^- formation in cell-free systems but competitively antagonizes the stimulatory effects of GTP[γS] (Seifert et al. 1986). Moreover, guanine nucleotides are hydrophilic molecules which are unlikely to cross the plasma membrane (Seifert et al. 1989d). These data suggest that the effects of guanine nucleotides on O_2^- formation in intact human neutrophils are medated through purinoceptors and those of guanine nucleotides in cell-free systems directly through G-proteins (see also Sect. 5.1.4).

With respect to the effects of guanine nucleotides in intact cells, there are apparently certain differences between human and rabbit neutrophils. In intact rabbit neutrophils, GTP[γS] has been reported to activate O_2^- formation, whereas GTP, ATP, and GDP[βS] are inactive (Elferink and Deierkauf 1989a). Polyarginine, which permeabilizes plasma membranes and is an activator of the respiratory burst, potentiates the effect of GTP[γS], and GTP[γS] prevents lactate dehydrogenase release caused by polyarginine (Elferink 1988; Elferink and Deierkauf 1989a; Ginsburg et al. 1989). Pertussis toxin does not inhibit GTP[γS]-induced O_2^- formation in rabbit neutrophils and partially inhibits that induced by GTP[γS]plus polyarginine (Elferink and Deierkauf 1989a). GTP[γS] has been suggested to permeate the plasma membrane of rabbit neutrophils and to activate NADPH oxidase directly through G-proteins (Elferink and Deierkauf 1989a; Elferink et al. 1990a).

3.3.2 Miscellaneous Stimulatory Agents

3.3.2.1 Lectins

Plant lectins are polypeptides which bind to specific sugar residues of plasma membrane glycoproteins and induce cell activation presumably by cross-linking and immobilizing cell surface receptors (Barondes 1981; Lis and Sharon 1986; Perez et al. 1986). In membranes of human neutrophils and mononuclear phagocytes, various lectin-binding glycoproteins have been identified, e.g., members of the adhesion glycoprotein family and the 183-kDa "mannose receptor" (Ozaki et al. 1984; Christiansen and Skubitz 1988; see also Sect. 3.3.2.12.1). The effects of ConA in U-937 cells are mediated through a glycoprotein with an apparent molecular mass of 140 kDa (Balsinde and Mollinedo 1990). Very recent results show that stimulatory effects of ConA on NADPH oxidase in phagocytes are mediated by the CD11c antigen (Lacal et al. 1990). There are certain similarities and dissimilarities between activations of NADPH oxidase by lectins and chemotactic peptides.

ConA is probably the most extensively studied plant lectin with respect to the effects on NADPH oxidase, but other lectins, e.g., wheat germ agglutinin and phytohemagglutinin, activate the respiratory burst as well. Lectins induce a sustained and reversible respiratory burst in phagocytes, and the effect of ConA is antagonized by α-D-glucopyranoside and α-methyl-mannoside (Romeo et al. 1973, 1974; Cohen et al. 1980; Pick and Keisari 1981; H.J. Cohen et al. 1982, 1984; Lambeth 1988). The ConA-induced respiratory burst is not substantially inhibited by pertussis toxin (Verghese et al. 1985a; Rossi et al. 1986; Lad et al. 1986; Lu and Grinstein 1989). As GDP and GDP[βS] inhibit ConA-induced oxygen consumption in electropermeabilized human neutrophils, it has been suggested that pertussis toxin-insensitive G-proteins are involved in the signal transduction pathway (Lu and Grinstein 1989; see also Sect. 3.3.1.8.1).

Similar to fMet-Leu-Phe, ConA induces phosphoinositide degradation and an increase in cytoplasmic Ca^{2+} in human neutrophils (Rossi et al. 1986). The ConA-induced O_2^- formation is inhibited by removal of extracellular Ca^{2+} and is potentiated by CB (Cohen et al. 1980, 1984; Scully et al. 1986). We observed that ConA, unlike fMet-Leu-Phe, only marginally induces O_2^- formation in human neutrophils in the absence of CB (unpublished results). ConA interacts synergistically with other activators of the respiratory burst, e.g., with PMA and chemotactic peptides (Kitagawa et al. 1980b; Cohen et al. 1980; Dorio et al. 1987). In Ca^{2+}-depleted neutrophils, fMet-Leu-Phe plus ConA induce a respiratory burst without phosphoinositide degradation, suggesting that phospholipase C-independent processes are involved in the activation of NADPH oxidase by lectins (Rossi et al. 1986; see also Sects. 3.2.2, 3.2.3). Additionally, activation of phospholipase C, release of arachidonic acid, and an increase in cytoplasmic Ca^{2+} are apparently not critically involved in the activation of NADPH oxidase by ConA-opsonized particles (Rossi et al. 1989). In analogy to fMet-Leu-Phe, activation of the respiratory burst by ConA in electropermeabilized human neutrophils depends on ATP and Mg^{2+}, pointing to the importance of phosphorylation reactions in the signal transduction pathway (Lu and Grinstein 1989; Grinstein et al. 1989; see also Sects. 3.2.7, 5.1.4.3).

Unlike fMet-Leu-Phe, ConA increases cytoplasmic Ca^{2+} primarily via influx from the extracellular space and induces diacylglycerol release from a phospholipid pool which is different from that mobilized by chemotactic peptides (Korchak et al. 1988a). Somewhat unexpectedly, ConA has been reported to induce translocation of protein kinase C from the plasma membrane to the cytosol (Costa-Casnellie et al. 1986; see also Sect. 3.2.2.1). Moreover, ConA specifically activates the respiratory burst in PMA-differentiated U-937 cells (Balsinde and Mollinedo 1988).

3.3.2.2 Lipopolysaccharides

LPS are glycolipids present in the outer cell wall of gram-negative bacteria (Lüderitz et al. 1978; Braun 1975). LPS plays a role in the induction of the pathopysiological changes following infection with gram-negative bacteria, e.g., hypotensive shock, disseminated intravascular coagulation, and metabolic changes (Lüderitz et al. 1978; Ulevitch et al. 1984; Kitagawa and Johnston 1985; Goldman et al. 1986; Hamilton and Adams 1987; Worthen et al. 1988) The "lipid A" component of LPS is responsible for most of the biological effects of LPS (Hall and Munford 1983; Munford and Hall 1986). In phagocytes, LPS binds to the plasma membrane and stimulates phagocytosis, adherence, and release of arachidonic acid (Dahinden et al. 1983a; Cooper et al. 1984; Leslie and Detty 1986; Aderem et al. 1986b; Hamilton and Adams 1987). With regard to the respiratory burst, lack of effects, stimulatory and inhibitory effects of LPS have been reported.

On one hand, LPS has been reported to activate the respiratory burst in adherent but not in suspended neutrophils (Dahinden et al. 1983a,b; Seifert et al. 1990; see also Sect. 3.4.3). On the other, Nathan (1987) reported that LPS does not substantially activate H_2O_2 formation in adherent human neutrophils. Exposure of neutrophils to LPS for 30–60 min primes the cells for enhanced $O_2^{\bullet-}$ formation upon subsequent stimulation with PMA or fMet-Leu-Phe (Guthrie et al. 1984). In addition, LPS primes mononuclear phagocytes for an enhanced respiratory burst (Johnston et al. 1978; Sasada and Johnston 1980; Pabst and Johnston 1980; Pabst et al. 1982; Cooper et al. 1984; Kitagawa and Johnston 1985).

The molecular mechanism by which LPS potentiates $O_2^{\bullet-}$ formation is only incompletely understood, and the effects of LPS may be cell type-specific. In human neutrophils, priming of the respiratory burst by LPS is pertussis toxin-insensitive (Forehand et al. 1989). In contrast, pertussis toxin has been reported to inhibit LPS-induced activation of murine P388D₁ macrophages and LPS-induced cytokine production in U-937 cells (Jakway and DeFranco 1986; Daniel-Issakani et al. 1989). In U-937 cells, LPS may reduce or enhance pertussis toxin-catalyzed ADP ribosylation of G_{i2}, and LPS has been reported to induce phosphorylation of G_{i2} in these cells (Daniel-Issakani et al. 1989). These data suggest that pertussis toxin-sensitive and -insensitive signal transduction pathways are involved in the activation of phagocytes by LPS.

In macrophages and B-lymphocytes, LPS has been reported to induce phosphoinositide degradation, Ca^{2+} mobilization, release of arachidonic acid, alterations in gene expression and protein synthesis (Cooper et al. 1984; Rosoff and Cantley 1985; Leslie and Detty 1986; Hamilton and Adams 1987; Prpic et al. 1987). In addition, lipid A has been reported directly to activate protein kinase C in RAW 264.7 macrophages (Wightman and Raetz 1984). Interestingly, LPS induces myristoylation of a 68-kDa protein in macrophages (Aderem et al. 1986a). Myristoylation of the 68-kDa protein may augment subsequent phosphorylation of this protein by protein kinase C (Rosen et al. 1989). Finally, the LPS-induced generation of prostaglandins with subsequent increase in cAMP has been suggested to be involved in the activation of murine peritoneal macrophages (Benninghoff et al. 1989; see also Sects. 3.2.6.1, 4.1).

Alterations in the expression of formyl peptide receptors and enhanced synthesis of PAF are additional mechanisms to explain the stimulatory effects of LPS on neutrophils (Kitagawa and Johnston 1985; Goldman et al. 1986; Worthen et al. 1988). LPS enhances chemotactic peptide-induced actin polymerization in human neutrophils (Howard et al. 1990). The role of membrane potential changes in LPS-induced activation of phagocytes is controversial (Larsen et al. 1985; Forehand et al. 1989). Priming of human neutrophils by LPS depends on Ca^{2+} mobilization but apparently does not depend on alterations in protein kinase C activity or in the cellular content of cytochrome b_{-245} (Forehand et al. 1989). Moreover, priming with LPS is not associated with a change in the K_m for NADPH of NADPH oxidase (Forehand et al. 1989). Finally, activation of the respiratory burst in neutrophils by LPS has been reported to require the presence of serum (Wilson et al. 1982).

LPS prevents potentiation of the respiratory burst by IFN-γ and TNF-α in murine peritoneal macrophages (Ding and Nathan 1987; see also 3.4.1). Inhibitors of cyclooxygenase partially antagonize the inhibitory effect of LPS, whereas prostaglandins and dibutyryl cAMP mimic its effects (Ding and Nathan 1987). These data suggest that cyclooxygenase products formed in response to LPS increase cAMP and thus inhibit H_2O_2 formation (see also Sect. 4.1). LPS also inhibits immune complex-induced H_2O_2 formation in murine peritoneal macrophages treated with IFN-γ but not that in untreated cells (Johnston et al. 1985). Finally, in cultured human blood monocytes, LPS inhibits the respiratory burst induced by opsonized particles or by PMA, and glucocorticoids partially block the inhibitory effects of LPS (Rellstab and Schaffner 1989; see also Sect. 4.2.1).

During the past few months, some interesting new data concerning modulation of the respiratory burst by LPS have been published. Heiman et al. (1990) reported that a nontoxic derivative of lipid A, monophosphoryl lipid A, inhibits LPS-induced priming of neutrophils for enhanced O_2^- formation. Monophosphoryl lipid A may inhibit the binding of LPS to cellular binding sites. Kharazmi et al. (1990) showed that the various types of LPS isolated from *Pseudomanoas aeruginosa* strains are functionally nonequivalent with respect to their effects on the respiratory burst. These data suggest that the chemical composition of LPS critically determines its biological effects on NADPH oxidase. Aida and Pabst (1990) reported that plasma is required for priming by LPS of the respiratory burst in human neutrophils. These authors suggested that plasma prevents inactivation of LPS. Cassatella et al. (1990) studied the effect of LPS on gene expression in human neutrophils, suggesting that enhanced expression of the β-subunit of cytochrome b_{-245} accounts for the stimulatory effect of LPS on the respiratory burst. Like LPS, IFN-γ enhances expression of the β-subunit of the cytochrome as well (Cassatella et al. 1990; see also Sect. 3.3.1.3.2). In human neutrophils, LPS and TNF-α induce the synthesis and myristoylation of a 82-kDa protein (Thelen et al. 1990). This protein is apparently the neutrophil homologue of the *m*ristoylated, *a*lanine-*r*ich *C*-kinase substrate, referred to as "MARCKS", which is present in various other cell types. LPS and TNF-α do not induce phosphorylation of MARCKS but potentiate its phosphorylation induced by PMA and fMet-Leu-Phe (see also Sect. 3.3.1.3.4).

3.3.2.3 Muramyl Peptides

Muramyl peptides are the smallest active moieties of bacterial cell walls which can replace killed *Mycobacteria* in Freund's complete adjuvant (Adam et al. 1981; Karnovsky 1986; Kotani et al. 1986; Bahr and Chedid 1986). Muramyl peptides show a broad spectrum of biological effects, such as immunoadjuvant activity, pyrogenicity, antitumor activity, contraction of smooth muscle, nonspecific protection against infection, promotion of slow-wave sleep, and activation of macrophages (Adam et al. 1981; Karnovsky 1986; Kotani et al. 1986; Bahr and Chedid 1986). In macrophages, high-affinity binding sites for muramyl peptides have been identified (Silverman et al. 1986). Muramyl peptides potentiate the respiratory burst induced by PMA and opsonized zymosan in mononuclear phagocytes including human monocytes (Cummings et al. 1980; Pabst and Johnston 1980; Pabst et al. 1982; Silverman et al. 1985). Interestingly, serotonin also potentiates the respiratory burst in mononuclear phagocytes and inhibits binding of muramyl peptides (Silverman et al. 1985). In addition, serotonin antagonists inhibit the binding of serotonin and muramyl peptides and the respiratory burst induced by these stimuli (Silverman et al. 1985). These data suggest that muramyl peptides and sertonin act via the same receptor (see also Sect. 3.2.6.2).

In suspended human neutrophils, muramyl dipeptides per se do not activate $O_2^{\cdot-}$ formation, but they have been reported to act as primers for an enhanced respiratory burst (Wright and Mandell 1986; Seifert et al. 1990).

The *Bacillus anthracis* toxin, anthrax toxin, consists of three proteins, i.e., edema factor, lethal factor, and protective antigen, which act in binary combinations (Blaustein et al. 1989). Protective antigen plays a role in the penetration of lethal factor and edema factor into the cytosol (Blaustein et al. 1989). Edema factor is a calmodulin-dependent adenylyl cyclase, but the mode of action of lethal factor is not known (Blaustein et al. 1989). Protective antigen plus edema factor or lethal factor inhibit LPS- or muramyl dipeptide-induced potentiation of $O_2^{\cdot-}$ formation in human neutrophils (Wright and Mandell 1986). In contrast, anthrax toxin does not inhibit fMet-Leu-Phe- or PMA-induced $O_2^{\cdot-}$ formation in human neutrophils, suggesting that LPS and muramyl dipeptide activate neutrophils by mechanisms different from those of chemotactic peptides or phorbol esters (see also Sect. 3.3.2.2).

3.3.2.4 Retinoids

The role of retinoids in the regulation of NADPH oxidase is very controversial, and the results of studies performed with these drugs are difficult to interpret.

Protein kinase C may be an important intracellular target of action of retinoids, and these substances may modulate the activity of this enzyme in a very complex manner. Some authors reported that retinoids inhibit protein kinase C (Taffet et al. 1983; Cope 1986). In contrast, Lochner et al. (1986) reported that all-*trans* retinal does not substantially inhibit protein kinase C in neutrophils, and Ohkubo et al. (1984) showed that retinoic acid may activate protein kinase C under certain experimental conditions.

On one hand, retinoic acid and all-*trans* retinal have been reported to activate O_2^- formation in human and guinea pig neutrophils (Badwey et al. 1986, 1989b). The mechanism by which retinoids activate NADPH oxidase has been suggested to involve activation of phospholipase C, increase in membrane fluidity, and association of protein kinase C with the plasma membrane (Badwey et al. 1986, 1989b; Lochner et al. 1986). Seifert and Schächtele (1988) found that retinoic acid but not retinal activates O_2^- formation in human neutrophils, whereas retinoids fail to activate NADPH oxidase in dibutyryl cAMP-differentiated HL-60 cells. Retinoids have also been reported to potentiate the fMet-Leu-Phe-induced respiratory burst in human neutrophils and HL-60 cells (Cooke and Hallett 1985; Seifert and Schächtele 1988). Unlike O_2^- formation induced by PMA, that induced by retinal is not substantially inhibited by H-7 or staurosporine, suggesting that these stimuli activate the respiratory burst through different mechanisms (Badwey et al. 1989b; see also Sect. 3.2.2.3).

On the other hand, inhibitory effects of retinoids on the respiratory burst have been repeatedly observed. Retinoids inhibit fMet-Leu-Phe-induced O_2^- formation in human neutrophils in the presence of CB (Camisa et al. 1982; Seifert and Schächtele 1988). In addition, certain retinoids inhibit PMA-induced O_2^- formation in neutrophils (Witz et al. 1980; Cooke and Hallett 1985; Seifert and Schächtele 1988). Paradoxically, in HL-60 cells retinoids have been found to potentiate PMA-induced O_2^- formation (Seifert and Schächtele 1988).

3.3.2.5 Digitonin

Digitonin and saponin are bulky detergents related to cholesterol and activate the respiratory burst in various types of phagocytes (Zatti and Rossi 1967; Cohen and Chovaniec 1978a,b; Yamashita et al. 1985). Activation of O_2^- formation by digitonin in guinea pig neutrophils is characterized by a lag phase and by reversibility (Cohen and Chovaniec 1978a,b). Digitonin activates O_2^- formation with a biphasic concentration-response function and in a pH- and temperature-dependent manner. NADPH has been reported to enhance O_2^- formation, suggesting that NADPH crosses the plasma membrane and serves as electron donor for O_2^- formation. NEM and EGTA inhibit digitonin-induced O_2^- formation when added to cells prior to the stimulus, and Ca^{2+} is required for activation of O_2^- formation by digitonin (Cohen and Chovaniec 1978a,b). These data suggest that activation of O_2^- formation by digitonin is a Ca^{2+}-dependent and NEM-sensitive process (Cohen and Chovaniec 1978b; see also Sect. 4.3.3).

3.3.2.6 Hexachlorocyclohexanes

Hexachlorocyclohexanes are a group of isomeres and show some structural similarity to inositol. γ-Hexachlorocyclohexane, also referred to as γ-benzene hexachloride or lindane, is used as insecticide and ectoparasiticide. Hexachlorocyclohexanes are very lipophilic, accumulate in plasma membranes, and have complex effects on phosphoinositide metabolism (Hokin and Brown 1969; Fisher and Mueller 1971; Omann and Lakowicz 1982; Meade et al. 1984; Parries and Hokin-Neaverson 1985). Interestingly, activations of the respiratory burst by hexachlorocyclohexanes and receptor agonists show some properties that they have in common.

The α-, γ- and δ-isomers of hexachlorocyclohexane but not the β-isomer activate the respiratory burst in alveolar macrophages, human neutrophils, and differentiated HL-60 cells (Holian et al. 1984; Kuhns et al. 1986; English et al. 1986; Seifert et al. 1989c, 1991). γ-Hexachlorocyclohexane is a similarly effective activator of O_2^- formation as PMA, but the insecticide is several orders of magnitude less potent than the phorbol ester (English et al. 1986; Seifert et al. 1989c, 1991). In contrast to O_2^- formation, hexachlorocyclohexanes inhibit chemotaxis and actin polymerization (Kaplan et al. 1988).

Similar to chemotactic peptides, hexachlorocyclohexanes activate phosphoinositide degradation and induce an increase in cytoplasmic Ca^{2+} (Holian et al. 1984; English et al. 1986). Hexachlorocyclohexane-induced O_2^- formation is a reversible process and is reactivated by chemotactic peptides (Holian et al. 1984). In contrast to activation of NADPH oxidase by PMA, that induced by hexachlorocyclohexanes is terminated by removal of the stimulus, and particulate fractions of PMA- but not of hexachlorocyclohexane-treated cells generate O_2^- (English et al. 1986). These data suggest that the permanent presence of hexachlorocyclohexanes is required for NADPH oxidase activation, and that these agents activate O_2^- generation by a mechanism distinct from that of PMA. Interestingly, γ-hexachlorocyclohexane has a weak stimulatory effect on protein kinase purified from rat brain (Seifert, unpublished results).

3.3.2.7 Alcohols

The role of short-chain aliphatic alcohols in the regulation of the respiratory burst is controversial, and several mechanisms may be involved in their effects. On one hand, hypertonic glycerol has been reported to induce $O_2^{\bullet-}$ formation in various types of phagocytes (Kaneda and Kakinuma 1986). Glycerol-induced $O_2^{\bullet-}$ formation is reversible and is associated neither with cytotoxicity nor with exocytosis but with marked changes in morphology (Kaneda and Kakinuma 1986). In addition, ethanol at concentrations of 0.1–0.5 M has been reported to be a weak activator of $O_2^{\bullet-}$ formation in rat alveolar macrophages (Dorio et al. 1988). On the other hand, aliphatic alcohols have been reported to inhibit receptor agonist- and PMA-induced $O_2^{\bullet-}$ formation (Yuli et al. 1982; Dorio et al. 1988; Bonser et al. 1989). The mechanism by which alcohols modulate the respiratory burst may involve alteration of the affinity state of formyl peptide receptors and of the activity of G-proteins, of phospholipases C and D, and of protein kinase C (Hoek et al. 1987; Rubin and Hoek 1988; Dorio et al. 1988; Rooney et al. 1989; Bonser et al. 1989; see also Sects. 3.2.2.1, 3.3.1.1.1).

3.3.2.8 Thymol

Thymol is used as antiseptic and antifungal agent. Thymol activates $O_2^{\bullet-}$ formation in neutrophils of various species including guinea pig, primates, and man (Suzuki et al. 1987; Suzuki and Furuta 1988). In guinea pig neutrophils, thymol induces $O_2^{\bullet-}$ formation with a lag time of 30 s (Suzuki and Furuta 1988). The precise mode of action of thymol is not known. Activation of NADPH oxidase by thymol does not depend on extracellular Ca^{2+}, is associated with a decrease in the cellular content of ATP, is inhibited by trifluoperazine, and is subject to homologous desensitization. In addition, exposure of cells to thymol potentiates PMA-induced $O_2^{\bullet-}$ formation.

3.3.2.9 Bleomycin

The induction of pulmonary fibrosis is an important unwanted effect of the antineoplastic agent bleomycin. Interestingly, bleomycin has been shown to enhance $O_2^{\bullet-}$ formation in alveolar macrophages of guinea pigs, suggesting that activation of the respiratory burst in mononuclear phagocytes may contribute to bleomycin-induced fibrosis (Conley et al. 1986). Glucocorticoids partially inhibit this effect of bleomycin (see also Sect. 4.2.1), but the molecular mode of action of bleomycin is still undefined.

3.3.2.10 Neuraminidase

Exogenous neuraminidase induces the release of sialic acid from human neutrophils (Henricks et al. 1982). Upon stimulation with opsonized staphylococci, neuraminidase-treated neutrophils generate larger amounts of O_2^- than control cells, suggesting that Fc receptor-mediated activation of NADPH oxidase is facilitated by removal of sialic acid (Henricks et al. 1982; see also Sect. 3.3.1.5). Exogenous neuraminidase also enhances O_2^- formation in phagocytosing neutrophils (Suzuki et al. 1982). In contrast, exogenous neuraminidase does not affect binding of fMet-Leu-Phe to formyl peptide receptors and fMet-Leu-Phe-induced O_2^- formation in neutrophils. The results of these studies suggest that enhancement of O_2^- formation by neuraminidase is stimulus dependent.

3.3.2.11 1,25-Dihydroxyvitamin D_3

1,25-Dihydroxyvitamin D_3 primes murine peritoneal macrophages and human monocyte-derived macrophages for enhanced O_2^- and H_2O_2 formation (M.S. Cohen et al. 1986; Gluck and Weinberg 1987). In addition, peritoneal macrophages from vitamin D_3-deficient mice show an impaired respiratory burst, and incubation of the vitamin D_3-deficient phagocytes with the hormone in vitro partially restores the respiratory burst (Gavison and Bar-Shavit 1989). These data suggest that 1,25-dihydroxyvitamin D_3 plays a role in the functional maturation of macrophages (Gavison and Bar-Shavit 1989; see also Sects. 3.4.4.1.2, 3.4.4.1.3).

3.3.2.12 Particulate Stimuli

In addition to opsonized particles (see also Sect. 3.3.1.5.5), unopsonized particles, e.g., zymosan, bacteria and latex beads, may activate the respiratory burst and phagocytosis in various types of phagocytes.

3.3.2.12.1 Unopsonized Fungal and Bacterial Components

Phagocytes possess mannose/fucose-specific plasma membrane "receptors," and ingestion of particles through mannose-specific mechanisms has also been referred to as lectinophagocytosis (Stahl et al. 1978, 1980; Warr 1980; Shepherd et al. 1982; Largent et al. 1984; Gbarah et al. 1989; see also Sect. 3.3.2.1).

Unopsonized zymosan has been reported effectively to activate the respiratory burst in primed murine macrophages (Berton and Gordon 1983b). Culture of primed murine peritoneal macrophages in the presence of IgG causes desensitization to zymosan-induced O_2^- formation (Valletta and Berton 1987). Unopsonized zymosan also activates O_2^- formation in suspended or adherent murine alveolar macrophages (Sugar and Field 1988). Opsonization of zymosan with complement does not enhance the effect of zymosan, suggesting that functional complement receptors are not present in these cells (Sugar and Field 1988). Various unopsonized fungi, zymosan, and the polysaccharide mannan have been shown to stimulate H_2O_2 formation in murine neutrophils (Danley and Hilger 1981). In addition, mannan enhances zymosan-induced H_2O_2 formation. Furthermore, the primary constituent of mannan, D-mannose, but not other monosaccharides, inhibits the stimulatory effects of zymosan or mannan. The inhibitory effect of 2-deoxy-D-glucose on the respiratory burst may be explained by inhibition of glycolysis (Danley and Hilger 1981). Finally, mannose does not inhibit H_2O_2 formation induced by PMA or opsonized Sephadex beads. These data suggest that activation of the respiratory burst in murine neutrophils by unopsonized fungal components involves mannose-specific mechanisms.

In human neutrophils and macrophages, unopsonized zymosan is not an effective activator of the respiratory burst (Goldstein et al. 1975; Roos et al. 1981; Ezekowitz et al. 1985; Meshulam et al. 1988). In contrast, opsonized and unopsonized *Candida albicans* hyphae have been shown to be similarly effective activators of the respiratory burst in human neutrophils (Meshulam et al. 1988). Activation of O_2^- formation by unopsonized hyphae is accompanied by phosphoinositide degradation and Ca^{2+} mobilization. Unlike the respiratory burst induced fMet-Leu-Phe, that induced by hyphae is not accompanied by plasma membrane depolarization and is partially pertussis toxin-insensitive (Meshulam et al. 1988). In hamster alveolar macrophages, phagocytosis of unopsonized particles is associated with an inhibition of the respiratory burst (Kobzik et al. 1990).

Recently, unopsonized type 1 fimbriated *Escherichia coli* has been reported to induce a respiratory burst in human neutrophils and peritoneal macrophages through a mannose-specific mechanism (Gbarah et al. 1989). Chemiluminescence induced by these bacteria is abrogated by prior exposure to PMA and is inhibited by sphingosine, H-7, and K-252a (Gbarah et al. 1989). The authors interpreted their results in such a way that protein kinase C is involved in the mannose-specific activation of the respiratory burst (see also Sects. 3.2.2.3, 3.2.2.5). Positive charges may play a role in the activation of the respiratory burst by fimbrinated *E. coli* strains (Steadman et al. 1990).

3.3.2.12.2 Latex Particles

Latex particles induce oxygen consumption and H_2O_2 formation in human neutrophils after a short lag time (Segal and Coade 1978; Hallett and Campbell 1983; Curnutte and Tauber 1983; Cooke and Hallett 1985). In contrast to opsonized zymosan, latex particles do not induce substantial release of O_2^- into the extracellular space (Curnutte and Tauber 1983). In addition, CB inhibits the latex bead-induced respiratory burst (Hallett and Campbell 1983; see also Sect. 3.2.5). Furthermore, chemotactic peptides but not latex beads induce myeloperoxidase release (Hallett and Campbell 1983). The kinetics of oxygen consumption induced by PMA and that induced by latex beads are similar, and latex beads but not chemotactic peptides compete with PMA in a simple common-target manner (Cooke and Hallett 1985). Finally, retinal inhibits the respiratory burst induced by latex beads but not that induced by fMet-Leu-Phe (Cooke and Hallett 1985; see also Sect. 3.3.2.4). These data suggest that protein kinase C plays a central role in the latex bead-induced respiratory burst, and that these particles and chemotactic peptides activate NADPH oxidase by different mechanisms (Cooke and Hallett 1985).

3.3.2.12.3 Crystals

Chronic inhalation of mineral dusts, e.g., of quartz or of asbestos, may lead to pulmonary diseases characterized by accumulation of mononuclear cells and neutrophils (Craighead and Mossman 1982; Mossman et al. 1983; Kamp et al. 1989). In addition, asbestos induces epithelial damage, fibrosis, and malignant tumors (Craighead et al. 1982; Mossman et al. 1983). Besides lysosomal enzyme release, activation of the respiratory burst in mononuclear phagocytes and neutrophils may be a mechanism by which asbestos and quartz induce tissue damage (Davies et al. 1974; Donaldson and Cullen 1984; Elferink and Ebbenhout 1988; Cantin et al. 1988; Kamp et al. 1989). Somewhat unexpectedly, activation of NADPH oxidase by asbestos has been shown to be pertussis toxin sensitive (Elferink and Ebbenhout 1988). Similar to chemotactic peptides, the respiratory burst induced by asbestos depends on extracellular Ca^{2+}, but it is not known whether asbestos binds to specific receptor proteins (Elferink and Ebbenhout 1988). The long- or short-term application of asbestos or quartz to sheep potentiates PMA-induced O_2^- formation in alveolar macrophages (Cantin et al. 1988). These data suggest that mineral dust inhalation primes macrophages for enhanced O_2^- formation in vivo and contributes to the pathogenesis of pulmonary fibrosis.

Urate crystals play a major role in the pathogenesis of acute and chronic gouty arthritis (Woolf and Dieppe 1987). The inflammatory potential of urate crystals is modulated by surface charge and protein coating. Upon exposure to monosodium urate crystals, human neutrophils undergo a respiratory burst (Simchowitz et al. 1982). Coating of monosodium urate with IgG potentiates $O_2^{\cdot-}$ formation in human neutrophils (Abramson et al. 1982; see also Sect. 3.3.1.5). The mechanism by which uncotated urate crystals activate NADPH oxidase may involve the synthesis of leukotrienes and Ca^{2+} mobilization (Poubelle et al. 1987; Terkeltaub et al. 1990). In addition, recent data suggest that both pertussis toxin-sensitive and -insensitive mechanisms are involved in urate crystal-induced neutrophil activation (Terkeltaub et al. 1990). In contrast, the induction of pleuritis by calcium pyrophosphate crystals in rats has been reported to depress the respiratory burst of the corresponding peritoneal macrophages (Bird et al. 1985).

3.4 Miscellaneous Aspects of NADPH Oxidase Activation

3.4.1 Relation of the Respiratory Burst to the Synthesis of Reactive Nitrogen Oxide Intermediates: Role of Arginine

R-NO, e.g., hydroxylamine and nitric oxide, has been suggested to play a role in phagocyte-induced cytotoxicity, carcinogenesis, vasodilation, and inhibition of platelet aggregation (Hibbs et al. 1987a,b, 1988; Iyengar et al. 1987; Ding et al. 1988; Rimele et al. 1988; Stuehr et al. 1989; Stuehr and Nathan 1989; Salvemini et al. 1989). IFN-γ and LPS induce the parallel synthesis of reactive oxygen species and R-NO in murine macrophages, and combinations of certain cytokines or of cytokines with LPS synergistically induce R-NO synthesis (Iyengar et al. 1987; Ding et al. 1988; see also Sects. 3.3.1.3, 3.3.2.2). In addition to macrophages, human neutrophils and dibutyryl cAMP-differentiated HL-60 cells have been shown to release R-NO in the absence or presence of chemoattractants (Wright et al. 1989; Schmidt et al. 1989). The effectiveness order of fMet-Leu-Phe, PAF, and LTB$_4$ to induce the formation of R-NO and $O_2^{\cdot-}$ in these phagocytes is the same, suggesting that the initial signal transduction steps for both processes are identical (Schmidt et al. 1989). In contrast to $O_2^{\cdot-}$ formation, the fMet-Leu-Phe-induced release of R-NO in suspended human neutrophils is delayed in onset, is long lasting, and is not potentiated by CB (Schmidt et al. 1989 and unpublished results). Superoxide dismutase potentiates chemoattractant-induced release of R-NO and neutrophil-induced inhibition of platelet aggregation and counteracts neutrophil-induced contraction of smooth vascular muscle, probably by preventing $O_2^{\cdot-}$-induced degradation of R-NO (Gryglewski et al. 1986; Schmidt et al. 1989; Salvemini et al. 1989; Ohlstein and Nichols 1989). $O_2^{\cdot-}$ and nitric oxide react to form peroxynitrite which rapidly decomposes and generates a potent oxidant (Beckman et al. 1990). It has been suggested that the parallel synthesis and release of $O_2^{\cdot-}$ and nitric oxide from phagocytes through formation of peroxynitrite represents an additional mechanism by which these cells exert their cytotoxic actions (Beckman et al. 1990).

The inhibitors of arginine-metabolizing enzymes L-canavanine and N^G-monomethyl-L-arginine prevent synthesis of R-NO, suggesting that the terminal guanidino nitrogen atoms of L-arginine are physiological precursors for R-NO synthesis (Ding et al. 1988; Marletta et al. 1988; Hibbs et al. 1988; Schmidt et al. 1989). In addition, N^G-monomethyl-L-arginine inhibits fMet-Leu-Phe-induced chemotaxis, and arginine or dibutyryl cGMP antagonize this effect (Kaplan et al. 1989). These data suggest that R-NO through activation of soluble guanylyl cyclase and subsequent formation of cGMP is involved in the regulation of chemotaxis (Kaplan et al. 1989; see also Sect. 3.2.6.2). In contrast, bacterial killing and activation of the respiratory burst are not affected by L-canavanine or N^G-monomethyl-L-arginine (Ding et al. 1988; Schmidt et al. 1989, and unpublished results). J774.C3C macrophages, which cannot undergo a respiratory burst, release R-NO, and the combination of LPS plus IFN-γ enhances R-NO release but decreases H_2O_2 production (Iyengar et al. 1987; Ding et al. 1988; see also Sects. 3.3.2.2, 3.4.4.1.5). These data suggest that activation of NADPH oxidase and the synthetisis of R-NO are independently regulated processes.

L-Arginine, however, plays a role in the regulation of the respiratory burst. The extracellular fluid in injured tissue contains L-arginine only at very low concentrations, probably due to degradation of the amino acid by macrophage-derived arginase (Albina et al. 1989a,b). In resident and primed rat peritoneal macrophages, $O_2^{\cdot-}$ formation is enhanced following incubation in L-arginine-deficient medium, whereas L-arginine at concentrations above 0.1 mM inhibits $O_2^{\cdot-}$ formation (Albina et al. 1989a,b). In contrast, N^G-monomethyl-L-arginine counteracts inhibition of the respiratory burst by L-arginine (Albina et al. 1989b). These data suggest that depletion of L-arginine primes macrophages for enhanced $O_2^{\cdot-}$ formation, and that oxidative metabolism of L-arginine through L-arginine deiminase inhibits the respiratory burst (Albina et al. 1989a,b). Finally, N^G-monomethyl-L-arginine slightly enhances $O_2^{\cdot-}$ formation in rat mast cells, possibly due to inhibition of release of R-NO which may react with O_2^- (Salvemini et al. 1990; see also Sect. 3.4.4.2.4).

3.4.2 Relation of the Respiratory Burst to Ion Fluxes

3.4.2.1 Na$^+$/H$^+$ Exchange

The Na$^+$/H$^+$ exchange is involved in the regulation of the cytosolic H$^+$ and Na$^+$ concentrations in many cell types including phagocytes (Grinstein and Furuya 1984; Grinstein et al. 1984; Seifter and Aronson 1986; Grinstein and Rothstein 1986). The intracellular Na$^+$ concentration is maintained at much lower concentrations than in the extracellular space by active extrusion of Na$^+$ through Na$^+$/K$^+$-ATPase. The driving force for H$^+$ extrusion through Na$^+$/H$^+$ exchange is the energy of the inwardly directed electrochemical gradient of Na$^+$.

Upon exposure to fMet-Leu-Phe or PMA, neutrophils undergo a transient acidification followed by alkalinization (Grinstein et al. 1985, 1986, 1988; Grinstein and Furuya 1986a,b). The acidification may be attributable to the generation of H$^+$ through activation of the respiratory burst, i.e., hexose monophosphate shunt and NADPH oxidase, as neutrophils of CGD patients do not undergo acidification (Grinstein et al. 1986, 1988; Grinstein and Furuya 1986a,b; Wright et al. 1986). In contrast, Naccache et al. (1989) suggested that acidification is not directly linked to activation of NADPH oxidase. The delayed alkalinization is discussed to be due to activation of the amiloride-sensitive Na$^+$/H$^+$ exchange by protein kinase C (Besterman and Cuatrecasas 1984; Simchowitz 1985a,b; Grinstein et al. 1985, 1986, 1988; Grinstein and Furuya 1986a,b). Amiloride has been used as a pharmacological tool in order to clarify the role of the Na$^+$/H$^+$ exchange in the regulation of NADPH oxidase.

The role of the Na$^+$/H$^+$ exchange in the regulation of O_2^- formation is controversial. In the absence of extracellular monovalent cations, fMet-Leu-Phe does not induce O_2^- formation in human neutrophils, and extracellular Na$^+$ restores their capacity to generate O_2^- in a concentration-dependent manner (Simchowitz and Spilberg 1979). Activation of the respiratory burst by opsonized zymosan, ConA, and PMA depends, at least in part, on extracellular Na$^+$ (Wright et al. 1986, 1988; Nasmith and Grinstein 1986). In the absence of extracellular Na$^+$, PMA does not cause alkalinization but only acidification in human neutrophils (Nasmith and Grinstein 1986). Relieving acidification restores the ability of PMA maximally to stimulate the respiratory burst (Nasmith and Grinstein 1986; Wright et al. 1988). These data suggest that Na$^+$ is not obligatorily involved in the activation of NADPH oxidase but indirectly inhibits the activation process by inhibiting Na^+/H^+ exchange.

Berkow et al. (1987a) reported that amiloride inhibits Na^+ influx in human neutrophils without affecting PMA-induced O_2^- formation, and in cultured rat Kupffer's cells, activation of the respiratory burst does not depend on extracellular Na^+ or on intracellular pH changes (Dieter et al. 1987). Amiloride prevents fMet-Leu-Phe-induced alkalinization and partially inhibits O_2^- formation (Simchowitz 1985a; Berkow et al. 1987a). The amount of O_2^- generated correlates with the extent of intracellular alkalinization, but alkalinization per se is not sufficient to activate the respiratory burst, as fMet-Leu-Phe may induce alkalinization without a respiratory burst (Simchowitz 1985a). The interpretation of the results of studies reporting on the effects of amiloride is hampered by the fact that amiloride does not only inhibit Na^+/H^+ exchange but among others inhibits protein kinases including protein kinase C and modulates the activity of G-proteins (Davis and Czech 1985; Besterman et al. 1985; Anand-Srivastava 1989). Molski et al. (1986) showed that most of the fMet-Leu-Phe-stimulated Na^+ influx is not coupled to H^+ efflux. In addition, H-7 blocks fMet-Leu-Phe-induced cell alkalinization without inhibiting O_2^- formation, indicating that an increase of the intracellular pH is not obligatorily required for activation of O_2^- formation.

3.4.2.2 Membrane Depolarization

Neutrophils show a resting potential of -30 to -75 mV which is maintained primarily by K^+ conductance (Korchak and Weissmann 1978; Kuroki et al. 1982; Majander and Wikström 1989). Upon exposure to various stimuli such as PMA, ConA, A 23187, zymosan, and fMet-Leu-Phe, the membrane rapidly depolarizes and slowly repolarizes (Korchak and Weissmann 1978; Jones et al. 1981; Kuroki et al. 1982; Cameron et al. 1983; Sullivan et al. 1984). The question of how far Na^+ influx contributes to depolarization is matter of debate (Korchak and Weissmann 1980; Jones et al. 1981; Kuroki et al. 1982; Majander and Wilkström 1989).

Membrane depolarization precedes O_2^- formation, and fMet-Leu-Phe-induced membrane potential changes correlate with the respiratory burst (Korchak and Weissmann 1978; Whitin et al. 1980; Seligmann and Gallin 1980; Jones et al. 1981; Korchak et al. 1983; Cameron et al. 1983; Fletcher and Seligmann 1986). In addition, neutrophils of CGD patients do not undergo membrane depolarization and do not generate O_2^- upon exposure to PMA, fMet-Leu-Phe, or ConA (Whitin et al. 1980; Seligmann and Gallin 1980; Castranova et al. 1981). Moreover, under certain experimental conditions depolarization may be sufficient to induce a respiratory burst (Rossi et al. 1981a). These data support a role of membrane depolarization is the regulation of NADPH oxidase.

In contrast, the membrane depolarization induced by A 23187 has been reported to be not impaired in neutrophils of CGD patients, and inhibition of membrane depolarization by EGTA has no inhibitory effect on the PMA-induced respiratory burst in murine peritoneal macrophages (Seligmann and Gallin 1980; Castranova et al. 1981; Lepoivre et al. 1982; Sullivan et al. 1984). On one hand, partial depolarization of neutrophils by a high K^+ buffer has been reported to inhibit fMet-Leu-Phe- and PMA-induced O_2^- formation (Martin et al. 1988). On the other hand, depolarization by increasing the extracellular K^+ concentration has been shown to have no effect on O_2^- formation in the presence of fMet-Leu-Phe (Kuroki et al. 1982). Moreover, the concentration of PMA required to induce membrane depolarization is lower than that to activate the respiratory burst (Seeds et al. 1985). Finally, spontaneous depolarization of neutrophils without a respiratory burst and activation of H_2O_2 formation by fMet-Leu-Phe without membrane depolarization have been observed (Seeds et al. 1985). These data suggest that membrane depolarization may precede activation of NADPH oxidase but is not sufficient to induce its activation.

NADPH oxidase has been suggested to be electrogenic (L.M. Henderson et al. 1987, 1988a,b). The release of O_2^- from phagocytes into the extracellular space would be expected to be associated with substantial membrane potential changes if no compensating ions crossed the plasma membrane (L.M. Henderson et al. 1987, 1988a,b). In neutrophil cytoplasts, the PMA-induced respiratory burst is associated with membrane depolarization, and the electroneutral amiloride-sensitive Na^+/H^+ exchange cannot contribute to the compensation for charge translocated by NADPH oxidase (see also Sect. 3.2.5). The inhibitor of NADPH oxidase diphenylene iodonium prevents membrane depolarization in neutrophil cytoplasts and acidification in stimulated neutrophils. The extent of depolarization is modulated by the pH gradient across the plasma membrane, and blockers of H^+ efflux enhance PMA-induced membrane depolarization and inhibit acidification of the extracellular medium and O_2^- formation. Extracellular NH_4Cl restores the ability to generate O_2^-, possibly due to the movement of NH_4^+ across the plasma membrane. These data suggest that NADPH oxidase is electrogenic, that NADPH oxidase activity is limited by the movement of ions as charge compensators, and that H^+ efflux compensates for the release of O_2^- and membrane depolarization (L.M. Henderson et al. 1987, 1988a,b).

3.4.2.3 Anion Transport

The stilbene sulfonic acids 4'4-diisothiocyanostilbene-2'2'-disulfonic acid (DIDS) and 4-acetamido-4'-isothiocyanostilbene-2,2'-disulfonic acid (SITS), have been suggested to block anion transport processes (Korchak et al. 1980, 1982). In addition, these agents may inhibit neutrophil Mg^{2+}-ATPase and may react with receptor agonists (Korchak et al. 1982; Vostal et al. 1989). The results of studies concerning the effects of these compounds on the respiratory burst are controversial. DIDS and SITS have been reported to inhibit O_2^- formation induced by zymosan opsonized in fresh serum but not that induced by zymosan opsonized in heat-decomplemented serum or by PMA (Tauber and Goetzl 1981). DIDS does not inhibit O_2^- formation induced by PMA and fMet-Leu-Phe but that induced by LTB_4 and PAF (Smith et al. 1984a). DIDS and SITS inhibit exocytosis and aggregation but not O_2^- formation induced by immune complexes and A 23187 in human neutrophils (Korchak et al. 1980; Kaplan et al. 1982).

3.4.3 Adherence

Most of the in vitro studies on the regulation of the respiratory burst in neutrophils have been carried out with suspended cells. However, in vivo, neutrophils may adhere to surrounding tissues upon activation (Hoffstein et al. 1985; Nathan 1987; Kownatzki and Uhrich 1987). Therefore, studies with adherent neutrophils are of considerable interest with respect to the activation of the respiratory burst in vivo (see also Sect. 1). Certain extracellular matrix proteins show stimulatory effects on the respiratory burst in human neutrophils (see also Sect. 3.3.1.4).

The results of various authors show that adherence may result in an augmentation of the respiratory burst in neutrophils. In comparison to suspended cells, fMet-Leu-Phe induces a greatly prolonged respiratory burst in neutrophils which adhere to petri dishes or polystyrene surfaces coated with serum, fibronectin, or human umbilical vein endothelial cells (Dahinden et al. 1983b; Nathan 1987). A variety of stimuli including LPS and cytokines have been reported to activate NADPH oxidase only in adherent but not in suspended neutrophils (Dahinden and Fehr 1983; Dahinden et al. 1983a; Nathan 1987; see also Sects. 3.3.1.3, 3.3.2.2). The onset of the respiratory burst in adherent human neutrophils is delayed (Nathan 1987). The CR3 receptor may play an important role in mediating the adherence-dependent respiratory burst in human neutrophils (Entman et al. 1990; Shappell et al. 1990; see also Sects. 3.3.1.5.2 and 6.2.1). Nylon-adherent human neutrophils generate considerably higher amounts of

O_2^- than nonadherent cells upon stimulation with fMet-Leu-Phe, C5a, PAF, or A 23187 but not exposure to PMA (Kownatzki and Uhrich 1987). Adherence of neutrophils to nylon fibers is also associated with increased chemiluminescence (Clifford et al. 1984). Interestingly, adherence of neutrophils to plastic surfaces is accompanied by actin polymerization, the extent of which is additively enhanced by fMet-Leu-Phe (Southwick et al. 1989). In contrast to actin polymerization induced by fMet-Leu-Phe, that induced by adherence is slow in onset, long lasting, and not inhibited by pertussis toxin (Southwick et al. 1989). These data suggest that the mechanism by which chemotactic peptides and adherence induce actin polymerization is different (see also Sect. 3.2.5). Activation of the respiratory burst in plastic-adherent neutrophils may involve Ca^{2+}-dependent processes and activation of protein kinase C (Ginis and Tauber 1990).

In contrast to the above studies, Hoffstein et al. (1985) reported that neutrophils which adhere to protein-coated surfaces, show a reduced respiratory burst in comparison to suspended cells upon exposure to fMet-Leu-Phe and PMA but not upon exposure to opsonized zymosan. This inhibition of O_2^- formation by surface contact has been reported to be rapid in onset and to be reversible upon resuspension of the cells (Hoffstein et al. 1985).

The respiratory burst in mononuclear phagocytes is also affected by the state of adherence. In primed mouse peritoneal macrophages, surface contact per se has been shown to induce a respiratory burst which depends on extracellular Mg^{2+} or Ca^{2+} and is prevented by the local anesthetic lidocaine (Berton and Gordon 1983a). Prolonged maintenance of macrophages as monolayer cultures is associated with a progressive loss of their ability to undergo a respiratory burst (Berton and Gordon 1983a). This decrease in respiratory burst activity is prevented by maintaining the phagocytes in a nonadherent state (Berton and Gordon 1983a). Primed murine peritoneal macrophages generate only low amounts of O_2^- in suspension, but when the macrophages are allowed to adhere to a glass surface, PMA is a very effective activator of the respiratory burst (M.S. Cohen et al. 1981). Interestingly, adherence of J774.1 macrophages to a glass surface is associated with a transient activation of phosphoinositide degradation (Zabrenetzky and Gallin 1988). Suspended murine alveolar macrophages do not generate O_2^- upon exposure to PMA, but PMA induces a massive respiratory burst in adherent macrophages cultured for 48 h (Sugar and Field 1988). Zymosan and *Blastomyces dermatitidis* conida induce O_2^- formation both in suspended and in adherent macrophages, but both stimuli are more effective in adherent than in suspended cells (Sugar and Field 1988).

Human monocytes cultured in vitro differentiate into macrophages (Nakagawara et al. 1981; Sasada et al. 1987). This differentiation process has been reported to be associated with an initial increase and a delayed decrease in the ability to undergo a respiratory burst (Nakagawara et al. 1981; Sasada et al. 1987). Human monocytes cultured under conditions preventing adherence retain their ability to generate O_2^- upon stimulation with PMA (Zeller et al. 1988).

3.4.4 Properties of the Respiratory Burst in some Specialized Cell Types

Most of the studies concerning the respiratory burst have been carried out with neutrophils or with mononuclear phagocytes. The magnitude and the stimulus-specificity of the respiratory burst among neutrophils of various species varies considerably (Badwey et al. 1980, 1983; Styrt 1989). In addition, macrophages primed by various agents show great differences in their ability to undergo a respiratory burst (Badwey et al. 1980, 1983). The regulation of the respiratory burst in neutrophils and mononuclear phagocytes in general is dealt with in other sections of this review.

This section focuses on characteristics of the respiratory burst in some specialized cell types, which are summarized in Table 10. It is generally assumed that the respiratory burst is a specialized property of phagocytes, but an increasing number of studies indicates that "respiratory burst-like processes" occur also in nonphagocytic cell types. This field is still in its beginnings, but we anticipate that research in this area will reveal many interesting and unexpected results.

3.4.4.1 Phagocytes

3.4.4.1.1 Eosinophils

Eosinophils comprise 2%–3% of the circulating leukocytes in healthy subjects. Eosinophils may accumulate in the skin and in the respiratory and gastrointestinal tract, and eosinophilia may be associated with allergic, neoplastic, or parasitic disease (Butterworth and David 1981). Thus, eosinophils are assumed to play a role in the pathogenesis of allergic reactions and inflammatory tissue injury, in the killing of helminths, and in host defense against bacteria (Butterworth and David 1981; Pincus et al. 1981; Yazdanbakhsh et al. 1987; Petreccia et al. 1987).

Early studies suggested that there are quantitative differences in the respiratory burst activity between human neutrophils and eosinophils (De-

Table 10. Properties of the respiratory burst in some cell types

Cell type	Properties	Selected references
Eosinophils	Quantitative/qualitative differences with respect to activation, kinetics, regulation and stimulus specificity in comparison to neutrophils. Eosinophilia may be associated with increased or reduced respiratory burst activity. Role in pathogenesis of inflammatory tissue injury, allergic reactions (type I), host defense against helminth parasites.	Pincus et al. (1981), Prin et al. (1984), Yamashita et al. (1985), Yazdanbakhsh et al. (1985, 1987a,b), Petreccia et al. (1987), Koenderman and Bruijnzeel (1989)
Mast cells/basophils	Activation of the respiratory burst upon exposure to anti-IgE and compound 48/80. Involvement of proteases in the activation process? Physiological role not known.	Henderson and Kaliner (1978), Kitagawa et al. (1980c), Schinetti et al. (1984)
HL-60 cells	Human promyelocytic cell line, may be differentiated into monocyte- or neutrophil-like cells by various agents. Differentiation-dependent expression of various signal transduction components (e.g., formyl peptide receptors, protein kinase C, cytochrome b_{-245}, cytosolic activation factors). Functional similarities and dissimilarities with human neutrophils and monocytes, respectively. Functional nonequivalence of the various differentiations.	Collins et al. (1978, 1979), Newburger et al. (1979), Breitman et al. (1980), Chaplinski and Niedel (1982), Harris and Ralph (1985), Collins (1987), Parkinson et al. (1987), Seifert and Schultz (1987b), Thompson et al. (1988), Rao et al. (1989), Seifert et al. (1989b,c)
K-562 cells	Human erythroleukemia cell line, may be differentiated into neutrophil-like cells by the inhibitor of DNA topoisomerase I, camptothecin, and acquires the ability to reduce NBT.	Chou et al. (1990)
Karpas 120 cells	Derived from a patient with acute myeloblastic leukemia. Activation of respiratory burst upon stimulation with PMA and latex beads but not with opsonized zymosan.	Newburger et al. (1980a)
U-937 cells	Human promonocytic cell line, acquires the ability to generate O_2^{-} upon differentiation with a variety of agents. Stimulus-specific expression of signal transduction components.	Clement and Lehmeyer (1983), Harris and Ralph (1985), Balsinde and Mollinedo (1988)

Table 10. (continued)

Cell type	Properties	Selected references
Kupffer's cells	Principally similar to peritoneal macrophages but quantitative differences. Priming with LPS, muramyl dipeptide, IFN-γ.	Laskin et al. (1988), Rieder et al. (1988a)
J774 Macrophages	Derived from a murine reticulum sarcoma. Defect of NADPH oxidase in J774.C3C cells (structural or functional abnormality of cytochrome b_{-245}?).	Damiani et al. (1980), Y. Tanaka et al. (1982), Kiyotaki et al. (1984)
B-Lymphocytes	Human tonsillar B-lymphocytes reduce NBT upon stimulation with PMA or IgG. Cytochrome b_{-245} present. Cytotoxic role of B-lymphocytes? Certain Ebstein-Barr virus-transformed B-lymphocytes also show a respiratory burst; K_m for NADPH ranges from 30 to 250 μM.	Volkman et al. (1984), Maly et al. (1988, 1989), Hancock et al. (1989)
Isolated glomeruli, mesangial cells	Chemiluminescence, O_2^- and H_2O_2 formation upon exposure to PMA, proteases, opsonized zymosan, TNF-α, IL-1α and complement components.	Shah and Naum-Bedigiam (1981), Baud et al. (1983, 1986), Adler et al. (1986), Basci and Shah (1987), Miyanoshita et al. (1989), Radeke et al. (1990)
Renal tubular epithelial cells	O_2^- and H_2O_2 formation upon stimulation with opsonized zymosan or heat-aggregated IgG. Role in the pathogenesis of renal injury?	Rovin et al. (1990)
Glia cells	NBT reduction, chemiluminescence, and O_2^- formation upon stimulation with PMA, zymosan or antibody-coated erythrocytes. Priming by IFN-γ. Role in immunological reactions in the brain?	Sonderer et al. (1987), Woodroofe et al. (1989)
Thyroid cells	Particulate NADPH oxidase, generates O_2^- and H_2O_2, insensitive to inhibition by KCN. Ca^{2+} dependency, stimulation by ATP; Role in the synthesis of thyroid hormones.	Deme et al. (1985), Nakamura et al. (1987, 1989)
Carotid body	Absorbance spectrum similar to the reduced spectrum of neutrophil NADPH oxidase. Modulation of spectral changes by KCN and diphenylene iodonium; Sensor for oxygen tension in the blood?	Acker et al. (1989), Cross et al. (1990)

Table 10. (continued)

Cell type	Properties	Selected references
Human fibroblasts	Sustained O_2^- formation, insensitive to inhibition by KCN and NaN₃. Stimulation by IL-1α or TNF-α. Physiological role in regulation of fibroblast proliferation, pathogenesis of fibrosis.	Meier et al. (1989), Murrel et al. (1990)
Fat cells	Insulin-stimulated H_2O_2 formation; related to glucose metabolism. H_2O_2 role as intracellular signal molecule?	May and de Haen (1979a), Mukherjee and Mukherjee (1982)
Endothelial cells	Stimulation of O_2^- formation by PMA, A 23187, Il-1, IFN-α and bradykinin. Physiological role in vasoconstriction? functional inactivation of vasorelaxant nitric oxide? Formation of peroxynitrate from O_2^- and nitric oxide (cytotoxicity)?	Rosen and Freeman (1984), Wei et al. (1985), Matsubara and Ziff (1986a,b), Auch-Schwelk et al. (1989), Beckman et al. (1990), Holland et al. (1990)
Sea urchin eggs	"Respiratory burst of fertilization." Generation of H_2O_2 and O_2^-. Stimulation by fertilization or Ca^{2+} ionophores. Partial sensitivity to inhibition by KCN, activated by Ca^{2+} and ATP, possible role of protein kinases in the activation process. Inhibition by procaine, phenothiazines and SH reagents. Role in the formation of the fertilization membrane.	Foerder et al. (1978), Turner et al. (1985), Weidman et al. (1985), Heinicke and Shapiro (1989), Takahashi et al. (1989)
Radish	Plasmalemma-associated NAD(P)H oxidase, oxidation of NAD(P)H, production of O_2^-. Stimulation by ferulic acid, inhibition by KCN, EDTA and ascorbic acid. Physiological role not known.	Vianello and Macri (1989)

Chatelet et al. 1977). The results of studies on this subject are not consistent. Controversial results are explained, at least in part, by the difficulties in obtaining a sufficiently high number of eosinophils from healthy subjects and in separating these cells from neutrophils (Petreccia et al. 1987; Yazdanbakhsh et al. 1987a).

The chemotactic tetrapeptide Val-Gly-Ser-Glu has been shown to enhance zymosan-induced $O_2^{\cdot-}$ formation in eosinophils and to inhibit the one in human neutrophils (Beswick and Kay 1981). fMet-Leu-Phe has been reported to induce a respiratory burst in eosinophils, but in these cells the chemotactic peptide is apparently less effective than in neutrophils (Yazdanbakhsh et al. 1987a,b). In contrast, Yamashita et al. (1985) did not find a stimulatory effect of fMet-Leu-Phe on $O_2^{\cdot-}$ formation in eosinophils. Upon exposure to PMA, digitonin, or NaF eosinophils show a higher respiratory burst activity than neutrophils (Roos et al. 1984; Yamashita et al. 1985; Shult et al. 1985; Petreccia et al. 1987; Yazdanbakhsh et al. 1985, 1987a). In comparison to neutrophils, IgG- and complement-coated particles are less potent but similarly effective stimulators of H_2O_2 formation in human eosinophils, suggesting cell type differences in the plasma membrane receptors for these agonists (Yazdanbakhsh et al. 1985; see also Sect. 3.3.1.5). Activation of $O_2^{\cdot-}$ formation by opsonized zymosan and PMA in eosinophils and neutrophils differs with respect to the kinetics, and CB has been reported to potentiate A 23187- or ConA-induced $O_2^{\cdot-}$ formation in human neutrophils but not in eosinophils (Yamashita et al. 1985; Petreccia et al. 1987; see also Sect. 3.2.5). NADPH oxidases of neutrophils and eosinophils possess the same K_m values for substrates, but the enzyme of eosinophils shows a higher V_{max} (Yamashita et al. 1985). These data indicate that the respiratory burst in eosinophils and neutrophils is different with respect to the activation process, activity, stimulus specificity, and kinetics. These cell-type specific properties of the respiratory burst in eosinophils have been suggested to be of relevance in host defense against parasites and in inflammatory tissue injury (Petreccia et al. 1987).

With respect to the activity of the respiratory burst in eosinophilia, the results are controversial. Both increased and decreased activity of the respiratory burst in the patients' eosinophils has been observed (Bass et al. 1980; Pincus et al. 1981; Winquist et al. 1982; Prin et al. 1984). These results indicate that eosinophils from eosinophilic subjects are functionally heterogenous and that eosinophils from eosinophilic and normal subjects are not functionally equivalent (see also Sect. 6.2.3).

3.4.4.1.2 U-937 Cells

In comparison to HL-60 cells, only a relatively limited number of studies on the regulation of the respiratory burst in the human promonocytic cell line U-937 have been performed. Upon exposure to a variety of agents, including dibutyryl cAMP, PMA, 1,25-dihydroxyvitamin D_3, dimethyl sulfoxide, and IFN-γ, U-937 cells acquire the ability to undergo a respiratory burst upon exposure to various stimuli including PMA and opsonized zymosan (Clement and Lehmeyer 1983; Harris and Ralph 1985; Harris et al. 1985; Roux-Lombard et al. 1986; Balsinde and Mollinedo 1988; Banks et al. 1988; Barzaghi et al. 1989; Saussy et al. 1989). Differentiation of U-937 cells is associated with an increase in the cellular content of cytochrome b_{-245} (Balsinde and Mollinedo 1988). Interestingly, differentiation of U-937 cells with various agents may result in the differential expression of signal transduction components required for the activation of NADPH oxidase by specific stimuli (Harris and Ralph 1985; Balsinde and Mollinedo 1988; Polla et al. 1989; see also Sects. 3.2.3.1, 3.3.2.1). Finally, recent studies showed that pertussis toxin-insensitive G-proteins are involved in the activation of phospholipase C by PAF and leukotrienes in dimethyl sulfoxide-differentiated U-937 cells, but the regulation of NADPH oxidase was not analyzed in these investigations (Barzaghi et al. 1989; Saussy et al. 1989; see also Sects. 3.2.1, 3.3.1.6, 3.3.1.7).

3.4.4.1.3 HL-60 Cells

HL-60 cells are a popular model system to study signal transduction processes in human myeloid cells in general and regulation of NADPH oxidase in particular. Certain cytokines, 1,25-dihydroxyvitamin D_3, or PMA induce monocytic differentiation of HL-60 cells (Harris and Ralph 1985; Trinchieri et al. 1986; Collins 1987; Thompson et al. 1988). Dimethyl sulfoxide, dimethyl formamide, retinoic acid, and the cAMP-increasing agents PGE_1, cholera toxin, and dibutyryl cAMP induce neutrophilic differentiation of HL-60 cells (Collins et al. 1978, 1979; Newburger et al. 1979; Breitman et al. 1980; Chaplinski and Niedel 1982; Kitagawa et al. 1984; Harris and Ralph 1985; Collins 1987; Thompson et al. 1988).

Differentiated HL-60 cells generate $O_2^{\cdot-}$ upon exposure to various agents including chemotactic peptides, PMA, A 23187, arachidonic acid, and γ-hexachlorocyclohexane (Seifert et al. 1989c). The expression of formyl peptide receptors and of cytosolic activation factors and increases in the activity of protein kinase C and in the amount of cyctochrome b_{-245} may contribute to the induction of the respiratory burst during myeloid differentiation (Chaplinski et al. 1982; Roberts et al. 1982; Kitagawa et al. 1984; Newburger et al. 1984; Harris and Ralph 1985; Zylber-Katz and

Glazer 1985; Collins 1987; Makowske et al. 1988; see also Sects. 2.4.3, 3.1.1, 3.3.1.1, 5.1.6, 6.1).

In general, regulation of the respiratory burst in dimethyl sulfoxide-differentiated HL-60 cells and human neutrophils is assumed to be similar (Thompson et al. 1988; see also Sect. 3.2.5.1). In contrast, there are substantial differences in oxidative metabolism between 1,25-dihydroxyvitamin D_3-differentiated HL-60 cells and human monocytes (Thompson et al. 1988). With respect to the effects of protein kinase C inhibitors, retinoids, and purine and pyrimidine nucleotides on O_2^- formation in intact cells and to the regulation of NADPH oxidase in cell-free systems, there are substantial differences between dibutyryl cAMP-differentiated HL-60 cells and human neutrophils (Seifert and Schächtele 1988; Seifert et al. 1989b,c,d). In addition, there are differences in the regulation of NADPH oxidase in the cell-free system between dimethyl sulfoxide- and dibutyryl cAMP-differentiated HL-60 cells and dimethyl sulfoxide-differentiated HL-60 cells and human neutrophils, respectively (Seifert and Schultz 1987a,b; Seifert et al. 1988a, 1989a,b,c). Moreover, substantial differences in the activation of O_2^- formation by fMet-Leu-Phe and PMA between dibutyryl cAMP- and retinoic acid-differentiated HL-60 cells have been observed (Rao et al. 1989).

3.4.4.1.4 Kupffer's Cells

Kupffer's cells are tissue macrophages of the liver and are involved in the elimination of material taken up in the gastrointestinal tract (Bhatnagar et al. 1981; Matsuo et al. 1985; Laskin et al. 1988). Activation of the respiratory burst in Kupffer's cells has been suggested to play a role in hepatocyte damage in inflammatory processes of the liver (Arthur et al. 1986; Rieder et al. 1988a). The availability of techniques to isolate and to culture Kupffer's cells has facilitated studies on the respiratory burst in these cells (Bhatnagar et al. 1981; Matsuo et al. 1985; Lepay et al. 1985; Arthur et al. 1986; Laskin et al. 1988; Rieder et al. 1988a). In general, the respiratory burst in Kupffer's cells shows properties similar to those of peritoneal macrophages, but there are quantitative differences between both cell types (Laskin et al. 1988; Rieder et al. 1988a). Various stimuli, e.g., zymosan particles, C5a, and PMA, have been reported to activate the respiratory burst in Kupffer's cells (Bhatnagar et al. 1981; Arthur et al. 1986; Laskin et al. 1988; Rieder et al. 1988a). Treatment of the host with bacteria or exposure to LPS, muramyl dipeptide, or IFN-γ primes Kupffer's cells for an enhanced respiratory burst (Matsuo et al. 1985; Lepay et al. 1985; Arthur et al. 1986; Rieder et al. 1988a).

3.4.4.1.5 J774.16 Cells

The macrophagelike cell line J774.16 was established from a murine reticulum cell sarcoma (Damiani et al. 1980). J774 cells undergo a respiratory burst upon stimulation with PMA, aggregated immunoglobulin, or zymosan (Damiani et al. 1980; Tanaka Y. et al. 1982; Tosk et al. 1989). Bacteria or LPS prime J774 macrophages for an enhanced respiratory burst, as do IFN-α and IFN-β (Tosk et al. 1989; see also Sect. 3.3.1.3.3). The variant cell clone, J774.C3C, does not undergo a respiratory burst but still possesses the ability to generate R-NO (Iyengar et al. 1987; see also Sect. 3.4.1). J774.C3C cells have been reconstituted with an $O_2^{\cdot-}$-generating system using zymosan particles covalently coupled to glucose oxidase (Tanaka Y. et al. 1982). The molecular basis of the defective respiratory burst in J774.3C3 cells apparently does not involve alterations of glucose transport, changes in hexose monophosphate shunt activity, changes in protein kinase C-mediated phosphorylation reactions, or reduction in the cellular content of cytochrome b_{-245} (Kiyotaki et al. 1984). The defect of the respiratory burst in J774.C3C cells has been suggested to be associated with a structural or functional abnormality of cytochrome b_{-245} (Kiyotaki et al. 1984).

3.4.4.2 Other Cell Types

3.4.4.2.1 B-Lymphocyte Cell Lines

There is increasing evidence for the assumption that lymphocytic cell lines may undergo a respiratory burst. Recently, fMet-Leu-Phe has been shown to induce phosphoinositide degradation in human peripheral blood lymphocytes, but the question whether the chemotactic peptide induces $O_2^{\cdot-}$ formation in these cells, has not been investigated (Schubert and Müller 1989; see also Sect. 3.3.1.1). Activation of human natural killer cells has been suggested to be associated with enhanced chemiluminescence, and Sendai virus has been reported to induce a respiratory burst in rat thymocytes (Helfand et al. 1982; Kolbuch-Braddon et al. 1984). However, contaminating monocytes and neutrophils may have contributed, at least in part, to the respiratory burst (Maly et al. 1988). In addition, mitogen-stimulated human lymphocytes have been reported to reduce NBT but not to generate H_2O_2 or $O_2^{\cdot-}$ (Melinn and McLaughlin 1987).

Ebstein-Barr virus (EBV) transformed B-lymphoblasts may show toxicity against tumor cells (Bersani et al. 1987). Certain EBV-transformed human B-lymphocyte cell lines have been shown to generate $O_2^{\cdot-}$ upon exposure to PMA, a contamination with phagocytes being unlikely

(Volkman et al. 1984). In contrast, EBV-transformed lymphocytes from CGD patients do not generate O_2^- (Volkman et al. 1984). EBV-transformed F1 and HELL cells but not EBV-negative U-266 plasmocytoma cells or EBV-positive Burkitt lymphoma WIL-2 cells possess the ability to undergo a respiratory burst (Maly et al. 1988). F1 and HELL cells but not WIL-2 or U-266 cells contain cytochrome b_{-245} and the 45-kDa diphenylene iodonium-binding protein, and diphenylene iodonium inhibits PMA-stimu-lated O_2^- formation in F1 and HELL cells (Maly et al. 1988, 1989; see also Sect. 2.4.1). In addition, the lymphoma cell lines P3HR-1, Jijoye, and RPMI 1788 contain cytochrome b_{-245} and generate O_2^- upon exposure to PMA (Hancock et al. 1989). The K_m for NADPH of NADPH oxidase amounts to about 250 μM for P3HR-1 and RPMI 1788 cells and to about 30 μM for Jijoye cells (Hancock et al. 1989). Moreover, NADH is a much less effective substrate for NADPH oxidase in these cells than is NADPH (Hancock et al. 1989). Finally, nontransformed B-lymphocytes may generate reactive oxygen intermediates in vivo, as human tonsillar B-lymphocytes have been shown to contain cytochrome b_{-245} and to reduce NBT upon stimulation with PMA or immunoglobulins (Maly et al. 1989). Kobayashi et al. (1990) showed that most peripheral B-lymphocytes but not T-lymphocytes or natural killer cells possess cytochrome b_{-245} and reduce NBT upon stimula-tion. Interestingly, cytochrome b_{-245} is not found in pre-B-lymphocytes or pre-B cells and disappears during the terminal differentiation of B-lym-phocytes to plasma cells (Kobayashi et al. 1990). These data suggest that some human B-lymphocyte cell lines possess an electron transport chain closely related to that present in neutrophils or monocytes, and that human B-lym-phocytes may possess cytotoxic properties.

Recently described substances which activate NADPH oxidase in EBV-transformed lymphocyte cell lines include PAF (Leca et al. 1990), TNF-α, IL-1β, LPS, NaF, A 23187, and ionomycin (Hancock et al. 1990).

3.4.4.2.2 Mesangial Cells, Tubular Epithelial Cells, and Glia Cells

The presence of a respiratory burst in mesangial cells was suggested by the finding that PMA induces chemiluminescence in isolated rat glomeruli (Shah and Nauun-Bedigian 1981). Unlike in other cell types, the PMA-in-duced respiratory burst in rat glomeruli is sensitive to inhibition by various cAMP-increasing agents (Miyanoshita et al. 1989; see also Sect. 4.1). In cultured rat mesangial cells, opsonized zymosan induces O_2^- and H_2O_2 formation and eicosanoid release, and CB inhibits the respiratory burst (Baud et al. 1983; see also Sects. 3.2.5, 3.3.1.5). Inhibitors of lipoxygenases and glucocorticoids inhibit the respiratory burst in mesangial cells (Baud et al. 1983, 1986; see also Sects. 3.2.4, 4.2.1). In addition, human complement components and normal human serum have been reported to synergistically activate O_2^- and H_2O_2 formation in cultured rat mesangial cells (Adler et al. 1986). Furthermore, certain proteases have been reported to activate the respiratory burst in glomeruli (Basci and Shah 1987; see also Sect. 3.2.8). Finally, human mesangial cells were recently shown to generate O_2^- and H_2O_2 upon stimulation with opsonized zymosan, whereas unopsonized zymosan and PMA are only weakly active or inactive (Radeke et al. 1990). The cytokines, TNF-α and IL-1α, induce a very long-lasting release of reactive oxygen species in mesangial cells (Radeke et al. 1990). The respiratory burst in mesangial cells may play a role in the pathogenesis of renal inflammatory processes and glomerular damage (Baud et al. 1983; Adler et al. 1986; Radeke et al. 1990).

In addition to mesangial cells, tubular epithelial cells are discussed as playing a part in the pathogenesis of renal injury in various conditions (Rovin et al. 1990). This view is supported by the recent finding that epithelial cells from the proximal tubulus, the cortical collecting duct, and the papillary collecting duct from rabbit generate reactive oxygen species in the absence of chemical stimulation (Rovin et al. 1990). Opsonized zymosan and heat-aggregated IgG enhance this basal formation of O_2^- and H_2O_2 in a time- and concentration-dependent manner. The identity of the enzyme involved in the formation of reactive oxygen species in tubular epithelial cells is not yet known.

Upon exposure to PMA, microglia cells from neonatal and adult rats generate O_2^-, and the respiratory burst is enhanced by IFN-γ (Woodroofe et al. 1989). In addition, murine glia cells have been reported to reduce NBT, and these cells show enhanced chemiluminescence upon stimulation with PMA, zymosan, or antibody-coated bovine erythrocytes (Sonderer et al. 1987). These data suggest that the respiratory burst in glia cells plays a role in immunopathological reactions of the brain.

3.4.4.2.3 Thyroid Cells, Epidermis Cells, and Chondrocytes

The synthesis of thyroid hormones by thyroid peroxidase requires iodination of tyrosine residues in thyreoglobulin, and this reaction depends on H_2O_2 (Deme et al. 1985; Nakamura et al. 1987, 1989). Plasma membrane fractions of thyroid cells possess an H_2O_2-generating and NADPH-oxidizing enzyme system which shows some properties similar to NADPH oxidase of phagocytes. The K_m for NADPH of the thyroid enzyme amounts to 35 μM, and NADH is a much less effective electron donor. KCN does not inhibit H_2O_2 formation, whereas Ca^{2+} and ATP enhance H_2O_2 formation (Deme et al. 1985; Nakamura et al. 1987). The primary product of the enzyme reaction has been suggested to be $O_2^{\cdot-}$, and H_2O_2 may be provided by dismutation of $O_2^{\cdot-}$ (Nakamura et al. 1989).

Murine epidermal cells generate H_2O_2 upon stimulation with PMA, a process which may play a role in PMA-mediated tumor promotion in skin (Robertson et al. 1990).

Rabbit articular chondrocytes generate H_2O_2 upon exposure to ConA, and IFN-γ and TNF show priming effects (Tiku et al. 1990). In comparison to alveolar macrophages, chondrocytes generate larger amounts of H_2O_2. Production by chondrocytes of reactive oxygen intermediates may play a part in the cartilage matrix degradation that occurs in arthritis.

3.4.4.2.4 Carotid Body, Fibroblasts, Fat Cells, and Endothelial Cells

The rat carotid body possesses an NADPH oxidase which shows certain properties similar to the ones of NADPH oxidase in neutrophils (Acker et al. 1989). NADPH oxidase of the carotid body has been suggested to play a role as sensor for the oxygen concentration in the arterial blood (Acker et al. 1989). The presence of an NADPH oxidase in the rat carotid body is further supported by the finding that this tissue shows a typical spectrum of cytochrome b and that diphenylene iodonium inhibits H_2O_2 formation (Cross et al. 1990).

In adherent cultured human skin fibroblasts, IL-1 and TNF-α were found to induce a long-lasting respiratory burst (Meier et al. 1989; see also Sect. 3.4.3). $O_2^{\cdot-}$ formation in fibroblasts is not inhibited by NaN_3 or KCN and is enhanced by NADPH, suggesting that an NADPH oxidase similar to that in phagocytes is involved in the generation of reactive oxygen intermediates in these cells (Meier et al. 1989).

Murrell et al. (1990) reported that cultured human fibroblasts generate and release $O_2^{\cdot-}$, and that $O_2^{\cdot-}$ at the concentrations released may stimulate proliferation of the fibroblasts. Prolonged autocrine stimulation of fibroblast replication by $O_2^{\cdot-}$ may contribute to the pathogenesis of fibrosis (Murrell et al. 1990).

Several years ago, insulin was reported to stimulate H_2O_2 formation in rat epididymal fat cells (May and de Haen 1979). Nerve growth factor induces the formation of H_2O_2 in adipocytes as well (Mukherjee and Mukherjee 1982). The formation of H_2O_2 is linked to the metabolism of glucose, and H_2O_2 has been suggested to play a role as intracellular signal molecule for some effects of insulin (May and de Haen 1979). It remains to be determined whether an NADPH oxidase-related enzyme system is involved in the formation of H_2O_2.

In 1984, Rosen and Freeman reported that endothelial cells generate and release $O_2^{\cdot-}$. Subsequently, Matsubara and Ziff (1986a) showed that

PMA enhances O_2^- formation in cultured human umbilical vein endothelial cells. Similar to O_2^- formation in neutrophils, PMA-triggered
O_2^- formation in endothelial cells does not depend on the presence of
extracellular Ca^{2+} (Matsubara and Ziff 1986a; see also Sect. 3.1.1.1). The
calcium ionophore A 23187 also augments O_2^- formation in endothelial
cells. A 23187 and the phorbol ester interact synergistically to enhance
O_2^- release (Matsubara and Ziff 1986a; see also Sect. 3.1.1.1). In addition
to the above stimuli, bradykinin, IFN-γ, and IL-1 show stimulatory
effects on O_2^- formation in human endothelial cells (Matsubara and Ziff
1986b; Holland et al. 1990). The physiological role of endothelium-in-
duced O_2^- formation may be complex. O_2^- or O_2^--derived radicals may
inactivate endothelium-derived nitric oxide and related reactive nitrogen
oxide intermediates which induce vasodilation (see also Sect. 3.4.1). In
addition, O_2^- may induce vasoconstriction and has been suggested to play
a role in the pathogenesis of certain types of arterial hypertension (Wei
et al. 1985; Auch-Schwelk et al. 1989; see also Sect. 6.2.4).

3.4.4.2.5 Sea Urchin Eggs

Decades ago, Otto Warburg observed that fertilization of sea urchin eggs
is associated with increased oxygen consumption; this process is referred to
as "the respiratory burst of fertilization" (cited in Foerder et al. 1978). The
complex biochemical changes accompanying fertilization have been
reviewed (Shapiro et al. 1981; Garbers 1989).

A number of studies showed that fertilization is associated with the
formation of reactive oxygen intermediates, e.g., H_2O_2, and that H_2O_2
may prevent polyspermy (Foerder et al. 1978; Turner et al. 1985; Weidman et al. 1985; Heinicke and Shapiro 1989). H_2O_2 is generated by the
"respiratory burst oxidase," and the H_2O_2 formed by this enzyme
provides the substrate for an "ovoperoxidase." The latter enzyme crosslinks tyrosine residues of surface glycoproteins of the egg and forms an
impermeable fertilization membrane (Foerder and Shapiro 1977; Foerder et al. 1978; Shapiro et al. 1981; Turner et al. 1985; Weidman et al.
1985; Heinicke and Shapiro 1989). In addition, the H_2O_2 formed may be
toxic for sperm (Shapiro et al. 1981).

The respiratory burst oxidase is activated by fertilization or by A 23187
and is a partially KCN-sensitive NAPDH:O_2 oxidoreductase, and H_2O_2 is
the initial product (Foerder et al. 1978; Turner et al. 1985; Heinecke and
Shapiro 1989). Recent data indicate that O_2^- is also formed during fertilization (Takahashi et al. 1989). In a cell-free system, the respiratory burst
oxidase is activated by Ca^{2+} at physiologically relevant concentrations and
by ATP and is inhibited by H-7, staurosporine, phenothiazines, and NEM

(Heinicke and Shapiro 1989; see also Sects. 3.2.2.3, 4.3.3, 5.1). These data show that the regulation of the NADPH oxidase of sea urchin eggs and of phagocytes is similar in some respects (Shapiro et al. 1981), and that kinases may be involved in the activation process.

3.4.4.2.6 Plant Cells

Finally, plant cells, e.g., radish, possess enzymes related to NADPH oxidase of animals (Vianello and Macri 1989). The formation of reactive oxygen intermediates in plant cells may be involved in the regulation of various processes, such as resistance to plant pathogens, growth, seed germination, and biosynthesis of lignin (Vianello and Macri 1989). A detailed discussion of the structure, catalytical properties, regulation and function of plant NAD(P)H oxidases is, however, beyond the scope of this review.

4 Inhibition of NADPH Oxidase

The respiratory burst is inhibited by numerous agents. In many cases, the precise mechanisms underlying inhibition of the respiratory burst are not yet known and are a subject of controversial discussion. The effects of protein kinase C inhibitors on the respiratory burst are dealt with in Sect. 3.2.2.3 and in Table 3. The inhibitory effects of cAMP-increasing agents on NADPH oxidase are summarized in Table 11 and are discussed with in Sect. 4.1. Table 12 summarizes the effects of various anti-inflammatory drugs on the respiratory burst (see Sect. 4.2), and Table 13 deals with the inhibitory effects of microbial products on NADPH oxidase (see Sect. 4.3.1). Finally, Tables 14 and 15 summarize data on miscellaneous inhibitors of the respiratory burst. Some of these agents are described in more detail in Sects. 4.3.2 and 4.3.3 and in various other sections of this review. Recently, Cross (1990) presented a critical and extensive review on inhibitors of NADPH oxidase.

4.1 cAMP-Increasing Agents

Neutrophils and mononuclear phagocytes possess G_s and adenylyl cyclase. Similar to other cell types, adenylyl cyclase in phagocytes is activated by stable guanine nucleotides, NaF, forskolin, and various intercellular signal molecules, i.e., prostaglandins, β-adrenergic agonists, and histamine (Stolc 1977; Verghese and Snyderman 1983; Lad et al. 1984; Verghese et al. 1985b; Meurs et al. 1986; Bokoch 1987; Motulsky et al. 1987). Unexpectedly, pertussis toxin was recently found to inhibit the increase in cAMP in human monocytes induced by β-adrenergic agonists and PGE_1, whereas that stimulated by forskolin is not affected (Griese et al. 1990). The inhibitory effect of pertussis toxin cannot be explained by the involvement of G_s as this G-protein is not a substrate for pertussis toxin (Gilman 1987). Unexpected effects of cAMP-increasing receptor agonists on signal transduction processes were also recently observed in HL-60 cells (Mitsuhashi et al. 1989). In

Table 11. Inhibition of the respiratory burst by cAMP-increasing agents

Inhibitory agent	Respiratory burst stimulated by	Mechanisms discussed	Selected references
Prostaglandins (e.g., PGE_1, PGE_2, PGD_2 $PGF_{2\alpha}$, PGI_2)	Chemotactic peptides. zymosan; A23187 and PMA (controversial); arachidonic acid (no effect); OAG (potentiation)	Stimulation of prostaglandin receptors, activation of adenylyl cyclase, increase in cAMP	Weissmann et al. (1980), Fantone et al. (1983), Lim et al. (1983), Fantone and Kinnes (1983), Lad et al. (1985a), Penfield and Dale (1985), Gryglewski et al. (1987)
β-Adrenergic agonists (e.g., isoproterenol, fenoterol)	Chemotactic peptides, A23187, zymosan; OAG, PMA, NaF (no effect)	Stimulation of β_1/β_2 receptors, activation of adenylyl cyclase, increase in cAMP	Schopf and Lemmel (1983), Lad et al. (1985a), Tecoma et al. (1986), Mack et al. (1986), Mueller et al. (1988), Mueller and Sklar (1989), Tenner et al. (1989)
Histaminergic agonists (e.g., histamine, dimaprit, impromidine, arpromidine)	Chemotactic peptides; PMA (no effect)	Stimulation of (cell type-specific?) H_2 receptors, activation of adenylyl cyclase, increase in cAMP, cAMP-independent pathways (phosphoinositide degradation, Ca^{2+} mobilization)?	Gespach and Abita (1982), Gespach et al. (1982), Seligmann et al. (1983), Ozaki et al. (1984b), Burde et al. (1989, 1990), Buschauer (1989), Mitsuhashi et al. (1989)
Adenosine A_2 agonists (e.g., adenosine, NECA)	Chemotactic peptides	Stimulation of A_2 receptors, increase in cAMP, cAMP-independent pathways?	Pike and Snyderman (1982), Garcia-Castro et al. (1983), Cronstein et al. (1983, 1985, 1988), Roberts et al. (1985), Schrier and Imre (1986), Elliott et al. (1986), Elliott and Leonhard (1989), Nielson and Vestal (1989)
ATP, ADP	Chemotactic peptides	Conversion to adenosine, other mechanisms?	McGarrity et al. (1988a,b, 1989)

Table 11. (continued)

Inhibitory agent	Respiratory burst stimulated by	Mechanisms discussed	Selected references
Cholera toxin	Chemotactic peptides	ADP ribosylation of G_s, permanent activation of adenylyl cyclase, increase in cAMP, ADP ribosylation of G-proteins coupling to formly peptide receptors?	Bokoch and Gilman (1984), Feltner et al. (1986), Verghese et al. (1986c), Gilman (1987), McLeish et al. (1989a), Seifert et al. (1989b)
NaF	Chemotactic peptides	Activation of G_s, activation of adenylyl cyclase, increase in cAMP	Wong (1983), Saad et al. (1987)
Bordetella pertussis adenylyl cyclase	Zymosan, PMA	Penetration into host cells, activation by cal-modulin, increase in cAMP	Pearson et al. (1987), Hanski (1989)
Cis-unsaturated fatty acids	Chemotactic peptides; NaF, PMA (no effect)	Activation of adenylyl cyclase through perturbation of plasma membrane structure, increase in cAMP?	Houslay and Gordon (1983), Wong and Chew (1984)
Forskolin	Chemotactic peptides (neutrophils), PMA (mesangial cells)	Direct activation of adenylyl cyclase	Burde et al. (1989), Miyanoshita et al. (1989)
Methylxanthines (e.g., IBMX, theophylline), cAMP-specific PDE inhibitors (e.g., Ro 20-1724, rolipram)	Chemotactic peptides, opsonized zymosan (controversial, inhibition or no inhibition; neutrophils, mononuclear cells), PMA (mesangial cells)	Inhibition of phosphodiesterases, increase in cAMP (and cGMP)	Lad et al. (1985a), Bessler et al. (1986), Cronstein et al. (1988), Burde et al. (1989), Elliott and Leonhard (1989), Miyanoshita et al. (1989), Yukana et al. (1989)
Dibutyrol cAMP	Chemotactic peptides (neutrophils), PMA (mesangial cells)	Activation of cAMP-dependent protein kinase	Kitagawa and Takaku (1982, Lad et al. (1985a), Kramer et al. (1988a), Burde et al. (1989), Miyanoshita et al. (1989)

Table 12 Inhibition of the respiratory burst by anti-inflammatory agents

Inhibitory agent	Respiratory burst stimulated by	Mechanisms discussed	Selected references
Glucocorticoids	Controversial (no inhibition or partial inhibition), various stimuli and cell types, no inhibition of exocytosis	Interference with chemotactic peptide-binding to formyl peptide receptors, induction of de novo protein synthesis (lipocortin) and inhibition of release of arachidonic acid, direct inhibition of NADPH oxidase; other mechanisms?	Oyanagui (1978), Drath and Kahan (1983, 1984), Schultz et al. (1985), Baud et al. (1986), Dieter et al. 1986), Müller-Peddinghaus and Wurl (1987), Rieder et al. (1988b), Maridonneau-Parini et al. (1989), Schleimer et al. (1989), Umeki and Soejima (1990)
Lipocortin I	A 23187; PMA (no inhibition)	Reduced activity of phospholipases and release of arachidonic acid	Machoczek et al. (1989), Maridonneau-Parini et al. (1989)
Cyclosporin A	Controversial (inhibition or no inhibition), various stimuli and cell types	Inhibition of phospholipases and/or protein kinase C and calmodulin-dependent processes; other mechanisms?	Drath and Kahan (1983, 1984), Janco and English (1983), Kharazmi et al. (1985), Niwa et al. (1986), Gschwendt et al. (1988), Chiara et al. (1989), R.J. Walker et al. (1989)
Gold compounds (e.g., auranofin, sodium aurothiomalate)	Various stimuli	Interference with cytoskeleton organization, increase in affinity of formyl peptide receptors, inhibition of protein kinase C	Davis et al. (1983), Harth et al. (1983), Hafström et al. (1984), Sung et al. (1984), Schultz et al. (1985), Froscio et al. (1989), Mahoney et al. (1989), Parente et al. (1989)
Sulfasalazine, olsalazine	Immune complexes, chemotactic peptides	Interference with binding of chemotactic peptides to formyl peptide receptors; other mechanisms?	Stenson et al. (1984), Neal et al. (1987b)
Chloroquine, mepacrine, primaquine	Various stimuli, stimulus dependency	Inhibition of phospholipase A_2, inhibition of glucose uptake; other mechanisms, unspecific effects?	Tauber and Simons (1983), Schultz et al. (1985), Hurst et al. (1986), Tsunawaki and Nathan (1986), Panus and Jones (1987)

Table 12. (continued)

Inhibitory agent	Respiratory burst stimulated by	Mechanisms discussed	Selected references
p-Bromophena-cylbromide	Various stimuli	Inhibition of phospholipase A_2, reduced release of arachidonic acid, inhibition of glucose uptake, unspecific effects	Smolen and Weissmann (1980), Bromberg and Pick (1983), Schultz et al. (1985), Maridonneau-Parini et al. (1986), Tsunawaki and Nathan (1986), Sakata et al. (1987b)
ETYA, BW 755C	Various stimuli	Inhibition of lipoxygenases, inhibition of glucose uptake, unspecific effects	Bokoch and Reed (1979), Smolen and Weissmann (1980), Schultz et al. (1985), Maridonneau-Parini et al. (1986), Tsunawaki and Nathan (1986), Ozaki et al. (1986b)
Nonsteroidal anti-inflammatory drugs (e.g., phenylbutazone, diclofenac, piroxicam, indomethacin, mefeamic acid, salicylates)	Controversial (no inhibition or inhibition), various stimuli, stimulus dependency	Inhibition of phospholipases A_2 and C, interference with binding of chemotactic peptides to formyl peptide receptors, inhibition of Ca^{2+} mobilization	Bokoch and Reed (1979), Kaplan et al. (1984), Abramson et al. (1985), Skubitz and Hammerschmidt (1986), Bomalaski et al. (1986), Müller-Peddinghaus and Wurl (1987), Perianin et al. (1987), Neal et al. (1987), Shelly and Hoff (1989)

Table 13. Inhibition of the respiratory burst by microbial products and components and by microorganisms

Inhibitory agent	Respiratory burst stimulated by	Mechanisms discussed	Selected references
Cytochalasin B	Various agents (e.g., digitonin, substance P)	Inhibition of actin polymerization	Williams and Cole (1981), Serra et al. (1988)
Pertussis toxin	Chemotactic peptides, other agonists for receptors which couple to phospholipase C via G-proteins	ADP ribosylation of G-proteins, uncoupling of receptors from G-proteins	Bokoch and Gilman (1984), Molski et al. (1984), Okajima and Ui (1984), Verghese et al. (1985a), Naccache et al. (1986)
Anthrax toxin	Potentiation of chemotactic peptide-induced O_2^- formation by LPS or muramyl dipeptide; activation by PMA and chemotactic peptides (no effect)	Unknown	Wright and Mandell (1986)
Wortmannin	Chemotactic peptides, C5a, PAF LTB4; PMA (no effect)	Inhibition of signal transduction processes independent of Ca^{2+} mobilization and activation of protein kinase C; inhibition of phospholipase D activation	Dewald et al. (1988), Reinhold et al. (1990)
LPS	Priming by IFN-γ and TNF	Release of prostaglandins with subsequent activation of adenylyl cyclase and increase in cAMP	Ding and Nathan (1987)
	Immune complexes; PMA (no effect)	Unknown; no interference with binding of immune complexes to Fc receptors	Johnston et al. (1985)
	Opsonized particles, PMA	Partial dexamethasone sensitivity	Rellstab and Schaffner (1989)

Table 13. (continued)

Inhibitory agent	Respiratory burst stimulated by	Mechanisms discussed	Selected references
Chlamydia trachomatis	Chemotactic peptides, PMA	Unknown; direct inhibition of NADPH oxidase?	Tauber et al. (1989b)
Candida albicans hyphae	Chemotactic peptides, A23187, opsonized zymosan; PMA (no effect)	Unknown; no interference with Ca^{2+} mobilization or phosphoinositide degradation	Smail et al. (1988)
Histoplasma capsulatum	Controversial; inhibition (macrophages) or stimulation (neutrophils) of the respiratory burst	Inhibition: unknown; stimulation: involvement of complement receptors	Wolf et al. (1989), Schnur and Newman (1990)
Leishmania donovani tartrate-resistant acid phosphatase	Chemotactic peptides	Unknown	Remaley et al. (1984)

Table 14. Inhibition of the respiratory burst by agents interfering with redox components of NADPH oxidase, proteins, and chemotherapeutic agents

Inhibitory agent	Respiratory burst stimulated by	Mechanisms discussed	Selected references
Compounds interfering with redox components			
5-Deaza FAD	Particulate activated NADPH oxidase	Inhibition of electron transfer via flavoprotein component of NADPH oxidase	Light et al. (1981)
Diphenylene iodonium	Various stimuli (intact cells)	Inhibition of a 45-kDa component of NADPH oxidase (flavoprotein?)	Cross and Jones (1986), Hancock and Jones (1987), Ellis et al. (1988, 1989)
Pyrimidine, imidazole	PMA (intact cells) and particulate activated NADPH oxidase	Inhibition of cytochrome b_{-245} reduction	Iizuka et al. (1985b), Ellis et al. (1989)
NADPH dialdehyde	Fatty acids (cell-free system)	Inactivation of the NADPH-binding component of NADPH oxidase (66 kDa)	Takasugi et al. (1989), Smith et al. (1989a,b)
Cibacron blue	Particulate activated NADPH oxidase	Inhibition of the NADPH binding site of NADPH oxidase	Yamaguchi and Kakinmma (1982)
Proteins			
Fibrinogen degradation products	Chemotactic peptides, OAG, PMA	Interference with agonist binding to formyl peptide receptors and protein kinase C activation	Kazura et al. (1989)
α₂-Macroglobulin	PMA	Unknown	Hoffman et al. (1983)
Heat shock proteins	Opsonized zymosan, PMA	Synthesis of heat shock proteins subsequent to cell injury	Maridonneau-Parini et al. (1988)
C-reactive protein	Chemotactic peptides, PAF, PMA	Interference with binding of stimuli to the plasma membrane, increase in cAMP?	Buchta et al. (1987a,b), Filep and Földers-Filep (1989)

Table 14. (continued)

Inhibitory agent	Respiratory burst stimulated by	Mechanisms discussed	Selected references
Haptoglobin	Chemotactic peptides, arachidonic acid, opsonized zymosan	Modulation of interaction of ligands with receptors	Oh et al. (1990)
IL-4	IFN-γ, PMA	Unknown; alteration of protein biosynthesis?	Lehn et al. (1989), Abramson and Gallin (1990)
Chemotherapeutic agents			
Doxycyclin	Various stimuli	Chelation of divalent cations	Sinico-Durieux et al. (1986)
Clindamycin	Chemotactic peptides	Interference with agonist binding to formyl peptide receptors	Moon et al. (1986), Solomkin et al. (1986)
Aminoglycosides (e.g., neomycin, tobramycin)	C5a (intact cells), arachidonic acid (cell-free system)	Inhibition of phospholipases, interaction with phospholipids, interference with agonist binding to chemoattractant receptors; other mechanisms?	Moon et al. (1986), Banks et al. (1988), Aviram and Sharabani (1989), Herrmann et al. (1989)
Anthracyclines (idarubicin, epirubicin)	Various stimuli	Unknown	Cairo et al. (1990)

Table 15. Inhibition of the respiratory burst by miscellaneous agents

Inhibitory agent	Respiratory burst stimulated by	Mechanisms discussed	Selected references
Essential fatty acid deficiency	Chemotactic peptides; PMA (no effect)	Inhibition of arachidonate metabolism, alteration of membrane structure	Palmblad et al. (1988b), Gyllenhammar and Palmblad (1989)
Lipid thiobis ester	Activated NADPH oxidase, cell-free system	Reversible interference with components of NADPH oxidase involved in its deactivation	Eklund and Gabig (1990)
Oxygenated sterols (6-ketocholestanol)	PMA	Perturbation of the plasma membrane	Vasconcelles et al. (1990)
Local anesthetics (e.g., cocaine, tetracaine)	fMet-Leu-Phe, PMA, A 23187	Unknown, inhibition of O_2^- formation without inhibition of phosphorylation of the 47 kDa protein	Haines et al. (1990)
Phenol compounds	PMA, opsonized zymosan	Metabolic activation by reaction with products released from activated neutrophils; interference with the assembly of the functionally active NADPH oxidase?	'T Hart et al. (1990), Simons et al. (1990)
Phalloidin	Chemotactic peptides	Inhibition of depolymerization of actin	Al-Mohanna and Hallett (1990)
Paraquat, t-butyl peroxide	ConA	Depletion of NADPH	Forman et al. (1980), Sutherland et al. (1985)
Mannose	Various stimuli	Inhibition of hexose monophosphate shunt	Rest et al. (1988)
Levamisole	Controversial (inhibition or none)	Unknown	Schinetti et al. (1984), Müller-Peddinghaus and Wurl (1987)
Triphenyltin compounds	Various stimuli	Unknown; inhibition of Ca^{2+} mobilization	Matsui et al. (1983a,b), Miura and Matsui (1989)

Table 15. (continued)

Inhibitory agent	Respiratory burst stimulated by	Mechanisms discussed	Selected references
SH reagents (e.g., NEM, p-chloromer-curibenzoate, ethacrynic acid)	Various stimuli	Covalent modifications of components of NADPH oxidase and/or components involved in its activation	Yamashita (1983), Yamashita et al. (1984), Cross et al. (1984), Pick et al. (1987), Elferink (1987), Akard et al. (1988)
Positively charged alkylamines	Various stimuli	Functional antagonism with the stimulatory effects of negatively charged lipids	Cross et al. (1984), Miyahara et al. (1987, 1988)
Aliphatic alcohols	Controversial (inhibition or activation)	Interference with G-proteins, phospholipases C and D, Ca^{2+} mobilization, membrane fluidization, increased affinity of formyl peptide receptors	Yuli et al. (1982), Hoek et al. (1987), Dorio et al. (1988), Nilsson and Palmblad (1988), Rubin and Hoek (1988), Bonser et al. (1989)
Fructose 1,6-diphosphate	PMA	Interference with the regulation of the intracellular ATP concentration	Schinetti and Lazarino (1986)
Amiloride	Chemotactic peptides	Inhibition of Na^+/H^+-exchange, inhibition of protein kinase C and other kinases, interference with G-proteins?	Besterman et al. (1985), Davis and Czech (1985), Simchowitz (1985), Berkow et al. (1987), Anand-Srivastava (1989)
Stilbene sulfonic acids	Controversial (no inhibition or inhibition various stimuli)	Blockade of anion transport, other mechanisms?	Korchak et al. (1980), Tauber and Goetzl (1981), Kaplan et al. (1982), Smith et al. (1984a), Vostal et al. (1989)
Tiazofurine	NaF, chemotactic peptides; PMA (no effect)	Inhibition of inosine monophosphate dehydrogenase, GTP depletion leading to inhibition of G-proteins	English et al. (1989)
Histamine H_1 antagonists	Chemotactic peptides, PMA	Inhibition of membrane potential changes, local anesthetic effects?	Ozaki et al. (1984b), Taniguchi and Takanaka (1984)
Diazepam	Chemotactic peptides	Unknown	Moon et al. (1986)

Table 15. (continued)

Inhibitory agent	Respiratory burst stimulated by	Mechanisms discussed	Selected References
Protease inhibitors (e.g., PMSF, TPCK)	Controversial (inhibition or no inhibition)	Inhibition of proteases?, interaction with SH groups? inhibition of actin polymerization? interaction with components of NADPH oxidase?	Goldstein et al. (1979), Kitagawa et al. (1979, 1980a), Hoffman and Autor (1982), Duque et al. (1983), Tsuan (1983), Basci and Shah (1987), Rao and Castronowa (1988), Conseiller and Lederer (1989)
Various tumor cells or tumor cell products	Controversial (inhibition or stimulation)	Unknown	Spitalny, (1980), Szuro-Sudol and Nathan (1982), Ghezzi et al. (1987), Lichtenstein (1987)
8-kDa factor from K 562 cells	Opsonized Zymosan; PMA and fMet-Leu-Phe (no effect)	Interference with adherence-related functions	Amar et al. (1990)
H_2O_2	Controversial (no inhibition or inhibition, various stimuli)	Unknown	Rajkovic and Williams (1985b), Mege et al. (1986)
Cd^{2+}, Zn^{2+}	PMA	Blockade of H+ efflux	L.M. Henderson et al. (1987, 1988a,b)
Pb^{2+}	LPS	Inhibition of glucose transport	Buchmüller-Rouiller et al. (1989)
Co^{2+}	PMA, chemotactic peptides	Interaction with Ca^{2+}-dependent processes in the intracellular space	Elferink and Deierkauf (1989b)

agreement with these authors, we found that PGE_1, isoproterenol, and histamine increase cytoplasmic Ca^{2+} in undifferentiated HL-60 cells (unpublished results; see also 4.1.1).

4.1.1 Receptor Agonists

Prostaglandins, β-adrenergic agonists, and histamine induce an increase in intracellular cAMP in various myeloid cell types including neutrophils, HL-60 cells and U-937 cells, and phosphodiesterase inhibitors potentiate receptor-mediated cAMP accumulation (Bourne and Melmon 1971; Busse and Sosman 1976; Hamachi et al. 1984; Gespach and Abita 1982; Gespach et al. 1982, 1985, 1986; Ham et al. 1983; Rivkin et al. 1975; Lad et al. 1985a; Verghese et al. 1985a).

Prostaglandins, β-adrenergic agonists, and histamine are similarly effective inhibitors of the chemotactic peptide-induced $O_2^{\cdot-}$ formation in human neutrophils, and the maximum rate of $O_2^{\cdot-}$ formation somewhat is less sensitive to inhibition by cAMP-increasing agents than the absolute amount of $O_2^{\cdot-}$ generated (Burde et al. 1989; Seifert et al. 1991). Other cAMP-increasing intercellular signal molecules, e.g., dopamine, glucagon, and vasointestinal peptide do not inhibit fMet-Leu-Phe-induced $O_2^{\cdot-}$ formation in human neutrophils (Seifert, unpublished results).

4.1.1.1 Prostaglandins Including Prostacyclin

In parallel with the increase in cAMP, various prostaglandins, e.g., PGE_1, PGE_2, PG 6-keto-E_1, PGD_2, and $PGF_{2\alpha}$ inhibit stimulated $O_2^{\cdot-}$ formation, exocytosis, and LTB_4 release in neutrophils (Rivkin et al. 1975; Weissmann et al. 1980; Ham et al. 1983; Fantone and Kinnes 1983; Fantone et al. 1983; Lad et al. 1985a; Gryglewski et al. 1987). Inhibition of $O_2^{\cdot-}$ formation by prostaglandins in neutrophils shows stimulus specificity, as is the case of other cAMP-increasing agents. For example, PGE_2 has been reported to inhibit the respiratory burst induced by intercellular signal molecules such as fMet-Leu-Phe, PAF, ATP, UTP, and combinations of these agents, whereas that induced by arachidonic acid, PMA, or A 23187 is insensitive to inhibition (Lim et al. 1983; Fantone and Kinnes 1983; Fantone et al. 1984; Penfield and Dale 1985; Gryglewski et al. 1987; Channon et al. 1987; Seifert et al. 1989b; Burde et al. 1989). In contrast, PGE_2 has been reported to inhibit the PMA-induced respiratory burst in murine macrophages primed with LPS (Metzger et al. 1981).

Inhibition of the respiratory burst by prostaglandins may be of relevance in vivo. The bactericidal defect of neutrophils in thermally injured guinea pigs may be due, at least in part, to the increased formation of PGE_1 in the phagocytes (Bjornson et al. 1989). Exogenous PGE_1 mimics this defect, and inhibitors of cyclooxygenase antagonize inhibition of bactericidal activity (Bjornson et al. 1989; see also Sect. 4.2.2).

As far as the effects of prostacyclin on the respiratory burst are concerned, the results are inconsistent. For example, Gryglewski et al. (1987) did not find an inhibitory effect of prostacyclin and its stable analogue iloprost on receptor-mediated O_2^- formation in human neutrophils. In a recent study, Hecker et al. (1990) did not obtain positive evidence for an inhibitory role of prostacyclin and iloprost in the regulation of NADPH oxidase in human neutrophils either. We found that iloprost is a much less effective inhibitor of fMet-Leu-Phe-induced O_2^- formation in human neutrophils than PGE_1 (unpublished results). In contrast, Weissmann et al. (1980), Fantone and Kinnes (1983), Fantone et al. (1984), Mehta et al. (1988), and Werns and Lucchesi (1988) reported on inhibitory effects of prostacyclin and its stable analogues on neutrophil activation including the respiratory burst. Recently, another stable analogue of prostacyclin, beraprost, has been shown to inhibit fMet-Leu-Phe-induced phosphoinositide turnover and O_2^- formation in rat neutrophils (Kainoh et al. 1990). Inhibition of O_2^- formation by prostaglandins of the E-type series and by prostacyclin has been suggested to be of potential therapeutic value in the prevention of oxygen radical-induced cell damage in myocardial infarction, stroke, peripheral vascular disease, and inflammatory reactions (Weissmann et al. 1980; Werns and Lucchesi 1987, 1988; Gryglewski et al. 1987; see also Sect. 1).

4.1.1.2 β-Adrenergic Agonists

Rabbit mononuclear phagocytes have been shown to possess $β_1$- and $β_2$-adrenoceptors, and human neutrophils and HL-60 cells have been suggested to possess adrenoceptors of the $β_2$-subtype (Tecoma et al. 1986; Mueller et al. 1988; Sager et al. 1988; Tenner et al. 1989). Physiologically occurring and synthetic β-adrenergic agonists inhibit chemoattractant-induced O_2^- formation, which effect is abrogated by β-adrenergic antagonists (Schopf and Lemmel 1983; Lad et al. 1985a; Tecoma et al. 1986; Mueller et al. 1988). The inhibitory effects of β-adrenergic agonists on O_2^- formation and their stimulatory effects on intracellular cAMP are potentiated by inhibitors of phosphodiesterases (Lad et al. 1985a). About 40% of the plasma membrane β-adrenoceptors must be occupied with agonist for maximal inhibition of fMet-Leu-Phe-induced O_2^- formation, and the inhibitory effects of β-adrenergic agonists are very rapid in onset (Mueller et al. 1988). β-Adrenergic agonists have been reported not to inhibit substantially the respiratory burst induced by opsonized zymosan, PMA, OAG, and NaF (Mack et al. 1986; Mueller and Sklar 1989).

4.1.1.3 Histamine

Among other substances, histamine is released from mast cells and basophils during acute allergic and inflammatory reactions (Lagunoff et al. 1983; Warner et al. 1987). As far as the effects of histamine on phagocyte functions are concerned, the results are not consistent.

A number of studies conclusively showed that histamine induces cAMP accumulation in various types of phagocytes including human neutrophils and undifferentiated and differentiated HL-60 cells (Gespach and Abita 1982; Gespach et al. 1982, 1985, 1986; Mitsuhashi et al. 1989). As various H_2 agonists mimic the effects of histamine on cAMP accumulation, and as H_2 antagonists counteract their effects, it was suggested that neutrophils and HL-60 cells possess H_2 receptors which functionally couple to adenylyl cyclase (Gespach and Abita 1982; Gespach et al. 1982, 1985, 1986; Mitsuhashi et al. 1989).

With respect to cell functions, histamine has been reported to inhibit zymosan-induced exocytosis in human neutrophils (Busse and Sosman 1976). In addition, histamine inhibits chemotaxis and fMet-Leu-Phe-induced exocytosis, membrane depolarization, and $O_2^{\cdot-}$ formation in human neutrophils, but, paradoxically, histamine stimulates chemokinesis (Seligmann et al. 1983). In analogy to prostaglandins and β-adrenergic agonists, histamine does not inhibit PMA-induced $O_2^{\cdot-}$ formation (Seligmann et al. 1983; Penfield and Dale 1985; Tecoma et al. 1986). Histamine does not interfere with the binding of fMet-Leu-Phe to formyl peptide receptors, as is the case for β-adrenergic agonists but apparently not for prostaglandins (Fantone et al. 1983; Seligmann et al. 1983; Tecoma et al. 1986; Gryglewski et al. 1987).

The H_2 agonist impromidine is a more potent inhibitor of $O_2^{\cdot-}$ formation than histamine, and the partial H_1 and H_2 agonist betahistine is less potent and effective than histamine (Burde et al. 1989). The inhibitor of the cAMP-specific phosphodiesterase Ro 20–1724 additively enhances the inhibitory effects of histamine on $O_2^{\cdot-}$ formation (Burde et al. 1989). In addition, the H_2 antagonist famotidine competitively antagonizes the effect of histamine with a pA_2 value of 7.5 (Burde et al. 1989). Furthermore, histamine and impromidine inhibit fMet-Leu-Phe-induced $O_2^{\cdot-}$ formation in dibutyryl cAMP- and dimethyl sulfoxide-differentiated HL-60 cells, although to a lesser extent than in neutrophils (Burde et al. 1989). Finally, certain H_1 and H_2 antagonists at relatively high concentrations inhibit the respiratory burst (Taniguchi and Takanaka 1984; Ozaki et al. 1984; Burde et al. 1989, 1990; Seifert, unpublished results).

Unexpectedly, histamine via H_2 receptors apparently not only activates adenylyl cyclase but also stimulates phospholipase C in HL-60 cells (Mitsuhashi et al. 1989). These two signal transduction processes would be expected to interact in an antagonistic manner (see also Sect. 3.2.2). Recently, the effects of a large series of guanidine-type H_2 agonists structurally derived from impromidine (Buschauer 1989) on $O_2^{\cdot-}$ formation in human neutrophils were studied in our laboratory (Burde et al. 1990). All compounds studied inhibit fMet-Leu-Phe-induced $O_2^{\cdot-}$ formation, but the structure/activity relationship for the neutrophil H_2 receptor differs substantially from that for the H_2 receptor in a standard model, the guinea pig atrium (Burde et al. 1990). These data suggest that the H_2 receptor in human myeloid cells shows cell type-specific properties. With respect to the postulated Ca^{2+}-mobilizing properties of H_2 receptors (Mitsuhashi et al. 1989), Burde et al. (1989, 1990) did not obtain positive evidence for a stimulatory effect of H_2 agonists on $O_2^{\cdot-}$ formation in human neutrophils.

The situation concerning the role of histamine in the regulation of phagocyte functions is even more complex. Human mononuclear phagocytes have been suggested to possess not only H_2 receptors but also H_1 receptors which functionally couple to phospholipase C (Gespach et al. 1985, 1986; Cameron et al. 1986; Driver et al. 1989). Activation of phospholipase C through H_1 receptors would be expected to be associated with an activation of NADPH oxidase. In agreement with this hypothesis, zymosan-bound histamine has been reported to activate $O_2^{\cdot-}$ formation in guinea pig alveolar macrophages via H_1 receptors, and histamine has been shown to induce release of eicosanoids in human neutrophils (Diaz et al. 1979; Puustinen and Uotila 1984; see also Sects. 3.2.2, 3.2.4). In contrast, we did not obtain positive evidence for a stimulatory role of H_1 receptors in the regulation of NADPH oxidase in human neutrophils under various experimental conditions using betahistine (Burde et al. 1989) and other more selective H_1 agonists (unpublished results).

4.1.1.4 Adenosine

Adenosine is a degradation product of adenine nucleotides and is released into the extracellular space from various cell types (Gordon 1986; see also Sect. 3.3.1.8). The effects of adenosine on the respiratory burst are not consistent. On one hand, adenosine has been reported to inhibit the respiratory burst in various types of phagocytes including human neutrophils and HL-60 cells (Cronstein et al. 1983, 1985, 1988; Schrier and Imre 1986; Seifert et al. 1989b,d). Inhibition of $O_2^{\cdot-}$ formation by adenosine depends on its permanent presence, and the order of potency of various A_1 and A_2 agonists suggests that adenosine acts through A_2 receptors (Roberts et al. 1985; Cronstein et al. 1985, 1988; Nielson and Vestal 1989; Eppell et al. 1989). The adenosine antagonist 8-phenyltheophylline, antagonizes inhibition of the respiratory burst caused by adenosine (Roberts et al. 1985). On the other hand, adenosine has been reported to stimulate the respiratory burst in certain types of phagocytes and to enhance chemotaxis (Tritsch and Niswander 1983; Schrier and Imre 1986; Rose et al. 1988; Ward et al. 1988c). In human mononuclear phagocytes, the inhibitory effects of adenosine on phagocytosis depend on the differentiation state of the cells (Eppell et al. 1989; see also Sect. 3.4.3). Stimulatory effects of adenosine on neutrophil chemotaxis are explained by the fact that these phagocytes do not only possess A_2 receptors but also A_1 receptors which latter promote chemotaxis (Cronstein et al. 1990). Interestingly, A_1 receptors may couple to pertussis toxin-sensitive G-proteins in neutrophils and may enhance the IgG-mediated respiratory burst (Salmon and Cronstein 1990). In addition to activation of adenosine receptors, adenosine may act through inhibition of methylation reactions (Pike and Snyderman 1982; Garcia-Castro et al. 1983).

The question of the mechanism by which adenosine inhibits O_2^- formation is a subject of current discussion. A_2 receptors interact with G_s, resulting in activation of adenylyl cyclase and cAMP accumulation (Nielson and Vestal 1989). In human monocytes, adenosine has been reported to induce a small and variable increase in cAMP, which is enhanced by a phosphodiesterase inhibitor (Elliott et al. 1986). In human neutrophils, neither adenosine nor the A_2 agonist 5'-N-ethylcarboxamidoadenosine (NECA) has been found to increase cAMP (Cronstein et al. 1988). In the presence of the phosphodiesterase inhibitor Ro-20 1724, A_2 agonists increase cAMP in human neutrophils, but Ro-20 1724 does not potentiate the inhibitory effect of NECA on O_2^- formation (Cronstein et al. 1988). These data suggest that inhibitory effects of adenosine on O_2^- formation do not necessarily depend on adenylyl cyclase activation and cAMP accumulation.

4.1.2 Other cAMP-Increasing Agents

The diterpene forskolin directly activates adenylyl cyclase and shows a weak inhibitory effect on chemotactic peptide-induced O_2^- formation in human neutrophils and diminishes the PMA-induced respiratory burst in rat glomeruli (Burde et al. 1989; Miyanoshita et al. 1989). In addition to their stimulatory effects on O_2^- formation, *cis*-unsaturated fatty acids may also inhibit O_2^- formation induced by fMet-Leu-Phe and NaF (Wong and

Chew 1984; see Sects. 3.1.2, 3.2.4). This effect may be due to an increase in membrane fluidity, which phenomenon is associated with activation of adenylyl cyclase (Houslay and Gordon 1983). Cholera toxin ADP-ribosylates G_s and leads to its peristent activation through inhibition of its intrinsic GTPase activity (Gilman 1987). Treatment of phagocytes with cholera toxin leads to cAMP accumulation and results in inhibition of chemotaxis, arachidonic acid release, exocytosis, and O_2^- formation (Bourne et al. 1973; Rivkin et al. 1975; Bokoch and Gilman 1984; Seifert et al. 1989b). Unexpectedly, cholera toxin shows no inhibitory effect on fMet-Leu-Phe-induced O_2^- formation in human neutrophils, whereas the effects of the chemotactic peptide and UTP in dibutyryl cAMP-differentiated HL-60 cells and the effect of a synthetic lipopeptide in neutrophils are partially inhibited by the toxin (Gabler et al. 1989; Seifert et al. 1989b, 1991).

The inhibitor of cAMP-specific phosphodiesterases, Ro 20–1724, and unspecific inhibitors of phosphodiesterases, e.g., the methylxanthines 3-isobutyl-1-methylxanthine, theophylline, and pentoxifylline inhibit the respiratory burst (Lad et al. 1985a; Bessler et al. 1986; Burde et al. 1989). In contrast, the cAMP-specific phosphodiesterase inhibitor rolipram has been reported to increase cAMP in human neutrophils without significantly inhibiting O_2^- formation (Elliott and Leonard 1989). The effects of methylxanthines on O_2^- formation are complex. Methylxanthines above 100 μM inhibit phosphodiesterases and inhibit the respiratory burst presumably via an increase in cAMP (Schmeichel and Thomas 1987; Yukawa et al. 1989). In contrast, methylxanthines below 100 μM act as competitive antagonists at A_2 receptors and potentiate fMet-Leu-Phe-induced O_2^- formation in human neutrophils and eosinophils (Schmeichel and Thomas 1987; Yukawa et al. 1989). Adenosine desaminase mimics the stimulatory effects of methylxanthines, and adenosine counteracts the stimulatory effects of these agents, suggesting that endogenous adenosine plays an inhibitory role in the regulation of NADPH oxidase (Schmeichel and Thomas 1987; see also Sect. 4.1.1.4). Finally, the cell-permeant analogue of cAMP, dibutyryl cAMP, inhibits chemotactic peptide-induced O_2^- formation (Kitagawa and Takaku 1982; Lad et al. 1985a; Bessler et al. 1986; Kramer et al. 1988a; Burde et al. 1989).

4.1.3 Mechanistic Aspects

The molecular mechanism by which cAMP-increasing agents inhibit the respiratory burst is not yet known exactly. Most but not all reports show that O_2^- formation induced by intercellular signal molecules is inhibited by cAMP-increasing agents, whereas O_2^- formation induced by stimuli which circumvent receptor stimulation, e.g., NaF, PMA, and A 23187, is not affected (Kitagawa and Takaku 1982; Seligmann et al. 1983; Penfield and Dale 1985; Mack et al. 1986; Miyanoshita et al. 1989). O_2^- formation induced by various classes of receptor agonists shows differential sensitivity to inhibition by cAMP-increasing agents (Gryglewski et al. 1987; Seifert et al. 1989b; Burde et al. 1989). Interestingly, the same is apparently true for protein kinase C-mediated desensitization of receptor agonist-mediated O_2^- formation (Bender et al. 1987; see also Sect. 3.2.2.5). These data raise the possibility that receptors for intercellular signal molecules, e.g., formyl peptide receptors, are targets for phosphorylation by cAMP-dependent protein kinase (Mueller and Sklar 1989). Phosphorylated formyl peptide receptors may be uncoupled from the G-proteins, a process referred to as heterologous desensitization (Lefkowitz and Caron 1986; Sibley et al. 1987; see also Sect. 3.2.2.5). Heterologous desensitization has been observed in numerous cellular systems (Lefkowitz and Caron 1986; Sibley et al. 1987). In addition, prostaglandins have been suggested to inhibit neutrophil functions by interference with the binding of formyl peptides to the plasma membrane (Fantone et al. 1983; Seligmann et al. 1983; Tecoma et al. 1986; Gryglewski et al. 1987).

G-proteins and phospholipase C may be additional targets for cAMP-mediated inhibited of NADPH oxidase. cAMP-dependent protein kinase has been suggested to phosphorylate G-proteins of the G_i family in differentiated HL-60 cells, and in the phosphorylated state these G-proteins may be uncoupled from effector systems, e.g., phospholipase C and/or NADPH oxidase (Misaki et al. 1989; see also Sect. 3.2.2.5). In fact, cAMP-increasing agents have been reported to inhibit phosphoinositide degradation, resynthesis of degraded phosphoinositides, and Ca^{2+} influx from the extracellular space in phagocytes (Farkas et al. 1984; Della Bianca et al. 1986b; Kato et al. 1986; Takenawa et al. 1986; Misaki et al. 1989). Moreover, the γ-isoenzyme of phospholipase C has recently been shown to be phosphorylated by cAMP-dependent protein kinase in vitro (Kim et al. 1989). In contrast, adenosine and theophylline apparently do not inhibit phosphoinositide degradation in human monocytes (Elliott and Leonard 1989).

Isoproterenol has only little effect on actin polymerization induced by chemotactic peptides (Tecoma et al. 1986). With respect to Ca^{2+} mobilization, the results are controversial. On one hand, the inhibitory effect of cAMP-increasing agents on phagocyte activation has been suggested to be not due to interference with Ca^{2+} mobilization (De Togni et al. 1984; Kato et al. 1986; Takenawa et al. 1986; Cronstein et al. 1988). Paradoxically, β-adrenergic agonists, prostaglandins of the E series and histamine have been shown to induce an increase in cytoplasmic Ca^{2+} in differentiated HL-60 cells (Mitsuhashi et al. 1989). On the other hand, inhibitory effects of isoproterenol and adenosine on Ca^{2+} mobilization have been observed (Tecoma et al. 1986; Nielson and Vestal 1989).

A number of proteins is phosphorylated by cAMP-dependent protein kinase in neutrophils and HL-60 cells (Helfman et al. 1983; Farkas et al. 1984; Misaki et al. 1989). Activation of NADPH oxidase by various stimuli is associated with the phosphorylation of a 47-kDa protein (see also Sects. 3.1.1, 3.2.2, 5.1.5, 6.1). Interestingly, the 47-kDa protein is apparently also a substrate for cAMP-dependent protein kinase (Kramer et al. 1988a). In contrast to the protein kinase C-induced phosphorylation of the 47-kDa protein, that induced by cAMP-dependent protein kinase is associated with an inhibition of $O_2^{\cdot-}$ formation (Kramer et al. 1988a). It is not known whether protein kinase C and cAMP-dependent protein kinase phosphorylate the 47-kDa protein at the same site, but it has been shown in other systems that phosphorylations catalyzed by different protein kinases may be functionally nonequivalent (Naka et al. 1983, Nastainczyk et al. 1987; Jahn et al. 1988).

We studied the role of cAMP-dependent protein kinase in the regulation of $O_2^{\cdot-}$ formation in a cell-free system from dimethyl sulfoxide-differentiated HL-60 cells (see also Sect. 5.1). Neither cAMP nor the catalytic subunit of cAMP-dependent protein kinase shows an inhibitory effect on arachidonic acid-induced $O_2^{\cdot-}$ formation both in the absence and presence of guanine nucleotides (unpublished results).

4.1.4 cAMP-Decreasing Agents

The role of cAMP-decreasing agents in the regulation of the respiratory burst is obscure, and there are only few experimental data (see also Sect. 3.2.6.1). Human neutrophils possess α_2-adrenergic receptors, whose activation causes inhibition of adenylyl cyclase and decrease in the cAMP concentration via G-proteins of the G_i family, i.e., presumably through the same G-proteins which mediate phagocyte activation by chemoattractants (Panosian and Marinetti 1983; Verghese and Snyderman 1983; Verghese et al. 1985a; see also Sect. 3.2.1.1). Apparently, the primary structure of G_i-proteins alone is not sufficient to determine which effector system, i.e., phospholipase C in the case of chemoattractants and adenylyl cyclase in the case of α_2-adrenergic agonists is regulated by a given subtype of G_i (Gierschik et al. 1989b). The α_2-adrenergic agonists B-HT 933 and B-HT 920 (Hammer et al. 1980; Starke 1987) at concentrations up to 100 μM, do not activate $O_2^{\cdot-}$ formation in human neutrophils in the presence or absence of CB (Seifert, unpublished results). In addition, B-HT 933 and B-HT 920 neither enhance nor inhibit fMet-Leu-Phe-induced $O_2^{\cdot-}$ formation in these cells. Furthermore, B-HT 933 and B-HT 920 fail to antagonize inhibition of fMet-Leu-Phe-induced O_2^- formation caused by PGE_1 or histamine at submaximally or maximally effective concentrations (Seifert, unpublished results).

Calcitonin has been reported to inhibit cAMP accumulation in human monocytes in a pertussis toxin-sensitive manner (Stock and Coderre 1982, 1987). With respect to human neutrophils, thyrocalcitonin does not activate O_2^- formation and does not affect the fMet-Leu-Phe-induced respiratory burst (Seifert, unpublished results).

In adherent but not in suspended human neutrophils, TNF-α induces a sustained decrease in the intracellular cAMP concentration (Nathan and Sanchez 1990). It has been suggested that this effect of TNF-α depends on the expression of integrins and is involved in the prolonged activation of the respiratory burst induced by this cytokine (see also Sect. 3.3.1.3.4).

Finally, GM-CSF has been reported substantially to inhibit adenylyl cyclase in human neutrophils (Coffey et al. 1988). How far this effect is causally linked to the GM-CSF-induced potentiation of O_2^- formation, remains unknown (see also Sects. 3.2.6.1, 3.3.1.3.5.2).

4.2 Anti-inflammatory Drugs

4.2.1 Glucocorticoids and Cyclosporin A

Glucocorticoids are widely used as anti-inflammatory and immunosuppressive agents. The results of studies concerning the effects of glucocorticoids on phagocyte functions in general and on the respiratory burst in particular are not consistent. On one hand, dexamethasone has been reported to inhibit prostaglandin release but not O_2^- formation in cultured rat Kupffer's cells (Dieter et al. 1986). Müller-Peddinghaus and Wurl (1987) also did not observe inhibitory effects of glucocorticoids on the respiratory burst in various types of phagocytes. Moreover, a recent study has shown that dexamethasone does not inhibit exocytosis in human neutrophils induced by fMet-Leu-Phe (Schleimer et al. 1989).

In contrast, other authors reported on inhibitory effects of glucocorticoids on the respiratory burst and phagocytosis in various types of phagocytes (Oyanagui et al. 1978; Lehmeyer and Johnston 1978; Drath and Kahan 1983, 1984; Rieder et al. 1988b). Long-term incubation with glucocorticoids has been reported to inhibit oxidative metabolism in cultured human monocytes, and priming with IFN-γ or LPS blocks the effects of glucocorticoids (Szefler et al. 1989). The mechanism by which LPS and IFN-γ antagonize inhibition of the respiratory burst caused by steroids may involve among others interference with the secretion of cytokines (Szefler et al. 1989). In guinea pig macrophages and in rat mesangial cells, dexamethasone inhibits $O_2^{\cdot-}$ formation induced by various stimuli (Baud et al. 1986; Maridonneau-Parini et al. 1989). Inhibition by dexamethasone of $O_2^{\cdot-}$ formation requires a lag time and depends on the stimulation of intracellular glucocorticoid receptors and on de novo protein synthesis (Baud et al. 1986; Maridonneau-Parini et al. 1989). In guinea pig macrophages, the effect of dexamethasone has been reported to be mimicked by lipocortin I, a glucocorticoid-induced protein which covers substrates of phospholipases (Maridonneau-Parini et al. 1989; Machoczek et al. 1989). In rat mesangial cells, exogenous arachidonic acid counteracts the inhibitory effect of dexamethasone on the respiratory burst, provided that conversion of the fatty acid to prostaglandins is prevented (Baud et al. 1986; see also Sects. 3.2.4, 3.4.4.2.2). Recently, hydrocortisone at very high concentrations was shown to inhibit PMA-induced $O_2^{\cdot-}$ formation in intact human neutrophils and $O_2^{\cdot-}$ formation in a cell-free system (Umeki and Soejima 1990). Apparently, these effects of glucocorticoids on the respiratory burst in human neutrophils are not related to modulation of protein synthesis.

The fungal cyclic undecapeptide cyclosporin A is used as immunosuppressive agent in patients following organ transplantation and suppresses antibody production and cell-mediated immunity (Bennett and Norman 1986). Cyclosporin A may interfere with the activation of phagocytes. The cyclic peptide binds with high affinity to human neutrophils and has been suggested to inhibit phospholipase A_2 (Kharazmi et al. 1985; Niwa et al. 1986). In vitro, cyclosporin partially inhibits $O_2^{\cdot-}$ formation in rat alveolar macrophages, but in vivo cyclosporin A does not inhibit $O_2^{\cdot-}$ formation in rat alveolar macrophages and neutrophils (Drath and Kahan 1983, 1984). Recently, cyclosporin A at therapeutically relevant concentrations has been reported to inhibit the PMA-induced respiratory burst in resident murine peritoneal macrophages in vitro, whereas activation of NADPH oxidase by ConA and receptor agonists is not affected by the undecapeptide (Chiara et al. 1989). Cyclosporin A apparently neither interferes directly with NADPH oxidase nor interferes with glucose transport (Chiara et al. 1989). At least in certain cellular systems, cyclosporines may inhibit protein kinase C- and calmodulin-dependent processes (Gschwendt et al. 1988; R.J. Walker et al. 1989). With respect to phagocytes of human origin, cyclosporin A has been reported to be without inhibitory effect on the respiratory burst in neutrophils and monocytes in vitro (Janco and English 1983; Kharazmi et al. 1985).

4.2.2 Nonsteroidal Anti-inflammatory Drugs

Nonsteroidal anti-inflammatory drugs at relatively high concentrations have been reported to inhibit various neutrophil functions such as aggregation, exocytosis, and O_2^- generation. The experimental data, however, are controversial. Among the nonsteroidal anti-inflammatory drugs which have been shown to inhibit the respiratory burst are phenylbutazone, diclofenac, acetyl salicylate, piroxicam, ibuprofen, indomethacin, and mefeamic acid, and the effects of these drugs show substantial stimulus and cell-type specificity (Bokoch and Reed 1979; Perianin et al. 1983; Kaplan et al. 1984; Abramson et al. 1985; Neal et al. 1987a; Weissmann 1987). In contrast, other authors did not find substantial inhibition of the respiratory burst by various anti-inflammatory drugs including piroxicam, phenylbutazone, and sulfinpyrazone (Müller-Peddinghaus and Wurl 1987).

Some authors reported that indomethacin inhibits agonist-induced O_2^- formation in human neutrophils (Bokoch and Reed 1979; Maridonneau-Parini et al. 1986; Neal et al. 1987a), whereas others did not find inhibitory effects of indomethacin on the respiratory burst (Bromberg and Pick 1983; Kaplan et al. 1984; Tsunawaki and Nathan 1986). Paradoxically, indomethacin per se has been reported to activate O_2^- formation in guinea pig macrophages and to enhance O_2^- formation induced by various stimuli (Bromberg and Pick 1983; Dale and Penfield 1985, 1987). The potentiating effect of indomethacin on O_2^- formation may be explained by inhibition of diacylglycerol kinase (Dale and Penfield 1985, 1987; see also Sect 3.2.2.4). Similar controversial results have been obtained for other anti-inflammatory drugs, e.g., for acetyl salicylate (Bokoch and Reed 1979; Kaplan et al. 1984; Maridonneau-Parini et al. 1986; Tsunawaki and Nathan 1986).

The mechanisms by which nonsteroidal anti-inflammatory agents inhibit the respiratory burst may be complex. Certain nonsteroidal anti-inflammatory drugs, e.g., diclofenac, ibuprofen, phenylbutazone, sulfinpyrazone, meclofenamate and tolmetin, have been suggested to interfere with the binding of chemotactic peptides to the plasma membrane (Dahinden and Fehr 1980; Perianin et al. 1987; Skubitz and Hammerschmidt 1986; Shelly and Hoff 1989). Acetyl salicylate and related compounds may inhibit phospholipase C in human monocytes and differentiated U-937 cells, and indomethacin and meclofenamate may inhibit phospholipase A_2 (Kaplan et al. 1978; Franson et al. 1980; Bomalaski et al. 1986; Muid et al 1988; see also Sects. 3.1.2, 3.2.2, 3.2.4). In addition, certain nonsteroidal anti-inflammatory drugs have been reported to interfere with early steps of cell activation, e.g., with Ca^{2+} mobilization (Northover 1985; Abramson et al. 1985; Kaplan et al. 1984; see also Sect. 4.1).

Sulfasalazine consists of 5-aminosalicyclic acid joined to sulfapyridine via an azo-linkage and is used in patients with colitis ulcerosa. Sulfasalazine has been reported to inhibit receptor agonist-induced $O_2^{\cdot-}$ formation in human neutrophils, whereas 5-aminosalicyclic acid and sulfapyridine are inactive (Neal et al. 1987b). These authors suggested that sulfasalzine not only serves as a precursor for 5-aminosalicyclic acid but also per se possesses anti-inflammatory properties. Sulfasalazine may inhibit neutrophil activation by interfering with the binding of formyl peptides to plasma membrane receptors (Stenson et al. 1984).

4.2.3 Chloroquine, Mepacrine, and Gold Compounds

Chloroquine and mepacrine possess anti-inflammatory and immunosuppressive properties, and mepacrine may improve pulmonary function in adult respiratory distress syndrome (Neal et al. 1987a; Panus and Jones 1987). Inhibition of the respiratory burst by mepacrine or chloroquine shows stimulus specificity (Hurst et al. 1986; Neal et al. 1987a). Mepacrine inhibits membrane depolarization at lower concentrations than $O_2^{\cdot-}$ formation, and phospholipid turnover is less sensitive to inhibition by mepacrine than $O_2^{\cdot-}$ formation (Tauber and Simons 1983) (see also Sect 3.4.2.2). The mechanism by which quinacrine and mepacrine inhibit the respiratory burst may involve inhibition of phospholipase A_2 and of glucose transport and more direct inhibition of NADPH oxidase (Cross et al. 1984; Schultz et al. 1985; Tsunawaki and Nathan 1986; Maridonneau-Parini et al. 1986; Henderson et al. 1989; see also Sects. 3.1.2, 3.2.4).

Gold compounds, e.g., triethylphosphine gold, sodium aurothiomalate, and auranofin, are used in the treatment of rheumatoid arthritis (Crooke et al. 1986). The mode of action of gold compounds may involve suppression of T-lymphocyte proliferation and inhibition of phagocyte functions, e.g., chemotaxis, exocytosis, phagocytosis, and production of leukotrienes (Davis et al. 1983; Hafström et al. 1984; Sung et al. 1984; Crooke et al. 1986; Parente et al. 1986b). Gold compounds are taken up by phagocytes in a time- and concentration-dependent manner which does not require metabolic energy (Snyder et al. 1986, 1987). Triethylphospine gold strongly inhibits fMet-Leu-Phe-induced $O_2^{\cdot-}$ formation (Davis et al. 1983; Sung et al. 1984). In analogy to the situation with the above-discussed anti-inflammatory agents, the effects of gold compounds on the respiratory burst are complex and depend on the specific gold compound, stimulus, and cell type studied. Sodium aurothiomalate inhibits chemiluminescence in blood monocytes from healthy volunteers and from patients with rheumatoid arthritis (Harth et al. 1983), and auranofin has been shown to inhibit the PMA-, NaF and fMet-Leu-Phe-induced respiratory burst in human neutrophils (Davis et al. 1982; Schultz et al. 1985). In contrast, auranofin does not inhibit O_2^- formation induced by A 23187 (Hafström et al. 1984), and sodium aurothiomalate does not affect PMA- and fMet-Leu-Phe-induced O_2^- formation in human neutrophils (Minta and Williams 1986). The mechanism by which gold compounds inhibit NADPH oxidase is currently under investigation. Auranofin has been suggested to interfere with formyl peptide receptors and with certain components of the cytoskeleton (Hafström et al. 1984). Interestingly, recent studies showed that gold compounds may modulate the activity of protein kinase C. Auranofin and sodium aurothiomalate reduce the cytosolic activity of protein kinase C, but auranofin induces neither degradation of the kinase nor its translocation to the plasma membrane (Parente et al. 1986a, 1989). In addition, auranofin has been shown to inhibit PMA-induced protein phosphorylation in human neutrophils (Parente et al. 1989). Furthermore, auranofin, sodium aurothiomalate, and gold thioglucose have been reported to inhibit partially purified protein kinase C in vitro (Parente et al. 1989; Froscio et al. 1989; Mahoney et al. 1989). Finally, gold compounds may directly inhibit NADPH oxidase and do not act as radical scavengers (Minta and Williams 1986).

4.3 Miscellaneous Inhibitory Agents

4.3.1 Protozoal, Fungal, and Bacterial Products

Various products or bacteria, fungi, and protozoa modulate the respiratory burst. The effects of pertussis toxin on the respiratory burst are dealt with in Sects. 3.2.1 and 3.3.1 and are summarized in Table 3. The effects of botulinum C2 toxin and CB and described in Sect. 3.2.5, those of anthrax toxin in Sect. 3.3.2.3, and those of cholera toxin in Sect. 4.1. The inhibitory properties of LPS, of polymyxin B from *Bacillus polymyxa*, of staurosporine from *Streptomyces species*, and of K-252a from *Nocardiopsis species* on the respiratory burst are described in Sects. 3.2.2.3 and 3.3.2.2. Some additional effects of infections on the respiratory burst are described in Sect 6.2.2.

In addition to pertussis toxin, *Bordetella species* release adenylyl cyclase as toxin into the extracellular space (Hanski 1989). Adenylyl cyclase enters target cells, e.g., neutrophils and monocytes, is activated by Ca^{2+}/calmodulin, and leads to a supraphysiological increase in cAMP (Pearson et al. 1987; Hanski 1989). This exogenous adenylyl cyclase effectively inhibits the respiratory burst induced by various particulate and soluble stimuli including opsonized zymosan and PMA without affecting cell viability (Pearson et al. 1987; see also Sect. 4.1).

Candida albicans hyphae release yet incompletely characterized factors which inhibit the fMet-Leu-Phe-, A 23187-, and opsonized zymosan- but not the PMA-induced respiratory burst in neutrophils (Smail et al. 1988). These factors do not substantially inhibit chemoattractant-induced chemotaxis or phosphoinositide degradation and increase in cytoplasmic Ca^{2+}. Apparently, these factors do not act at the level of formyl peptide receptors, of G-proteins, or of NADPH oxidase but at yet unknown steps on the signal transduction process (Smail et al. 1988; see also Sects. 3.2.2.1, 3.2.3).

Intact microorganisms may show inhibitory effects of NADPH oxidase. For example, *Histoplasma capsulatum* yeast inhibits priming of the respiratory burst by IFN-γ in murine macrophages, suggesting that ingestion of these yeast cells induces cellular deactivation (Wolf et al. 1989). In contrast, opsonized *Histoplasma capsulatum* shows stimulatory effects on the respiratory burst in human neutrophils (Schnur and Newman 1990). Moreover, *Chlamydia trachomatis* partially inhibits fMet-Leu-Phe- and PMA-induced $O_2^{\bullet-}$ formation in human neutrophils (Tauber et al. 1989b). The effect of *Chlamydia* is rapid in onset and has been suggested to take place at the level of NADPH oxidase (Tauber et al. 1989b).

The wortmannins are fungal products and are very potent and effective inhibitors of chemoattractant-induced exocytosis and $O_2^{\bullet-}$ formation in human neutrophils, whereas activation by PMA is unaffected (Dewald et al. 1988). Apparently, the wortmannins do not interfere with phospholipase C, protein kinase C, or increase in cytoplasmic Ca^{2+} and NADPH oxidase (Dewald et al. 1988). The wortmannins have been shown to be very useful pharmacological tools to analyze signal transduction sequences in neutrophils. Two signal transduction sequences have been postulated to be initiated by chemotactic peptides, and both processes are required for activation of NADPH oxidase. One process is Ca^{2+}/protein kinase C dependent and wortmannin resistant, and the other process is Ca^{2+} independent but wortmannin sensitive (Dewald et al. 1989; see also Sects. 3.1.1, 3.2.2, 3.2.3). The results of a recent study suggest that wortmannin may interfere with chemotactic peptide-induced activation of phospholipase D (Reinhold et al. 1990; see also Sect. 3.2.2.1).

Leishmania donovani causes kala azar, multiplies in macrophages, and is sensitive to inactivation by products of the respiratory burst (Murray 1981; Pearson and Steigbigel 1981; Murray and Cartelli 1983). One of the factors which contribute to the parasites' ability to circumvent destruction by the host may be a tartrate-resistant acid phosphatase from the external surface of the promastigotes which inhibits $O_2^{\bullet-}$ formation in human neutrophils

(Remaley et al. 1984). The mechanism by which the phosphatase inhibits the respiratory burst is obviously related to its catalytic activity, as a phosphatase inhibitor abolishes the effect of the enzyme on NADPH oxidase (Remaley et al. 1984). In addition, a lipophosphoglycan from *Leishmania donovanii* has been reported to inhibit protein kinase C, and this compound may play a role in the inhibition of the respiratory burst of the host cell as well (McNeely and Turco 1987). Related mechanisms to those described above may play a role in the suppression of the respiratory burst in the host cell by phagocytosed *Toxoplasma gondii* (Wilson et al. 1980).

4.3.2 Endogenous Proteins

When cells or intact organisms are stressed by elevated temperatures, trauma, or certain chemicals, they respond by synthesizing a number of proteins, referred to as heat-shock proteins (Currie and White 1981; Schlesinger et al. 1982; Lindquist 1986; Pelham 1986; Subjeck and Shyy 1986; Maridonneau-Parini et al. 1988). The heat-shock response is found in many cell types including neutrophils and monocytes and is assumed to be involved in the protection of the cell from injury (Polla et al. 1987; Deguchi et al. 1988; Maridonneau-Parini et al. 1988). Recent studies raised the interesting possibility that heat-shock proteins play a role in myeloid differentiation processes (Richards et al. 1988; Yufu et al. 1989). Exposure of human neutrophils to elevated temperatures or heavy metals is associated with the synthesis of a variety of heat shock proteins and reversible inhibition of O_2^- formation (Maridonneau-Parini et al. 1988). This inhibition of the respiratory burst apparently does not depend on cytosolic pH or thiol group oxidation (Maridonneau-Parini et al. 1988). Moreover, in neutrophil cytoplasts and in intact guinea pigs elevated temperatures inhibit the respiratory burst (Malawista and van Blaricom 1987; Bjornson et al. 1989). One possible interpretation of these findings is that the synthesis of heat-shock proteins represent an endogenous mechanism to inhibit the respiratory burst and hence to protect tissues from oxidative damage.

C-Reactive protein is composed of five 21-kDa subunits, which are arranged in cyclic symmetry (Müller and Fehr 1986). C-Reactive protein is synthesized in the liver and is present in serum at low concentrations (Buchta et al. 1987a,b). Following acute trauma or infection, the plasma concentration of C-reactive protein increases greatly, a phenomenon referred to as acute phase response, and C-reactive protein accumulates in inflamed and injured tissue (Buchta et al. 1987a,b). C-Reactive protein rapidly binds to high-affinity binding sites on the plasma membrane of human neutrophils (Buchta et al. 1987a,b). Aggregated human IgG prevents binding of C-reactive protein, suggesting its association with Fc receptors (Buchta et al. 1987b; see also Sect 3.3.1.5.3). C-Reactive protein may play a role in the regulation of various neutrophil functions including motility, phagocytosis, and $O_2^{\cdot-}$ formation (Kilpatrick and Volanakis 1985; Müller and Fehr 1986; Buchta et al. 1987a,b; Filep and Földes-Filep 1989). C-Reactive protein does not activate the hexose monophosphate shunt but at low concentrations enhances PMA-induced $O_2^{\cdot-}$ formation in human neutrophils (Müller and Fehr 1986; Buchta et al. 1987a). In contrast, C-reactive protein at higher concentrations inhibits the effect of PMA (Buchta et al. 1986). In addition, C-reactive protein has been shown to inhibit fMet-Leu-Phe- and PAF-induced $O_2^{\cdot-}$ formation in a time- and concentration-dependent manner (Filep and Földers-Filep 1989). C-Reactive protein reduces the binding of the chemotactic peptide and PAF to the plasma membrane (Filep and Földers-Filep 1989), and an increase in cAMP may also play a role in the inhibition of neutrophil activation by C-reactive protein (Buchta et al. 1987a; see also Sect 4.1). These data suggest that C-reactive protein plays a protective role against oxygen radical-induced tissue injury in the acute phase of inflammatory processes.

Haptoglobin is another acute phase protein (Oh et al. 1990). Haptoglobin has been reported to bind to specific sites on human neutrophils and to inhibit $O_2^{\cdot-}$ formation induced by fMet-Leu-Phe, opsonized zymosan, and arachidonic acid whereas the respiratory burst induced by PMA is not affected by the acute-phase protein (Oh et al. 1990). These authors suggested that haptoglobin interferes with the receptor ligand interaction in neutrophils. In addition, the acute phase reactant α_1-acid glycoprotein was shown to inhibit the respiratory burst stimulated by various agents (Laine et al. 1990).

The fibrinogen degradation product fragment E_3 is present in blood of patients with disseminated intravascular coagulation. Fragment E_3 has been reported to inhibit receptor agonist-, OAG-, and PMA-induced $O_2^{\cdot-}$ formation in neutrophils (Kazura et al. 1989). The mechanism by which fragment E_3 inhibits the respiratory burst may involve interference with agonist binding to plasma membrane receptors and inhibition of protein kinase C. Inhibition of the respiratory burst by fragment E_3 may contribute to the impaired host defense against bacterial infections in disseminated intravascular coagulation (Kazura et al. 1989).

The major surfactant-associated protein is a glycoprotein with an apparent molecular mass of 28-36 kDa and is involved in the metabolism of lung surfactant compounds (Weber et al. 1990). This protein inhibits the respiratory burst in canine phagocytes, which effect is counteracted by treatment with collagenase (Weber et al. 1990). These data suggest that the surfactant-associated protein plays a role in the regulation of the respiratory burts in alveolar macrophages.

Finally, the cytokine IL-4 has been recently reported to inhibit the stimulatory effect of IFN-γ on the respiratory burst in human monocytes (Lehn et al. 1989). The inhibitory effect of IL-4 is evident when added prior to or together with IFN-γ to the cells and is accompanied by a decrease in cytotoxic activity of the phagocytes (Lehn et al. 1989; see also Sect 3.3.1.3.2). Inhibitory effects of IL-4 on the respiratory burst in human mononuclear phagocytes were also reported by Abramson and Gallin (1990). IL-1 partially antagonizes the inhibitory effects of IL-4. In contrast, O_2^- formation in human neutrophils is not inhibited by IL-4 (Abramson and Gallin 1990). The molecular mode of action of IL-4 on the respiratory burst remains to be determined and may involve alterations in the biosynthesis of proteins (Abramson and Gallin 1990).

IL-4 has not only inhibitory but also stimulatory effects on the respiratory burst. In murine bone marrow-derived macrophages, IL-4 primes for an enhanced respiratory burst upon exposure to PMA or zymosan (Phillips et al. 1990). IL-4 and TNF-α interact in a synergistic manner to prime for enhanced O_2^- formation, whereas IL-4 and IFN-γ interact in an antagonistic manner.

4.3.3 SH Reagents

A number of studies concerning the effects of SH reagents on NADPH oxidase have been performed (see also Sects. 2.1, 2.2, 3.1.1). In intact guinea pig neutrophils, showdomycin, a very slowly penetrating SH reagent, does not inhibit O_2^- formation induced by various stimuli, suggesting that SH groups at the extracellular surface of the plasma membrane are not involved in NADPH oxidase activation. In contrast, the cell-permeant SH reagent NEM inhibits receptor agonist-, lectin-, digitonin-, and cytochalasin-induced O_2^- formation but not that induced by A 23187 or NaF, suggesting the existence of NEM-sensitive and -insensitive activation pathways (Tsan et al. 1976; Yamashita 1983; Yamashita et al. 1984). In addition, NEM terminates O_2^- formation induced by various stimuli, but NEM does not affect the activity of the particulate NADPH oxidase (Akard et al. 1989). In the cell-free system, NEM has been shown to inactivate cytosolic components but not membrane components of NADPH oxidase (Akard et al. 1988). These data suggest that NEM interferes with an activation step of

NADPH oxidase, and that continuous replenishment of cytosolic components is required for maintenance of O_2^- formation (Akard et al. 1988; see also Sects. 5.1.2, 5.1.5).

Ethacrynic acid and apomorphine have been suggested to inhibit the respiratory burst by reacting with SH groups (Elferink et al. 1982, 1987). In addition, ozone and certain unsaturated aldehydes, e.g., acrolein and crotonaldehyde, inhibit O_2^- formation (Witz et al. 1987). The latter two agents have been shown to decrease the cellular content of free SH groups (Witz et al. 1987).

Cross-linking agents, e.g., disuccinimidyl suberate and dithiobis (succinimidylpropionate), inhibit O_2^- formation in human neutrophils induced by a variety of agents. In contrast, monovalent analogues of the cross-linkers are inactive, and dithiothreitol counteracts the inhibitory effects of cross-linkers (Aviram and Henis 1984; Aviram et al. 1984). It has been suggested that cross-linkers interfere with the activation process of NADPH oxidase but not with its activity (Aviram et al. 1984).

5 Reconstitution and Regulation of NADPH Oxidase Activity in Cell-free Systems

Studies concerning the mechanisms of NADPH oxidase activation in intact cells rely primarily on correlations and/or dissociations between various parameters and on the use of drugs which are assumed to interfere with certain steps of signal transduction processes. The results of several studies with intact and electropermeabilized phagocytes suggest that receptor agonist-mediated activation of NADPH oxidase does not necessarily depend on activation of phospholipase C and protein kinase C and on increase in cytoplasmic Ca^{2+} (see also Sects. 3.2.2, 3.2.3). Unfortunately, the interpretation of studies with various drugs such as protein kinase C inhibitors may be hampered by their lack of specificity (see also Sects. 3.2.2.3, 3.2.2.4). In addition, the respiratory burst is activated by stimuli which circumvent G-proteins and mimic certain aspects of receptor-induced activation, e.g., phorbol esters, cell-permeant diacylglycerols, fatty acids, and Ca^{2+} ionophores. Thus, studies with intact phagocytes can hardly answer the question whether activation of the respiratory burst by receptor agonists is due to direct interaction of G-proteins or low molecular mass GTP-binding proteins with NADPH oxidase or due to indirect activation through the formation of intracellular signal molecules.

These limitations are overcome by the use of cell-free systems which allow very effective manipulation of the experimental conditions. As is pointed out below, the cell-free systems have certain limitations as well. For example, in comparison to intact cells, relatively few agents, i.e., fatty acids, guanine and adenine nucleotides, phorbol esters, and phosphatidic acid, activate NADPH oxidase in the cell-free system, but most other stimuli described in Sect. 3.3 are apparently inactive.

5.1 Reconstitution and Regulation of NADPH Oxidase Activity by Fatty Acids and Sodium Dodecyl Sulfate

5.1.1 Historical Remarks

The establishment of cell-free systems for the reconstitution of NADPH oxidase activity with components from resting phagocytes was a breakthrough for the understanding of NADPH oxidase regulation. Cell-free systems were first described by Heyneman and Vercauteren (1984) and Bromberg and Pick (1984). Heyneman and Vercauteren (1984) reported that oleic or linoleic acid activate O_2^- formation in postnuclear fractions of horse neutrophils. Bromberg and Pick (1984) obtained similar results with guinea pig macrophages and showed that particulate and cytosolic components are required for reconstitution of NADPH oxidase activity. Subsequently, analogous reconstitution systems have been established for human neutrophils (Curnutte 1985; McPhail et al. 1985; Seifert et al. 1986), human monocytes (Thelen and Baggiolini 1990), differentiated HL-60 cells (Seifert and Schultz 1987b; Nozawa et al. 1988), pig neutrophils (Fujita et al. 1987; Tanaka et al. 1988), and bovine neutrophils (Ligeti et al. 1988; Doussiere et al. 1988). In these systems, fatty acids or SDS, membranes, and the cytosolic fraction of phagocytes are all required to reconstitute O_2^- formation, and omission of one of these components abolishes enzyme activity.

5.1.2 Some General Aspects

As is the case for intact phagocytes, there are certain functional differences between the cell-free systems from various types of phagocytes, but principally their regulatory properties are similar. With respect to the kinetic and catalytic properties of NADPH oxidase, to the insensitivity to inhibition by KCN and NaN₃, to the pH optimum, and to the salt sensitivity, the cell-free systems for NADPH oxidase activation and NADPH oxidase preparations from activated cells show similar properties (Bromberg and Pick 1984, 1985; Heyneman and Vercauteren 1984; Curnutte 1985; McPhail et al. 1985; Curnutte et al. 1987b; Fujita et al. 1987; Clark et al. 1987; Pick et al. 1987; Ligeti et al. 1988; Pilloud et al. 1989b; Nozaki et al. 1990; see also Sect. 2.1).

Not only neutrophil plasma membranes but also the specific granules contain the membrane components of NADPH oxidase (Clark et al. 1987). The kinetic properties of the plasma membrane-bound and granule-associated enzyme are very similar, and both components interact additively to reconstitute O_2^- formation in the presence of cytosol (Clark et al. 1987). These data suggest that the enzyme associated to specific granules represents an intracellular storage pool for NADPH oxidase which is translocated to the plasma membrane upon stimulation (see also Sect. 2.5). Preliminary data indicate that neutrophil granules contain a protein which inhibits activation of NADPH oxidase in the cell-free system (Aviram and Faber 1990).

The membrane-associated components of NADPH oxidase from resting phagocytes have been solubilized using various detergents. Curnutte et al. (1987a) prepared deoxycholate extracts of human neutrophil plasma membranes and reconstituted NADPH oxidase activity by combining this extract with neutrophil cytosol and SDS. The solubilized enzyme shows kinetic properties similar to those of the plasma membrane-associated enzyme (Curnutte et al. 1987a). Pick et al. (1987) solubilized the membrane component of NADPH oxidase from resting guinea pig macrophages with the nonionic detergent, n-octylglucoside. Delipidation of the solubilized NADPH oxidase reduces its activity, and various phospholipids restore enzyme activity (Shpungin et al. 1989; see also Sect. 2.1). Activation of the solubilized NADPH oxidase of pig neutrophils in the cell-free system is also a phospholipid-dependent process (Nozaki et al. 1990). The membrane component of NADPH oxidase from differentiated HL-60 cells was solubilized with n-octylglucoside, whereas other detergents were much less effective in this regard (Seifert 1988, and unpublished results). As is the case for NADPH oxidase in native plasma membranes of HL-60 cells, the solubilized enzyme is reversibly activated by arachidonic acid and guanine nucleotides, suggesting that solubilization does not result in the loss of these regulatory properties, especially regulation by G-proteins (Seifert and Schultz 1987b; Seifert 1988; see also Sect. 5.1.4). Unfortunately, the solubilized NADPH oxidase of HL-60 membranes is very instable and loses its activity at 4°C within 3 h (Seifert, unpublished results), possibly due to delipidation or to loss of the quarternary structure of cytochrome b_{-245} (Shpungin et al. 1989; Nugent et al. 1989).

5.1.3 Activation by Fatty Acids

5.1.3.1 Lipid Specificity

Arachidonic acid, other cis-unsaturated fatty acids, trans-unsaturated fatty acids, and SDS activate NADPH oxidase in crude membrane preparations and in purified plasma membranes of neutrophils and macrophages from various sources (Bromberg and Pick 1984, 1985; Heyneman and Vercauteren 1984; Curnutte et al. 1987a,b; Fujita et al. 1987; Seifert and Schultz 1987a,b; Ligeti et al. 1988; Nozawa et al. 1988). In membranes of human neutrophils, saturated fatty acids, esters of unsaturated fatty acids and ETYA do not activate the enzyme (Seifert and Schultz 1987a). In addition, Triton X-100, Lubrol PX, and sodium cholate do not activate O_2^- formation in various systems (Bromberg and Pick 1985; Seifert and Schultz 1987a). In contrast, certain saturated fatty acids, e.g., lauric acid, activate NADPH oxidase in membranes of porcine neutrophils (Tanaka et al. 1987, 1988). The extent of O_2^- formation depends on the ratio of fatty acid to membrane phospholipids rather than on the concentration of the fatty acid (Ligeti et al. 1988). Fatty acids and SDS may activate NADPH oxidase due to their anionic amphiphilic character (Bromberg and Pick 1985). This assumption is supported by the finding that positively charged alkylamines but not neutral amphiphilic alkylalcohols inhibit fatty acid-induced O_2^- formation in cell-free systems of guinea pig neutrophils and in intact cells (Miyahara et al. 1987, 1988; see also Sect. 3.1.2.2). Fatty acids induce translocation of cytosolic components of NADPH oxidase to the plasma membrane (Tanaka et al. 1988). and treatment of intact phagocytes with various stimuli prior to cell disruption renders O_2^- formation in the cell-free system less dependent on the participation of cytosolic components, (McPhail et al. 1985; Bromberg and Pick 1985). In intact cells, activation of NADPH oxidase is accompanied by the association of the 47-kDa protein with the plasma membrane (Heyworth et al. 1989a; see also Sects. 3.1.1.1, 3.2.2, 5.1.5, 6.1.2). Recently, Clark et al. (1990) showed that activation of NADPH oxidase by phorbol esters in intact cells and by arachidonic acid in the cell-free system is associated not only with the translocation of the 47-kDa protein to the plasma membrane but also with translocation of the 66-kDa protein. Doussiere et al. (1990) reported on arachidonic acid-dependent translocation of proteins with apparent molecular masses of 17, 45, 53, and 65 kDa to the plasma membrane in the cell-free system.

5.1.3.2 Fatty Acids and the Role of Protein Kinase C

There has been a discussion on the question whether the effects of fatty acids on O_2^- formation in cell-free systems are mediated via protein kinase C or not (see also Sect. 3.1.2.2).

SDS has been shown to activate O_2^- formation in the cell-free system independently of phosphoinositide degradation (Traynor et al. 1989). In addition, solubilization of the membrane component of NADPH oxidase results in depletion of phosphoinositides from the enzyme preparation without loss of NADPH oxidase activity (Traynor et al. 1989). Furthermore, various phosphoinositides inhibit SDS-induced $O_2^{\bullet-}$ formation in a cell-free system from human neutrophils (Aviram and Sharabani 1989a).

It is well known the *cis*-unsaturated fatty acids activate protein kinase C in the absence or presence of Ca^{2+}, depending on the preparation of kinase studied (McPhail et al. 1984b; Murakami and Routtenberg 1985; Hansson et al. 1986; K. Murakami et al. 1986, 1987; Linden et al. 1986; Sekiguchi et al. 1987; Seifert et al. 1988c; Verkest et al. 1988) (see also 3.1.1). In addition, ETYA and *trans*-unsaturated fatty acids with the exception of elaidic acid activate protein kinase C (Seifert et al. 1988c). In contrast, saturated fatty acids and SDS are not effective activators of protein kinase C (McPhail et al. 1984b; Murakami and Routtenberg 1985; K. Murakami et al. 1986; Seifert et al. 1988c). Ca^{2+} enhances fatty acid-induced activation of protein kinase C (McPhail et al. 1984b; K. Murakami et al. 1986; Sekiguchi et al. 1987; Seifert et al. 1988c), possibly by increasing the hydrophobicity of protein kinase C (Walsh et al. 1984). In addition to Ca^{2+}, Zn^{2+} may stimulate or inhibit protein kinase C (Murakami et al. 1987; Csermely et al. 1988).

With respect to NADPH oxidase, ETYA does not activate $O_2^{\bullet-}$ formation but is an inhibitor of fatty acid-induced $O_2^{\bullet-}$ formation in the cell-free system (Seifert and Schultz 1987a). Saturated fatty acids, elaidic acid and SDS effectively activate $O_2^{\bullet-}$ formation in cell-free systems of different cell types, but they do not activate protein kinase C (Bromberg and Pick 1984, 1985; Curnutte et al. 1987b; Fujita et al. 1987; Tanaka et al. 1987; Seifert and Schultz 1987a; Pick et al. 1987; Tanaka et al. 1988; Babior et al. 1988; Ligeti et al. 1988). Fatty acids activate $O_2^{\bullet-}$ formation in the absence of exogenous Ca^{2+}, and chelation of endogenous Ca^{2+} does not prevent $O_2^{\bullet-}$ formation (Seifert and Schultz 1987a; Ligeti et al. 1988). In addition, Ca^{2+} and Zn^{2+} inhibit arachidonic acid-induced $O_2^{\bullet-}$ formation (Bromberg and Pick 1984; Fujita et al. 1987; Seifert et al. 1988a). $O_2^{\bullet-}$ formation induced by fatty acids and SDS is not inhibited by H-7 (Seifert and Schultz 1987a; Miyahara et al. 1987) or by staurosporine (Seifert, unpublished results), and purified protein kinase C is no substitute for neutrophil cytosol and does not enhance $O_2^{\bullet-}$ formation in the presence of neutrophil cytosol (Seifert and Schultz 1987a; see also Sects. 3.2.2.3, 5.2).

Neutrophils of patients with autosomal recessive CGD apparently possess normal protein kinase C activity, and undifferentiated HL-60 cells possess a functionally intact phospholipase C/protein kinase C system as well (Zylber-Katz and Glazer 1985; Stutchfield and Cockcroft 1988; Caldwell et al. 1988; Makowske et al. 1988; Wenzel-Seifert and Seifert 1990). In contrast, both types of phagocyte cytosol fail to reconstitute $O_2^{\bullet-}$ formation in the cell-free system (Curnutte 1985; Curnutte et al. 1987b, 1988; Seifert and Schultz 1987b; Parkinson et al. 1987; Nozawa et al. 1988; see also Sect. 5.1.5). In addition, there is a chromatographic dissociation between protein kinase C and the components which reconstitute O_2^- formation (Curnutte et al. 1986, 1987b). The removal of ATP from the reaction mixtures by preincubation with hexokinase and glucose or dialysis of cytosol does not abolish O_2^- formation in the cell-free system, suggesting that kinase-mediated reactions are not obligatorily or are only partially involved in the activation of NADPH oxidase (Gabig and English 1986; Clark et al. 1987; Seifert and Schultz 1987a; Fujita et al. 1987; Ligeti et al. 1988). However, ATP enhances O_2^- formation, indicating that kinase reactions play a facultative role in the regulation of O_2^- formation (Clark et al. 1987; Seifert and Schultz 1987a; see also Sect. 5.1.4.3). Finally, ATP[γS] is only a very poor substrate for protein kinase C, but this nucleotide effectively enhances O_2^- formation in membranes of HL-60 cells (Wise et al. 1982; Seifert et al. 1988b; see also Sect. 5.1.4.3). From all these data it was concluded that protein kinase C is not involved in the activation of NADPH oxidase by fatty acids in cell-free systems.

5.1.3.3 The Role of Calmodulin

Besides protein kinase C, calmodulin has been suggested to play a role in fatty acid-induced O_2^- formation, as certain calmodulin antagonists inhibit arachidonic acid-induced O_2^- formation (McPhail et al. 1986). In addition, calmodulin has been reported to enhance the activity of NADPH oxidase obtained from stimulated phagocytes (Jones et al. 1982). In contrast, chelation of Ca^{2+} by EGTA does not prevent activation of NADPH oxidase by fatty acids in cell-free systems (Seifert and Schultz 1987a; Sakata et al. 1987a; Nozawa et al. 1988). In addition, purified calmodulin does not enhance O_2^- formation in cell-free systems, and calmodulin antagonists inhibit arachidonic acid-induced O_2^- formation presumably by blocking hydrophobic interaction of fatty acids with NADPH oxidase rather than by inhibiting calmodulin-dependent processes (Sakata et al. 1987a; see also Sect. 3.2.3.1). Thus, an involvement of calmodulin in the regulation of NADPH oxidase in cell-free systems is not likely.

5.1.3.4 Other Mechanistic Aspects

ETYA is a potent inhibitor of lipoxygenases and cyclooxygenase and inhibits O_2^- formation induced by *cis*-polyunsaturated fatty acids, which are substrates for lipoxygenases and cyclooxygenase, and O_2^- formation induced by *cis*-monounsaturated and *trans*-unsaturated fatty acids, which are no substrates for these enzymes (Kinsella et al. 1981; Needleman et al. 1986; Seifert and Schultz 1987a). In addition, *bis*(*tert*-butyl)peroxide does not activate O_2^- formation in the cell-free system, and soybean lipoxygenase is no substitute for phagocyte cytosol (Seifert and Schultz 1987a). These results indicate that oxygenated metabolites of unsaturated fatty acids are not involved in activation of NADPH oxidase by fatty acids in cell-free systems. Inhibition by ETYA of O_2^- formation may be explained by unspecific effects of the fatty acid (see also Sects. 3.1.2.2, 3.2.4.1) or by competitive antagonism of ETYA with unsaturated fatty acids at sites which are not localized on arachidonic acid-metabolizing enzymes (Seifert and Schultz 1987a). It is also unlikely that fatty acids activate NADPH oxidase by increasing membrane fluidity, as saturated and *trans*-unsaturated fatty acids have been reported to be inactive in this respect (Klausner et al. 1980; Badwey et al. 1984; see also Sect. 3.1.2).

NADPH oxidase of human, bovine, and porcine neutrophils, once activated by arachidonic acid, apparently does not depend on the permanent presence of arachidonic acid and cytosol, suggesting that an activated complex consisting of membrane components, cytosolic components and fatty acid is formed (Clark et al. 1987; Curnutte et al. 1987b; Gabig et al. 1987; Fujita et al. 1987; Doussiere et al. 1988).

Bovine serum albumin, which binds fatty acids (Badwey et al. 1984), rapidly terminates arachidonic acid-induced O_2^- formation in HL-60 membranes, as is the case for intact human neutrophils (Badwey et al. 1984; Seifert and Schultz 1987b; see also Sect. 3.1.2). In a cell-free system from porcine neutrophils, removal of fatty acids by bovine serum albumin prevents O_2^- formation, and readdition of fatty acids restores enzyme activity in the presence of cytosol (Tanaka et al. 1988).

Recently, Fujimoto et al. (1990) suggested that SDS specifically modulates the functional state of the 66-kDa protein (see also Sect. 5.1.5). SDS may activate the 66-kDa protein in the presence of the membrane components of NADPH oxidase and may inactivate the 66-kDa protein when the membrane components of the oxidase are absent.

5.1.3.5 Physiological Relevance of Fatty Acid-Induced Activation of NADPH Oxidase in Cell-free Systems

As has been pointed out above, there are certain similarities between fatty acid-induced activation of NADPH oxidase in intact cells and in cell-free systems. With respect to the dependency on Ca^{2+}, there are substantial differences between the two systems (see Sects. 3.1.2, 3.2.4). Arachidonic acid induces arachidonic acid release in intact human neutrophils but not in neutrophil plasma membranes (Maridonneau-Parini and Tauber 1986). In addition, inhibitors of phospholipase A_2 block arachidonic acid-induced O_2^- formation in intact phagocytes but not in the cell-free system, and phospholipase A_2 does not activate NADPH oxidase in the cell-free system (Maridonneau-Parini and Tauber 1986). In addition, we found that mellitin, which activates phospholipase A_2 (Schoch and Sargent 1980), inhibits arachidonic acid-induced O_2^- formation in a cell-free system from HL-60 cells in the absence and in the presence of guanine nucleotides, and mellitin per se does not activate NADPH oxidase (unpublished results). These data suggest that activation of the respiratory burst by arachidonic acid in intact cells but not in cell-free systems may involve phospholipase A_2.

In most studies with cell-free systems, fatty acids are required at supraphysiological concentrations to activate the respiratory burst (Heyneman and Vercauteren 1984; McPhail et al. 1985; Seifert and Schultz 1987a,b). Arachidonic acid at concentrations which activate the respiratory burst in cell-free systems may be cytotoxic to intact neutrophils and macrophages (H.J. Cohen et al. 1986; Tsunawaki and Nathan 1986). However, arachidonic acid at a concentration as low as 16 μM has been shown to activate O_2^- formation in plasma membranes of human nuetrophils, and this concentration of the fatty acid may be within the physiological range in intact cells (Tsunawaki and Nathan 1986; Seifert et al. 1986; Seifert and Schultz 1987a). Thus, the question of the extent to which activation of NADPH oxidase by fatty acids in cell-free systems reflects a physiological process is still open (see also Sect. 5.1.1.4).

5.1.4 The Role of G-Proteins

The results of studies with cell-free systems discussed above suggest the existence of protein kinase C/Ca^{2+}-independent signal transduction pathways for the activation of NADPH oxidase. Thus, much work has been done to answer the question whether NADPH oxidase is regulated in a more direct way by G-proteins.

5.1.4.1 Guanine Nucleotides

The stable GTP-analogues GTP[γS] and [β,γ-NH]GTP are potent activators of G-proteins (Gilman 1987) and enhance fatty acid-induced O_2^- formation several-fold when added prior to or together with the fatty acid to the reaction mixture (Seifert et al. 1986, 1988b; Seifert and Schultz 1987a,b; Gabig et al. 1987; Ligeti et al. 1988; Doussiere et al. 1988). In addition, stable guanine nucleotides reinitiate O_2^- formation in membranes of HL-60 cells after the arachidonic acid-induced respiratory burst had ceased (Seifert and Schultz 1987b; Seifert et al. 1988b). Activation of NADPH oxidase by arachidonic acid follows a first-order reaction course (Aviram and Sharabani 1989b). GTP[γS] increases V_{max} of NADPH oxidase without affecting the K_m for NADPH, but the kinetic of O_2^- formation is sigmoid (Aviram and Sharabani 1989b). These data suggest that two processes are involved in the activation of NADPH oxidase, and that two separate pools of NADPH oxidase are present in neutrophil membranes (Aviram and Sharabani 1989b). In a cell-free system from resting macrophages, stable guanine nucleotides prevent loss of SDS-stimulated NADPH oxidase activity (Aharoni and Pick 1990).

G-proteins are assumed to bind guanine nucleotides at a nucleotide binding site in the presence of Mg^{2+} (Gilman 1987; Hingorani and Ho 1987; Yamazaki et al. 1987). In agreement with this suggestion, it has been reported that the stimulatory effects of guanine nucleotides on O_2^- formation require Mg^{2+} to be maximal (Seifert and Schultz 1987a; Gabig et al. 1987; Ligeti et al. 1988). However, even in the absence of Mg^{2+} and in the presence of EDTA, GTP[γS] stimulates O_2^- formation at least to some extent (Seifert and Schultz 1987b; Seifert et al. 1988b). This finding may be explained by the fact that Mg^{2+} is tightly bound to membranes, phospholipids and/or G-proteins and is only slowly removed by EDTA (Codina et al. 1984; Seifert and Schultz 1987b). In intact cells, Mg^{2+} also plays an important role in the activation process of NADPH oxidase (Gabler 1990).

GTP and [β,γ-NH]ATP do not potentiate fatty acid-induced O_2^- formation, and GDP and GDP[βS] inhibit O_2^- formation in the absence and presence of GTP[γS] (Seifert et al. 1986; Seifert and Schultz 1987b; Gabig et al. 1987; Ligeti et al. 1988; Doussiere et al. 1988). In addition, GDP and GDP[βS] terminate arachidonic acid- and GTP[γS]-stimulated O_2^- formation in membranes of HL-60 cells (Seifert and Schultz 1987b). In contrast, other nucleoside diphosphates do not ihibit O_2^- formation (Seifert and Schultz 1987b). These results indicate that GDP and GDP[βS] compete with G-protein-activating ligands, i.e., with endogenous GTP and less effectively with exogenous GTP[γS], and promote inactivation of G-proteins (Eckstein et al. 1979; Eckstein 1985; Gilman 1987) and subsequent deactivation of NADPH oxidase.

The stimulatory effects of stable GTP analogues on O_2^- formation in the cell-free system of human neutrophils are not inhibited by pertussis toxin or cholera toxin (Seifert et al. 1986; Gabig et al. 1987). In addition, the effects of GTP[γS], ATP[γS], and NaF on O_2^- formation in a cell-free systems from dimethyl sulfoxide-differentiated HL-60 cells are completely pertussis toxin insensitive (Seifert, unpublished results). Gabig et al. (1987) put forward the attractive hypothesis that cholera toxin- and pertussis toxin-insensitive G-proteins, distinct from either G_s or a G-protein of the G_i-family, are involved in the regulation of NADPH oxidase. Indeed, the

22-kDa GTP-binding protein, *rap*1, which is associated with cytochrome b_{-245}, is not a substrate for pertussis toxin or cholera toxin (Quinn et al. 1989; see also Sects. 2.4.3, 3.2.1.2, 5.1.5).

However, other possibilities to explain the toxin-insensitivity of the effects of GTP[γS] cannot be ruled out, as pertussis toxin insensitivity of effects of GTP[γS] has been reported for other effector systems regulated by G-proteins of the G_i family (Jakobs et al. 1984; Cockcroft and Stutchfield 1988). As far as inhibition of adenylyl cyclase is concerned, the onset of the effects of GTP[γS] is delayed in membranes of pertussis toxin-treated cells, but the maximal effect is unaffected (Jakobs et al. 1984). In membranes of pertussis toxin-treated human neutrophils, neither the onset nor V_{max} of GTP[γS]-stimulated O_2^- formation is affected by the toxin (Seifert et al. 1986). The same is true for HL-60 cells (Seifert, unpublished results). These results are in agreement with the assumption that ADP ribosylation of G-proteins by pertussis toxin impairs the interaction of G-proteins with agonist-occupied receptors but not the interaction of G-proteins with stable guanine nucleotides (Gilman 1987).

GTP[γS] has been suggested to interact with a cytosolic component prior to stimulation of membrane-bound G-proteins and formation of an active complex consisting of membrane components, cytosolic components, arachidonic acid, and GTP[γS] (Ligeti et al. 1988; Doussiere et al. 1988). Guanine nucleotides apparently promote translocation of a cytosolic component to the plasma membrane (Ligeti et al. 1989). Gabig et al. (1990) suggested that the cytosolic GTP-binding protein in its GTP[γS]-bound form is stabilized or activated by unsaturated fatty acids (see also Sect. 5.1.3). Seifert et al. (1989c) and Ishida et al. (1989) also obtained functional evidence for the participation of specific cytosolic components in the guanine nucleotide-dependent activation of NADPH oxidase. The identity of these cytosolic components is not yet known (see also Sects. 5.1.4.4, 5.1.5.2.3). Candidates are the 47-kDa protein (see Sect. 5.1.5.2.3) and a cytosolic 23-kDa GTP-binding protein (Ligeti et al. 1989; Stasia et al. 1989). Moreover, human neutrophil cytosol contains α-subunits of G-proteins of the G_i-family, i.e., G_{i2} (Rosenthal et al. 1987; Rudolph et al. 1989a,b; Volpp et al. 1989a). α-Subunits in neutrophil cytosol are regulated by GTP[γS] and Mg^{2+} as is suggested by changes of their hydrodynamic properties (Rudolph et al. 1989b).

5.1.4.2 NaF

Similar to intact cells, NaF, presumably as AlF_4^- (Sternweis and Gilman 1982; Bigay et al. 1985), stimulates O_2^- formation in cell-free systems of human neutrophils and HL-60 cells, i.e., the halide potentiates fatty acid-induced O_2^- formation (Seifert et al. 1986; Gabig et al. 1987; Seifert 1988; see also Sect. 3.2.1.3). NaF slightly reduces the activation rate of NADPH oxidase but increases V_{max} (Aviram and Sharabani 1989b).

5.1.4.3 Nucleoside Diphosphate Kinase

Nucleoside disphosphate kinase catalyzes the phosphorylation of GDP to GTP by ATP and may play a role in regulation of various G-protein-regulated effector systems such as adenylyl cyclase and phospholipase C (Kimura and Nagata 1979; Totsuka et al. 1982; Kimura and Johnson 1983; Anthes et al. 1987; Wieland and Jakobs 1989). ATP is not obligatorily required for reconstitution of NADPH oxidase activity but substantially enhances O_2^- formation (Seifert and Schultz 1987a; Clark et al. 1987; Seifert et al. 1988a,b; Ligeti et al. 1988; see also Sect. 5.1.3.2). In addition to GTP[γS], ATP[γS] potentiates O_2^- formation in HL-60 membranes, but ATP[γS] is less potent and effective than GTP[γS] (Seifert et al. 1988b). This finding may be explained by the following mechanism.

The thiophosphoryl group of ATP[γS] is transferred to phosphate acceptors by various kinases including nucleoside diphosphate kinase (Cassidy et al. 1979; Eckstein 1985). HL-60 membranes possess a nucleoside diphosphate kinase which catalyzes the thiophosphorylation of GDP by ATP[γS] to GTP[γS], resulting in activation of NADPH oxidase (Seifert et al. 1988b). This thiophosphorylation does not require added GDP, indicating that endogenous GDP in the cytosol and/or membrane is sufficient for serving as thiophosphoryl group acceptor (Seifert et al. 1988b). Indeed, it has been shown that GDP is tightly bound to G-proteins (Godchaux and Zimmerman 1979; Ferguson et al. 1986). In HL-60 cells, GDP bound to G-proteins may serve as acceptor for phosphate groups in nucleoside diphosphate kinase-mediated reactions as well. In HL-60 cells, G-proteins or low molecular mass GTP-binding proteins and nucleoside diphosphate kinase may be closely associated proteins, as is the case for other systems (Ohtsuki et al. 1986, 1987; Uesaka et al. 1987; Ohtsuki and Yokoyama 1987; Kimura and Shimada 1988; see also Sects. 2.4.3, 3.2.1.2, 5.1.5). The stimulatory effect of ATP[γS] but not that of GTP[γS] on O_2^- formation is abolished by the inhibitors of nucleoside diphosphate kinase, UDP, and ADP (Goffeau et al. 1967; Kimura and Shimada 1983; Seifert et al. 1988b).

The activity of nucleoside diphosphate kinase absolutely depends on Mg^{2+} (Parks and Agarwal 1973), and chelation of Mg^{2+} by EDTA abolishes the stimulatory effect of ATP[γS] but not that of GTP[γS] on NADPH oxidase (Seifert et al. 1988b). Moreover, phosphorylation of endogenous GDP to GTP by creatine kinase and creatine phosphate prevents thiophosphorylation of GDP to GTP[γS] by ATP[γS] and potentiation of O_2^- formation (Seifert et al. 1988b). In addition to HL-60 cells, thiophosphorylation of GDP to GTP[γS] by ATP[γS] has been shown to take place in atrial myocytes and in platelets, and this process is associated with persistent activation of the corresponding G-protein-regulated effector systems (Otero et al. 1988; Wieland and Jakobs 1989).

5.1.4.4 Some Open Questions

The studies described above provided substantial evidence for the assumption that NADPH oxidase is regulated by G-proteins and/or low molecular mass GTP-binding proteins. However, a number of problems remain to be resolved. For example, is NADPH oxidase, in analogy to adenylyl cyclase and retinal cGMP phosphodiesterase, under the *direct* control of G-proteins (Gilman 1987) or are additional, possibly cytosolic components required? The role of cytosolic G-proteins in the regulation of NADPH oxidase is not known, and the relation between the cytosolic 23-kDa GTP-binding protein identified by Ligeti et al. (1989) to NADPH oxidase activation and *rap*1 (Quinn et al. 1989) remains to be established (see also Sects. 2.4.3, 3.2.1.2, 5.1.5). The 23-kDa GTP-binding protein has been purified (Stasia et al. 1989). It is a substrate for protein kinase C in vitro, is apparently no substrate for *Clostridium botulinum* ADP-ribosyltransferase C3, and is not identical with calmodulin or the α-subunit of cytochrome b_{-245} (Stasia et al. 1989). In addition, the identity of several other putative low molecular mass components of NADPH oxidase remains to be clarified (see also Sect. 2.4.1, Table 1). Furthermore, the precise role of pertussis toxin-sensitive and/or -insensitive G-proteins and/or of low molecular mass GTP-binding proteins in the regulation of NADPH oxidase by various types of receptor agonists and stimuli circumventing receptor stimulation is not known. Finally, it remains to be established whether the 47-kDa protein which possesses a nucleotide-binding domain and shows homology to *ras* p21 GTPase-activating protein, interacts with *rap*1 (Quinn et al. 1989; Lomax et al. 1989; Volpp et al. 1989b).

NADPH oxidase preparations obtained from phagocytes treated with chemotactic peptides prior to cell disruption generate O_2^- (McPhail and Snyderman 1983). In contrast, in disrupted phagocytes, fMet-Leu-Phe does not stimulate O_2^- formation, whereas in electropermeabilized neutrophils, the chemotactic peptide stimulates a respiratory burst (McPhail et al. 1985; Grinstein and Furuya 1988; Nasmith et al. 1989). In our laboratory, we have undertaken many efforts during the past 4 years to demonstrate a stimulatory effect of fMet-Leu-Phe on O_2^- formation in cell-free systems from human neutrophils and HL-60 cells, but for unknown reasons all experiments gave negative results (unpublished results). In contrast, chemotactic peptides stimulate high-affinity GTPase and phospholipase C in plasma membranes of neutrophils and HL-60 cells (Hyslop et al. 1984; Okajima et al. 1985; C.D. Smith et al. 1985, 1986; Kikuchi et al. 1986; Williamson et al. 1988; Wilde et al. 1989). These data suggest that components required for receptor-mediated activation of NADPH oxidase are destroyed during cell disruption and/or that the integrity of cellular structures is required for this process.

Purified G-proteins of the G_i family or G_o reconstitute chemotactic peptide-stimulated GTPase and phospholipase C activity (Okajima et al. 1985; Kikuchi et al. 1986). So far, we have not found stimulatory effects of purified G_o or G_i from porcine brain (Rosenthal et al. 1986) on O_2^- formation in cell-free systems from HL-60 cells under a variety of experimental conditions (unpublished results). Recently, we studied the effects of various recombinant *ras*-proteins in cell-free systems. As in our experiments with purified G-proteins, we failed to detect stimulatory effects of recombinant *ras* proteins on O_2^- formation (unpublished results).

With respect to immunological studies, we did not observe any inhibitory effect of antibodies raised against the β-subunits of G-proteins or of an antibody raised against a highly conserved sequence of α-subunits of G-proteins (α_{common} peptide; Rosenthal et al. 1986; Rudolph et al. 1989a; Hinsch et al. 1988) on fatty acid-induced O_2^- formation in cell-free systems in the absence or the presence of GTP[γS] under various experimental conditions (unpublished results). In contrast, a number of reports show that antibodies raised against certain components of NADPH oxidase may inhibit O_2^- formation in intact cells or of the activated enzyme (Doussiere and Vignais 1988; Fukuhara et al. 1988; Berton et al. 1989; see also Sect. 2.4).

5.1.5 Cytosolic Activation Factors

Much progress has been made in the past 2 years with regard to the characterization of the cytosolic activation factors of NADPH oxidase. Initial studies suggested that this factor or one of these factors may be protein kinase C (McPhail et al. 1984b, 1985). However, subsequent studies provided convincing evidence for the assumption that protein kinase C is not involved in the activation of $O_2^{\cdot-}$ formation in this cell-free system (see also Sect. 5.1.3.2). Some of the properties of the cytosolic activation factors for NADPH oxidase are described in Sects. 2.4.1, 3.1.1.1, and 3.2.2.1.

5.1.5.1 Some General Properties

The cytosolic factor is heat labile and sensitive to proteolytic inactivation (Bromberg and Pick 1985; Seifert and Schultz 1987a; Seifert et al. 1989c; Fujita et al. 1987; Ligeti et al. 1988; Bolscher et al. 1989). Activation of intact human neutrophils with PMA is associated with the consumption and/or depletion of cytosolic components of NADPH oxidase (Umeki 1990; see also Sect. 3.1.1.1). The occurrence of the cytosolic activation factor is apparently cell type-specific. Crude cytosols of neutrophils and macrophages of various species and of HL-60 cells differentiated with vitamin D_3, dimethyl sulfoxide, or retinoic acid have been shown to reconstitute $O_2^{\cdot-}$ formation in the cell-free system (Bromberg and Pick 1984; Fujita et al. 1987; Seifert and Schultz 1987a,b; Parkinson et al. 1987; Ligeti et al. 1988; Nozawa et al. 1988; Seifert et al. 1989c).

In contrast to the above cells, cytosol of undifferentiated HL-60 cells and neutrophil cytosol of patients with autosomal recessive, cytochrome b_{-245}-positive CGD is inactive (Curnutte 1985; Curnutte et al. 1987b, 1988; Seifert and Schultz 1987b; Parkinson et al. 1987; Nozawa et al. 1988; see also Sect. 6.1.2). Neutrophil cytosol of autosomal recessive CGD patients does not inhibit the activity of control cytosol, indicating that the defect is due to the lack of stimulatory factors rather than to the presence of inhibitory factors (Curnutte et al. 1988). In contrast, neutrophils of these CGD patients possess a functionally intact membrane component of NADPH oxidase but do not generate $O_2^{\cdot-}$ upon stimulation with various agents (Curnutte et al. 1986, 1987a,b, 1988). These results underline the importance of the cytosolic activation factors for NADPH oxidase regulation in intact cells and the physiological relevance of the cell-free system.

Cytosols of brain, kidney, or liver of the rat, lymph node and thymus of the guinea pig, human lymphocytes and platelets as well as cytosols of the murine phagocytic cell lines $P338_1$ and MOPC 315 myeloma cells do not reconstitute $O_2^{\cdot-}$ formation in the cell-free system (Seifert and Schultz 1987a; Pick and Gadba 1988; Bolscher et al. 1989). Somewhat unexpectedly, cytosol of guinea pig thymus, lymph node lymphocytes, brain and mouse myeloma MOPC 315 cells has recently been reported to contain a factor with a molecular mass of 30–52 kDa, referred to as σ_1, which reconstitutes $O_2^{\cdot-}$ formation in a cooperative manner with the phagocyte-specific factor σ_2 (Pick et al. 1989).

5.1.5.2 Involvement of Multiple Cytosolic Activation Factors in the Regulation of NADPH Oxidase

The analysis of cytosolic activation factors by functional studies and protein purification studies revealed an unexpected complexity both within a given type of phagocyte and within various types of phagocytes. Table 16 summarizes some properties of the putative cytosolic activation factors in various cell types. Unfortunately, the nomenclature of cytosolic activation factors is still a matter of debate. At present, each group of authors prefers its own classification, and even within a given group the terms may change rapidly. In the following, we use the terms used by these authors, and we will attempt to compare the identity of the cytosolic activation factors.

5.1.5.2.1 Functional Studies

The results of functional studies suggest that more than one cytosolic activation factor is involved in the regulation of NADPH oxidase. The analysis of the kinetics of NADPH oxidase activation in a fully soluble system revealed that the enzyme is activated in a three-stage process (Babior et al. 1988). According to this model in the first step, the membrane component of NADPH oxidase (M) takes up a cytosolic factor (S) to form the complex [M•S]. In the second step, this complex is converted into the precatalytic species [M•S]*. In the third step, this complex takes up two additional, possibly identical cytosolic components, termed C_α and C_β. This process results in the formation of a low-activity (i.e., high K_m) NADPH oxidase ([M•S]*C_α) and subsequently in the formation of a high-activity (i.e., low K_m) NADPH oxidase ([M•S]*$C_\alpha C_\beta$).

Cytosol of dimethyl sulfoxide-differentiated HL-60 cells has been reported to reconstitute O_2^- formation in the absence and presence of GTP[γS], whereas cytosol of dibutyryl cAMP-differentiated HL-60 cells reconstitutes enzyme activity only in the presence of GTP[γS] (Seifert et al. 1989c). In addition, cytosolic proteins of dimethyl sulfoxide- and dibutyryl cAMP-differentiated HL-60 cells at submaximally stimulatory amounts synergistically stimulate O_2^- formation in the presence but not in the absence of GTP[γS]. These data suggest that two cytosolic activation factors are involved in the regulation of NADPH oxidase which are differently expressed in HL-60 cells (Seifert et al. 1989c). Apparently, one factor is involved in the maintenance of basal, fatty acid-induced O_2^- formation, the other factor mediates G-protein-mediated O_2^- formation, and the two factors interact synergistically to reconstitute G-protein-regulated O_2^- formation (Seifert et al. 1989c; see also Sects. 5.1.4.1, 5.1.4.4).

Table 16. Some properties of putative cytosolic activation factors of NADPH oxidase

Source	Properties of cytosolic activation factors	Selected references
Human neutrophils	Kinetic analysis of NADPH oxidase activation: with membrane component, factor S forms a precatalytic species. S may be defective in autosomal recessive CGD (47-kDa protein). Factors C_α and C_β convert NADPH oxidase in the high-activity (low K_m) form.	Babior et al. (1988), Curnutte et al. (1989b)
	Gel filtration: apparent molecular mass 10 kDa, single species.	Clark et al. (1987)
	Gel filtration: apparent molecular masses 59, 89, and 122 kDa.	Gabig et al. (1987)
	Heat-labile; functional evidence that the factor is not identical with protein kinase C.	Cox et al. (1987), Miyahara et al. (1987), Seifert and Schultz (1987a), Ligeti et al. (1988)
	Gel filtration: apparent molecular masses 30–40 and 250 kDa; chromatographic dissociation from protein kinase C.	Curnutte et al. (1987b)
	NADPH binding site; apparent molecular mass 66 kDa. Inactivated by NADPH-dialdehyde. May be identical with C2 and NCF-2.	Smith et al. (1989a,b)
	Isoelectric focusing: factor C1 (pI 3.1), C2 (pI 6.0), C3 (pI 7.0), C4 (pI 9.5); 5th component postulated. All five components are required for reconstitution. C2 (66-kDa protein?) and C4 (47-kDa protein?) may be defective in various types of autosomal recessive CGD.	Curnutte et al. (1989a)
	Binding to carboxymethyl Sepharose: Soluble oxidase component (SOC I) does not bind; SOC II binds to matrix. Heat-labile; for reconstitution of activity both SOC I and SOC II are required. SOC II defective in autosomal recessive CGD (47-kDa protein?).	Bolscher et al. (1989)
	Binding to GTP-agarose; apparent molecular masses 47 and 67 kDa. Both factors essential for reconstitution of activity.	Volpp et al. (1988)

Table 16. (continued)

Source	Properties of cytosolic activation factors	Selected references
Human Neutrophils	Binding to Mono-Q: Neutrophil cytosol factor (NCF)-1 = 47 kDa; NCF-2 = 65 kDa; NCF-3 (molecular mass?). All three components are required for reconstitution. Either NCF-1 or NCF-2 may be defective in autosomal recessive CGD.	Nunoi et al. (1988)
	Recombinant NCF-1 (NCF-47K) reconstitutes the defect in cytosol of autosomal recessive CGD patients. cDNA codes for a 41.4- to 41.9-kDa protein (pI 10.4). Arginine- and serine-rich COOH-terminal domain with potential protein kinase C phosphorylation sites. A 33 amino acid segment with 49% identity to *ras* p21 GTPase activating protein. N-terminal glycine (myristylation site?), nucleotide-binding domain. Homology to *src* protein kinases, phospholipase C and α-fodrin.	Lomax et al. (1989), Volpp et al. (1989b)
	Recombinant NCF-2 (r-p67) partially reconstitutes the defect in cytosol o autosomal recessive CGD patients. cDNA codes for a 526-amino acid protein. Acidic middle and COOH-terminal domains. Motif similarity to *scr*-related protein kinases, phospholipase C-γ, αfodrin, *ras* p21 GTPase activating protein and NCF-1. Functional similarities with NCF-1?	Leto et al. (1989, 1990)
HL-60 leukemic cells	Defect in undifferentiated HL-60 cells. Expressed in dimethyl sulfoxide-differentiated cells. Functional evidence for the involvement of two factors. One factor mediates basal, arachidonic acid-induced $O_2^{\bullet-}$ formation and is present in dimethyl sulfoxide-differentiated cells. The other factor mediates GTP-dependent $O_2^{\bullet-}$ formation and is present in dimethyl sulfoxide and dibutyryl cAMP-differentiated cells. Both factors interact synergistically to reconstitute GTP-dependent $O_2^{\bullet-}$ formation.	Seifert and Schultz (1987b), Parkinson et al. (1987), Nozawa et al. (1988), Nunoi et al. (1988), Volpp et al. (1988), Seifert et al. (1989c)

Table 16. (continued)

Source	Properties of cytosolic activation factors	Selected references
Guinea pig macrophages	Heat-labile; binds to 2′, 5′-ADP-agarose (NADPH binding site?), molecular mass < 100 kDa.	Bromberg and Pick (1985), Sha'ag and Pick (1988), Pick et al. (1989)
	Separation of σ_1 and σ_2 by various purification procedures. Both σ_1 and σ_2 are required for $O_2^{\cdot-}$ formation. σ_1 inactivated by proteases and heat but not by NEM. Gel filtration: 30–52 kDa. Occurs in cytosol of guinea pig thymus, lymph node lymphocytes, brain and mouse myeloma cell line MOPC 315. May be identical with NCF-3, SOC-1, and C1. σ_2 inactivated by proteases, heat and NEM. Gel filtration: 150–440 kDa. May contain the 47- and 66-kDa proteins; is phagocyte-specific.	
Bovine neutrophils	Proteolytic inactivation, interacts with GTP[γS], G-protein? Low molecular mass GTP-binding protein (23 kDa)?	Doussiere et al. (1988), Ligeti et al. (1988, 1989), Pilloud et al. (1989b), Stasia et al. (1989)
	Chromatography on Mono-Q and Mono-S: separation of two factors which are both required for reconstitution of $O_2^{\cdot-}$ formation.	
Pig neutrophils	Gel filtration: factor C_1 (300 kDa), C_2 (50 kDa), and C_3 (1.3 kDa). Both C_1 and C_2 are required for reconstitution of enzyme activity; C_1 may contain two compounds binding to 2′,5′-ADP agarose. One component of C_1 mediates GTP-dependent and another GTP-independent $O_2^{\cdot-}$ formation.	Fujita et al. (1987), Ishida et al. (1989)
	NADPH binding site, apparent molecular mass 66-kDa; 45 kDa protein may be proteolytic product of 66-kDa protein. Inactivated by NADPH-dialdehyde.	Takasugi et al. (1989)

The functional defect of a cytosolic activation factor in dibutyryl cAMP-differentiated HL-60 cells raises the question of the extent to which $O_2^{\cdot-}$ formation in cell-free systems reflects $O_2^{\cdot-}$ formation in intact cells. Intact dimethyl sulfoxide-differentiated HL-60 cells generate $O_2^{\cdot-}$ at substantially higher rates than dibutyryl cAMP-differentiated HL-60 cells upon stimulation with A 23187, PMA, arachidonic acid, and γ-hexachlorocyclohexane, whereas the chemotactic peptide is a more effective activator of $O_2^{\cdot-}$ formation in dibutyryl cAMP-differentiated HL-60 cells (Seifert et al. 1989c). Thus, HL-60 cells may be a useful model system to study the roles of cytosolic activation factors in the activation of NADPH oxidase by various stimuli in intact phagocytes which are yet incompletely understood (see also Sect. 3.4.4.1.3).

Recent data obtained by Levy et al. (1990a,b) show that, in fact, cytosolic activation factors for NADPH oxidase are differentially expressed in human myeloid cells. Specifically, cytosol from monocytes cultured for 6 days is virtually depleted of the 47-kDa protein but not of the 66-kDa protein (Levy et al. 1990a; see also Sect. 3.4.3). Conversely, cytosol from HL-60 cells cultured for 3 days in the presence of retinoic acid is devoid of the 66-kDa protein, whereas the 47-kDa protein is present. These authors suggested that these cytosols provide suitable model systems to study the defects in cytosolic activation factors present in autosomal recessive CGD. During the differentiation of HL-60 cells, the 47-kDa protein is detected earlier than the 66-kDa protein, and the latter protein is apparently the limiting cytosolic component for NADPH oxidase activation (Levy et al. 1990b).

Preincubation of neutrophil cytosol with 2'3'-dialdehyde NADPH prevents activation of NADPH oxidase in the cell-free system, apparently by covalently reacting with a 66-kDa protein (Smith et al. 1989a,b; Takasugi et al. 1989). Neutrophil cytosol treated with 2'3'-dialdehyde NADPH loses its ability to reduce the lag time for NADPH oxidase activation and to convert the enzyme from the high K_m form to the low K_m form (Smith et al. 1989b). 2'3'-Dialdehyde NADPH-treated neutrophil cytosol plus neutrophil cytosol from CGD patients with a defect in the 47-kDa protein reconstitute this functional abnormality (Smith et al. 1988b). These results suggest that the 66-kDa protein carries the NADPH-binding site of NADPH oxidase, and that its translocation from the cytosol to the plasma membrane is an early step in the activation of NADPH oxidase (see also Sect. 2.4.1 and below).

Data from Kleinberg et al. (1990) show that the 47-kDa protein but not the 66-kDa protein shortens the lag time of NADPH oxidase activation in the cell-free system. Additionally, experiments with a peptide that corresponds to a cytoplasmic carboxy-terminal domain of the β-subunit of cytochrome b_{-245} indicate that the 47-kDa protein is required early in the activation of NADPH oxidase, whereas the 66-kDa protein is essential for subsequent reactions resulting in the formation of the catalytically active NADPH oxidase.

5.1.5.2.2 Protein Purification Studies

The concept that multiple cytosolic activation factors are involved in the regulation of NADPH oxidase is also supported by the results of protein purification studies. Initial studies suggested that in cytosol of human neutrophils a single peptide with an apparent molecular mass of 10 kDa reconstitutes O_2^- formation (Clark et al. 1987). Other authors suggested that various peptides and/or proteins with apparent molecular masses of 40–250 kDa reconstitute enzyme activity in cell-free systems of human neutrophils (Curnutte et al. 1987b; Gabig et al. 1987). We also found that various molecules with apparent molecular masses of 40–300 kDa, as revealed by gel filtration, support O_2^- formation in cell-free systems of dimethyl sulfoxide-differentiated HL-60 cells (unpublished results).

The cytosolic activation factor from dimethyl sulfoxide-differentiated HL-60 cells is stable at 4°C for at least 2 weeks and is recovered in a functionally active state by ammonium sulfate precipitation at 35%–45% saturation (Seifert, unpublished results). In addition, the cytosolic activation factor from these cells binds to the dye orange A and to Heparin-Sepharose CL-6B and is eluted from these matrixes with 1-M KCl. The above procedures lead to approximately 5-fold, 10-fold, and 20-fold increases in specific activity of the cytosolic activation factor, and the combination of ammonium sulfate precipitation with subsequent chromatography on orange A results in a 25-fold enrichment in specific activity (unpublished results). The dyes blue B, red A, and green B are considerably less effective in binding the cytosolic activation factor than orange A (unpublished results). The cytosolic activation factor also binds to fast-flow phenyl Sepharose CL-4B, but we failed to elute the factor in a functionally active state from this matrix (unpublished results).

The cytosolic activation factor from guinea pig macrophages has been purified by chromatography on 2'5'-ADP-agarose and has been suggested to carry the NADPH binding site of NADPH oxidase (Sha'ag and Pick 1988; see also Sect. 2.4.1 and above). Further analysis suggested that two components, referred to as σ_1 and σ_2, are both required for reconstitution of O_2^- formation (Pick et al. 1989). The σ_1 factor is inactivated by proteases and by heat but not by NEM, may possess a molecular mass of 30–52 kDa, and is present not only in phagocytes but also in nonphagocytic cell types (Pick et al. 1989; see also Sect. 5.1.5.1). Unlike the σ_1 factor, the σ_2 is inactivated by NEM (see also Sect. 4.3.3), may possess a molecular mass of 150–440 kDa, and is apparently phagocyte-specific (Pick et al. 1989). Recently, Sha'ag and Pick (1990) characterized the nucleotide-binding properties of the σ_2 factor and showed that this protein contains a domain which recognizes the phosphate group at the ribose 2' position in adenosine and another domain which recognizes purine nucleoside triphosphates.

Porcine neutrophil cytosol was analyzed by gel filtration chromatography (Fujita et al. 1987; Ishida et al. 1989). Two components, operationally termed C_1 and C_2, have been reported to be involved in the reconstitution of NADPH oxidase activity. C_1 alone is not very effective in reconstituting O_2^- formation, but its effectiveness is potentiated either by GTP[γS] or by C_2. C_2 alone is inactive, and the effects of C_2 and GTP[γS] in the presence of C_1 are additive. C_1 has an apparent molecular mass of 300 kDa, and C_2 shows a molecular mass of 50 kDa as revealed by gel filtration. Analysis of C_1 by affinity chromatography on 2'5'-ADP agarose revealed that C_1 consists of at least two components, one mediating GTP-dependent and the other mediating GTP-independent regulation of NADPH oxidase (Ishida et al. 1989; see also Sects. 5.1.4.1, 5.1.5.2.1).

Chromatography of bovine neutrophil cytosol on Mono-Q and Mono-S columns resulted in the separation of two factors neither of which alone reconstitute O_2^- formation (Pilloud et al. 1989b). However, upon recombination both factors support O_2^- formation (Pilloud et al. 1989b). In a subsequent study, these authors provided evidence for the assumption that several proteins with apparent molecular masses ranging from 17 to 65 kDa may be involved in the reconstitution of O_2^- formation in the bovine neutrophil-derived cell-free system (Doussiere et al. 1990).

Using GTP-agarose affinity chromatography, a cytosolic activation complex for NADPH oxidase has been purified (Volpp et al. 1988). Polyclonal antibodies against this complex recognize a 47- and a 67-kDa protein, and there is a close correlation between the occurrence of the 47- and 67-kDa proteins and the amount of cytosolic activation factor in various cell types and column fractions (Volpp et al. 1988). Using anion exchange chromatography, Nunoi et al. (1988) identified a 47- and a 65-kDa protein in neutrophil cytosol, both of which are required for the reconstitution of O_2^- formation. The 47-kDa protein has operationally been termed neutrophil cytosol factor 1 (NCF-1), and the 65-kDa protein NCF-2. Interestingly, a third yet unknown factor, termed NCF-3, is required for the reconstitution of NADPH oxidase activity (Nunoi et al. 1988). Autosomal recessive CGD is associated with the more common defect of NCF-1 or with the less common defect of NCF-2 but apparently not with a defect of NCF-3 (Nunoi et al. 1988; Clark et al. 1989; see also Sect. 6.1.2).

By isoelectric focusing, four cytosolic activation factors in human neutrophil cytosol have been identified (Curnutte et al. 1989a). These factors have been operationally termed C1, C2, C3, and C4 and possess pI values of 3.1, 6.0, 7.0 and 9.1, respectively. As combinations of these four factors do not support O_2^- formation, a hitherto unknown fifth component has been suggested to be required for reconstitution of NADPH oxidase activity (Curnutte et al. 1989a). Autosomal recessive CGD may be associated with a defect in C2 or in C4 (Curnutte et al. 1989a).

Cytosolic activation factors from human neutrophils have been characterized by chromatography on carboxymethyl Sepharose (Bolscher et al. 1989). When tested separately, neither the column-bound protein nor the unbound protein reconstitute O_2^- formation, but the combination of the wash fraction with a fraction eluting at 125 mM NaCl restore enzyme activity. These results suggest that O_2^- formation depends on the presence of at least two cytosolic activation factors, one of which binds to carboxymethyl Sepharose. The component which does not bind to this matrix is referred to as soluble oxidase component I (SOC I), and the component binding to the column is termed SOC II. SOC II copurifies with a 47-kDa protein which is missing in autosomal-recessive CGD, and SOC II from control neutrophils reconstitutes the defect in neutrophil cytosol of these patients (Bolscher et al. 1989). This group also isolated a cytosolic factor which may specifically participate in GTP[γS]-dependent activation of NADPH oxidase (Bolscher et al. 1990).

A protein with an apparent molecular mass of 63-kDa was purified from cytosol of porcine neutrophils (Tanaka et al. 1990). Partial amino acid sequence analysis showed that it corresponds to the 66-kDa protein. Tanaka et al. (1990) detected neither heme nor flavin in the purified protein, suggesting that it acts as a regulatory component of NADPH oxidase and not as an electron transport component. Additionally, an antibody raised against this purified protein cross-reacts with a 65-kDa protein in human neutrophils and reduces the effectiveness of cytosol to reconstitute NADPH oxidase activity in the cell-free system. Precipitating the 47- and 66-kDa proteins with anionic amphiphiles, Chiba et al. (1990) did not obtain positive evidence for the presence of FAD or FMN in them.

Recently, Teahan et al. (1990) reported on the purification of the phosphorylated form of the 47-kDa protein from human neutrophils by chromatography on ion-exchange and hydroxyapatite columns. In addition, polyclonal antibodies against this protein were raised (Teahan et al. 1990).

Apparently, the 47-kDa protein identified by Volpp et al. (1988) corresponds to NCF-1 (Nunoi et al. 1988), C4 (Curnutte et al. 1989a), and SOC II (Bolscher et al. 1989). The cytosolic factor S (Babior et al. 1988) has been suggested to be a nonphosphorylated form of the 47-kDa protein (Curnutte et al. 1989b). In addition, one of the two C_1 components described by Ishida et al. (1989) may be identical with the 47-kDa protein. Thus, one of the cytosolic activation factors for NADPH oxidase represents the 47-kDa protein which is defective in most cases of autosomal recessive CGD (see also Sects. 3.1.1.1, 3.2.2.1, 5.1.5.2.3, 6.1.2).

The 67-kDa protein characterized by Volpp et al. (1988) is apparently identical with NCF-2 (Nunoi et al. 1988), C2 (Curnutte et al. 1989a), and possibly with one of the C_1 components (Ishida et al. 1989). The similarity of the molecular mass of NCF-2 with a cytosolic 65- to 67-kDa protein which is labeled by NADPH analogues (Smith et al. 1989a,b; Takasugi et al. 1989), suggests that NCF-2 may carry the NADPH-binding site of NADPH oxidase (see also Sect. 2.4.1). The σ_1 factor identified by Pick et al. (1989) may be identical with NCF-3 (Nunoi et al. 1988), SOC-1 (Bolscher et al. 1989), and C1 (Curnutte et al. 1989a), and σ_2 (Pick et al. 1989) may be composed of the 47- and 66-kDa proteins. The identity of other cytosolic components, e.g., the fifth cytosolic component postulated by Curnutte et al. (1989a), and the defect of the cytosolic activation factor(s) mediating GTP-dependent activation of O_2^- formation in dibutyryl cAMP-differentiated HL-60 cells (Seifert et al. 1989c) remain to be clarified (see also Sect. 5.1.4.4).

5.1.5.2.3 Molecular Cloning and Expression of Recombinant Proteins

Recently, NCF-1, also termed NCF-47K, has been cloned and functionally expressed in bacteria (Lomax et al. 1989; Volpp et al. 1989b). The cDNA for NCF-47K codes for a 41.4- to 41.9-kDa protein with a calculated pI value of 10.4. The protein possesses an arginine- and serine-rich COOH-terminal domain with putative phosphorylation sites for protein kinases and an N-terminal glycine. The protein shows homologies to phospholipase C, *src* protein kinases, and α-fodrin and possesses a nucleotide-binding domain. In addition, the protein carries a segment consisting of 33 amino acids with about 50% identity to *ras* p21 GTPase-activating protein. These properties of NCF-47K suggest that this protein participates in GTP-dependent regulation of NADPH oxidase, but the precise mode of interaction of NCF-47K with other components of NADPH oxidase remains to be determined (see also Sects. 2.4.3, 3.2.1, 5.1.4). Finally, recombinant NCF-47K has been shown functionally to reconstitute the defect of the 47-kDa protein in neutrophil cytosol of autosomal recessive CGD patients in the cell-free system (Lomax et al. 1989; Volpp et al. 1989b).

The cDNA for NCF-2 has also been cloned and recombinant NCF-2 (presently also referred to as r-p67) partially restores the functional defect of neutrophil cytosol of CGD patients with a defect of the 67-kDa protein (Leto et al. 1989, 1990). The cDNA for NCF-2 encodes a protein with 526 amino acids and possesses acidic middle and COOH-terminal regions (Leto et al. 1990). These regions share homology to sequence motifs present in the non-catalytic region of *src*-related protein kinases (Leto et al. 1990). This sequence motif was also found in a specific isoenzyme of phospholipase C, α-fodrin, *ras* p21 GTPase-activating protein and NCF-1 (Lomax et al. 1989; Volpp et al. 1989b; Leto et al. 1990). These structural similarities suggest that NCF-1 and NCF-2 share common functions in the regulation of NADPH oxidase (Leto et al. 1990).

5.2 Reconstitution and Regulation of NADPH Oxidase Activity by Phorbol Esters

Activation of NADPH oxidase by fatty acids in the cell-free system is independent of protein kinase C and Ca^{2+} (see Sects. 5.1.3.2, 5.1.3.3). Tsunawaki and Nathan (1986), Seifert and Schultz (1987a), and Traynor et al. (1989) did not find a stimulatory effect of PMA, diacylglycerol, or inositol 1,4,5-trisphosphate on $O_2^{\cdot-}$ formation in cell-free systems derived from human neutrophils and murine macrophages. In contrast, other authors succeeded in establishing a protein kinase C-dependent cell-free activation system for NADPH oxidase (Cox et al. 1985, 1987; Tauber et al. 1989a).

Protein kinase C-mediated activation of NADPH oxidase in neutrophil membranes requires not only the presence of PMA but also the addition of Ca^{2+} and exogenous phospholipids, e.g., phosphatidylserine (Cox et al. 1985). Protein kinase C present in neutrophil cytosol as well as purified protein kinase C from rat brain reconstitute $O_2^{\cdot-}$ formation (Cox et al. 1985). However, the system reconstituted by combination of neutrophil membranes, PMA, phospholipids, Ca^{2+}, and protein kinase C is considerably less effective in catalyzing $O_2^{\cdot-}$ formation than the system consisting of neutrophil membranes, neutrophil cytosol plus fatty acids or SDS (Cox et al. 1987). The pH optimum and K_m for NADPH of NADPH oxidase activated via protein kinase C are in agreement with the values obtained for NADPH oxidase activated through the protein kinase C-independent pathway in a cell-free system (Cox et al. 1987).

In addition to native protein kinase C, the proteolytically activated, Ca^{2+}/phospholipid-independent protein kinase C has been reported to stimulate $O_2^{\cdot-}$ formation in plasma membranes from resting human neutrophils in the presence of ATP and Mg^{2+} (Tauber et al. 1989a). The proteolytically activated protein kinase C is more effective in activating O_2^- formation than native protein kinase C, and neither PMA nor Ca^{2+} is required for reconstitution of O_2^- formation with the former kinase. Unexpectedly, activation of O_2^- formation by proteolytically activated protein kinase C depends on phosphatidylserine (Tauber et al. 1989a). Phosphatidylserine has been suggested to interact directly with a component of NADPH oxidase rather than with protein kinase C (see also Sects. 2.1, 5.1.2). These results show that protein kinase C-dependent and -independent pathways for the activation of NADPH oxidase exist not only in intact cells but also in cell-free systems.

Very recently, Burnham et al. (1990) showed that short chain diacylglycerols such as dioctanoylglycerol potentiate SDS-induced O_2^- formation in a cell-free system from human neutrophils. Apparently, diacylglycerols do not increase the sensitivity of SDS towards cytosolic components, and they do not mimic the effects of GTP[γS]. By contrast, PMA and mezerein do not substantially enhance SDS-induced O_2^- formation. Although diacylglycerols potentiate SDS-induced phosphorylation of the 47-kDa protein, various experimental data suggest that their effects are apparently not mediated through protein kinase C.

5.3 Reconstitution and Regulation of NADPH Oxidase Activity by Phosphatidic Acid

In intact neutrophils, chemoattractants induce the release of phosphatidic acid through activation of phospholipase D, and phosphatidic acid has recently been shown to stimulate the respiratory burst in intact phagocytes (Anthes et al. 1989; Billah et al. 1989; Ohtsuka et al. 1989). Moreover, there is a correlation between the chemotactic peptide-induced activations of phospholipase D and NADPH oxidase in human neutrophils (Bonser et al. 1989; see also Sect. 3.2.2.1).

The role of phosphatidic acid in the regulation of NADPH oxidase in cell-free systems is controversial. Bellavite et al. (1988) made the very interesting observation that phosphatidic acid activates NADPH oxidase in detergent extracts from membranes of resting pig neutrophils. Unlike fatty acid-induced O_2^- formation, that induced by phosphatidic acid has been reported not to depend on the presence of neutrophil cytosol. The phosphatidic acid-activated NADPH oxidase shows structural and catalytic properties similar to NADPH oxidase from activated cells or to NADPH oxidase activated by fatty acids plus cytosol (Bellavite et al. 1988; see also

Sects. 2.1, 5.1.2). In contrast, phosphatidic acid has been reported to inhibit SDS-induced O_2^- formation in a cell-free system from human neutrophils (Aviram and Sharabani 1989a). We did not find stimulatory effects of phosphatidic acid on O_2^- formation in cell-free systems from dimethyl sulfoxide-differentiated HL-60 cells under various experimental conditions (unpublished results). These data suggest that the effects of phosphatidic acid on NADPH oxidase in cell-free systems are species and/or cell type specific.

6 Pathology of NADPH Oxidase

6.1 Chronic Granulomatous Disease

Much information on the structure and regulation of NADPH oxidase is derived from studies on neutrophils from CGD patients (see also Sects. 2.4.3, 3.1.1.1, 5.1.5). CGD is a rare inherited disease; it occurs with a frequency of about 1:1 000 000 and may be divided into X-chromosomal and autosomal-recessive forms (Tauber et al. 1983). CGD usually becomes apparent in childhood and is characterized by recurrent infections with granuloma formation (Babior 1978b; Tauber et al. 1983). The clinical manifestations of CGD have been reviewed by Tauber et al. (1983). The patients' symptoms are the result of a defect of NADPH oxidase in their phagocytes, i.e., neutrophils and mononuclear phagocytes (Baehner and Karnovsky 1968; Hohn and Lehrer 1975; Curnutte et al. 1975). Upon stimulation, phagocytes of CGD patients do not undergo a respiratory burst, as revealed by hexose monophosphate shunt activity, oxygen consumption, O_2^{-} formation, and NBT reduction (Baehner and Nathan 1967; Nathan et al. 1969; Curnutte et al. 1974; Musson et al. 1982). Surprisingly, PMA has been reported to induce H_2O_2-dependent oxidation of 2'7'-dichlorofluorescein in neutrophils of CGD patients, suggesting that some PMA-activable oxidase is present in these phagocytes (Hassan et al. 1988). Table 17 summarizes some of the characteristics of NADPH oxidase in the various CGD forms.

6.1.1 Defect of Cytochrome b_{-245}

Most cases of X-chromosomal CGD are characterized by a defect of cytochrome b_{-245}, whereas most cases of autosomal-recessive CGD do not show apparent defects of the cytochrome (Segal et al. 1983; Royer-Pokora et al. 1986; Segal 1987; Teahan et al. 1987; Dinauer et al. 1987). Hybridization of monocytes from a cytochrome b_{-245}-negative, X-chromosomal CGD patient with monocytes from a cytochrome b_{-245}-positive patient resulted

Table 17. Chronic granulomatous disease

Mode of inheritance	Some characteristics of NADPH oxidase	Selected references
X-Chromosomal	No expression of the α- and β-subunits of cytochrome b-245; defect of the gene for the β-subunit; no defect of the cytosolic activation factors; abnormalities in the phosphorylation of the 47-kDa protein.	Segal et al. (1983), Curnutte (1985), Curnutte et al. (1987a,b), Royer-Pokora et al. (1986), Teahan et al. (1987), Dinauer et al. (1987), Okamura et al. (1988a,b), Clark et al. (1989), Parkos et al. (1989)
X-Chromosomal	Very rare; no defect of cytochrome b-245; normal phosphorylation of the 47-kDa protein.	Okamura et al. (1988b)
Autosomal recessive	No defect of cytochrome b-245; defect of phosphorylation of the 47-kDa protein; defect of the cytosolic activation factors. Most patients show a defect in the 47-kDa protein; few show a defect in the 66-kDa protein.	Segal et al. (1985), Hayakawa et al. (1986), Curnutte et al. (1986, 1987b, 1989a,b), Heyworth and Segal (1986), Kramer et al. (1988b), Okamura et al. (1988a,b), Nunoi et al. (1988), Volpp et al. (1988), Bolscher et al. (1989), Clark et al. (1989)
Autosomal recessive	Very rare; defect of expression of both subunits of cytochrome b-245. Defect of the gene for the α-subunit of the cytochrome. Abnormalities in the phosphorylation of the 47-kDa protein; no defect of cytosolic activation factors.	Weening et al. (1985), Okamura et al. (1988b), Parkos et al. (1989), Dinauer et al. (1990)
Autosomal recessive or X-chromosomal	Very rare; so-called "variant CGD." In contrast to other CGD forms low but detectable respiratory burst activity. Kinetics of NADPH oxidase are altered (reduction of V_{max}, decreased affinity for NADPH). IFN-γ may enhance O_2^- formation in phagocytes of certain patients with variant CGD.	Lew et al. (1981), Shurin et al. (1983), Tauber et al. (1983), Newburger et al. (1986), Ezekowitz et al. (1987)

in functional reconstitution of the respiratory burst (Hamers et al. 1984). In addition to the above CGD forms, very rare cases of autosomal-recessive, cytochrome b_{-245}-negative, and X-chromosomal cytochrome b_{-245}-positive CGD have been described (Weening et al. 1985; Okamura et al. 1988b; Dinauer et al. 1989).

Both the α- and the β-subunits of cytochrome b_{-245} are absent in neutrophils of patients with X-chromosomal and autosomal-recessive, cytochrome b_{-245}-negative CGD (Verhoeven et al. 1989; Parkos et al. 1989). The absence of both subunits of cytochrome b_{-245} in these CGD patients may be explained by the fact that stable expression of either subunit of the cytochrome depends on the presence of the other subunit (Parkos et al. 1989; Dinauer et al. 1989; see also Sect. 2.4.3). The β-subunit of cytochrome b_{-245} is encoded by the X-chromosome, and mutations affecting expression or structure of this gene result in the former type of CGD (Royer-Pokora et al. 1986; Teahan et al. 1987; Parkos et al. 1989; Verhoeven et al. 1989; Dinauer et al. 1989). It has been suggested that the gene for the α-subunit of cytochrome b_{-245} is defective in the corresponding autosomal-recessive form of CGD (Parkos et al. 1989). Recent data show that, in fact, autosomal recessive CGD may be due to defects in the gene encoding the α-subunit of cytochrome b_{-245} (Dinauer et al. 1990). Finally, the membrane-associated phosphorylated 47-kDa protein is missing in neutrophils of X-chromosomal cytochrome b_{-245}-negative CGD patients, suggesting that activation of NADPH oxidase depends on the phosphorylation of this protein and its subsequent association with cytochrome b_{-245} in the plasma membrane (Heyworth et al. 1989a).

6.1.2 Defect of Cytosolic Activation Factors

In neutrophils of healthy subjects, the 47-kDa protein is phosphorylated upon stimulation with a variety of agents including phorbol esters and chemotactic peptides, supporting a key role of this protein in the activation of NADPH oxidase (see also Sects. 3.1.1.1, 3.2.2.1, 5.1.5). In contrast, the 47-kDa protein is not phosphorylated in patients with autosomal-recessive cytochrome b_{-245}-positive CGD (Segal et al. 1985; Hayakawa et al. 1986; Heyworth and Segal 1986; Okamura et al. 1988a,b). In addition, purified protein kinase C does not phosphorylate the 47-kDa protein of autosomal-recessive CGD patients in vitro (Kramer et al. 1988b). The 47-kDa protein is localized both in the cytosol and in the membrane fraction of stimulated neutrophils, and recent studies have shown that the 47-kDa protein is one of the cytosolic activation factors for NADPH oxidase (Kramer et al. 1988b; Heyworth et al. 1989a; see also Sects. 5.1.5.2.2, 5.1.5.2.3). Most patients with autosomal recessive CGD show a defect of the 47-kDa protein, and few patients show a defect of the 66-kDa protein (Nunoi et al. 1988; Clark et al. 1989).

6.1.3 Variant Chronic Granulomatous Disease

In some CGD patients, the kinetics of NADPH oxidase, i.e., V_{max} and K_m for NADPH, are altered (Lew et al. 1981; Shurin et al. 1983; Newburger et al. 1986). Patients with this type of CGD, also referred to as variant CGD, generate low but detectable amounts of $O_2^{\cdot-}$ upon exposure to various stimuli, and the clinical symptoms are less severe than in patients with the other forms of CGD (Newburger et al. 1986; Ezekowitz et al. 1987). Variant CGD is inherited in an autosomal-recessive or an X-chromosomal manner, and severe infections may be associated with a further decrease in their neutrophils' capacity to generate $O_2^{\cdot-}$ (Newburger et al. 1986; see also Sect. 6.2.2). The role of IFN-γ in the treatment of CGD is described in Sect. 3.3.1.3.8.

6.2 Other Pathological States

Quantitative and/or qualitative alterations of the respiratory burst have been observed in various diseases, but in many cases the results are controversial. Some of the reasons which may explain these conflicting results are dealt with in Sect. 1. Table 18 summarizes some pathological states in which the activity of the respiratory burst is assumed to be altered.

6.2.1 Hematological Disorders

Neutrophils possess a myeloperoxidase which is located in the azurophilic granules and catalyzes the formation of HOCl with H_2O_2 and Cl^- as substrates (Roos 1980; Rossi 1986; Edwards and Swan 1986; Sandborg and Smolen 1988). An antibody raised against human myeloperoxidase has been reported to enhance fMet-Leu-Phe-induced $O_2^{\cdot-}$ formation in human neutrophils, and inhibition of myeloperoxidase partially inhibits inactivation of NADPH oxidase (Jandl et al. 1978; Edwards and Swan 1986). In addition, neutrophils of certain patients with myeloperoxidase deficiency have been reported to show enhanced phagocytosis and prolonged activation of the respiratory burst, as assessed by oxygen consumption, hexose monophosphate shunt activity, and $O_2^{\cdot-}$ and H_2O_2 formation (Klebanoff and

Table 18. Alterations of the respiratory burst activity in various pathological states

Pathological state	Respiratory burst activity (cell type)	Selected references
Human disorders		
Hematological disorders		
Myeloperoxidase deficiency	Enhanced or unaltered O_2^- formation; kinetics may be altered (neutrophils).	Rosen and Klebanoff (1976), Nauseef et al. (1983b), Stendahl et al. (1984)
Glucose-6-phosphate dehydrogenase deficiency	Defect of H_2O_2 formation and NBT reduction (neutrophils).	Baehner et al. (1972), Roos (1980)
Glutathione synthetase deficiency	Decreased O_2 consumption; H_2O_2 formation and hexose monophosphate shunt activity; O_2^- formation not substantially decreased (neutrophils).	Roos et al. (1979)
Glutathione peroxidase deficiency	Decreased O_2^- formation (neutrophils).	Matsuda et al. (1976)
Chediak-Higashi syndrome	Enhanced O_2^- formation (Epstein Barr virus-transformed B-lymphocyte cell lines, neutrophils).	Volkman et al. (1984)
Lactoferrin deficiency	Decreased O_2^- formation and O_2 consumption (neutrophils).	Boxer et al. (1982)
Paroxysmal nocturnal hemoglobinuria	Decreased PMA-stimulated O_2^- formation, no defect upon stimulation with opsonized zymosan or IgG (neutrophils).	Tauber et al. (1983), Huizinga et al. (1989)
Deficiency in leukocyte cell-adhesion molecules (CR3 receptor)	Defect of C3bi- and adhesion-mediated respiratory burst (neutrophils).	Hoogerwerf et al. (1990), Shappell et al. (1990)
Infections		
Bacterial infections	Priming for enhanced respiratory burst (neutrophils).	Bass et al. (1986), Briheim et al. (1989)
Virus infections	Controversial: decreased or increased respiratory burst activity (various parameter and various stimuli) (neutrophils, macrophages; various species).	Jones (1982), Abramson et al. (1984), Cassidy et al. (1989), Engels et al. (1989), Iglesias et al. (1989), Roberts et al. (1989)

Table 18. (continued)

Pathological state	Respiratory burst activity (cell type)	Selected references
Pulmonary and allergic disorders		
Adult respiratory distress syndrome	Increased $O_2^{\bullet-}$ formation and chemiluminescence (neutrophils).	Zimmerman et al. (1983)
Allergic asthma	Increased $O_2^{\bullet-}$ formation, priming by lipid mediators (alveolar macrophages)?	Damon et al. (1988)
Atopia	Controversial: unaltered, increased or decreased $O_2^{\bullet-}$ formation (eosinophils, macrophages).	Mrowietz et al. (1988), Koenderman and Bruijnzeel (1989)
Obstructive airway disease	Increased $O_2^{\bullet-}$ formation (neutrophils).	Renkema et al. (1989)
Sarcoidosis, pneumoconiosis, idiopathic pulmonary fibrosis	Increased $O_2^{\bullet-}$ and H_2O_2 formation, priming by cytokines? (alveolar macrophages).	Wallaert et al. (1990), Fels et al. (1987), Cassatella et al. (1989a), Strausz et al. (1990)
Cigarette smoking	Controversial: unaltered, increased or decreased respiratory burst activity (various parameters) (alveolar macrophages, neutrophils).	Greening and Lowrie (1983), Totti et al. (1984), Sasagawa et al. (1985), Thomassen et al. (1988)
Micsellaneous		
Neonatal stress	Enhanced PMA-stimulated $O_2^{\bullet-}$ formation (neutrophils).	Shigeoka et al. (1981)
Essential hypertension	Controversial: unaltered or increased respiratory burst activity (various stimuli, intact cells and cell-free system) (neutrophils).	Pontremoli et al. (1989), Seifert et al. (1990a)
Myotonic dystrophy	Decreased $O_2^{\bullet-}$ formation (neutrophils).	Mege et al. (1988)
Diabetes mellitus	Controversial: decreased or increased respiratory burst activity (various parameters) (neutrophils, monocytes).	Kitahara et al. (1980), Shah et al. (1983), Wierusz-Wysocka et al. (1987)
Acute alcohol intoxication	Transient and moderate decrease of fMet-Leu-Phe- and PMA-induced $O_2^{\bullet-}$ formation (neutrophils).	Sachs et al. (1990)

Table 18. (continued)

Pathological state	Respiratory burst activity (cell type)	Selected references
Crohn's disease and ulcerative colitis	Controversial: decreased or unaltered $O_2^{\cdot-}$ formation and H_2O_2 formation (neutrophils).	Verspaget et al. (1984, 1986)
Systematic lupus erythematosus and Felty's syndrome	Patients' serum contains factors which potentiate $O_2^{\cdot-}$ formation in healthy subjects.	Hashimoto et al. (1982)
Systemic necrotizing vasculitis and crescentic glomerulonephritis	Patients' serum contains anti-neutrophil cytoplasmic antibodies which stimulate chemiluminescence (neutrophils).	Falk et al. (1990)
Behçet's disease	Enhanced $O_2^{\cdot-}$ formation due to excessive cytokine production in lymphocytes (neutrophils).	Niwa et al. (1982), Niwa and Mizushima (1990)
Chronic renal failure	Controversial: unaltered, increased or decreased respiratory burst activity (various parameters, various stimuli). Patients undergoing or not undergoing dialysis (neutrophils, monocytes).	M.S. Cohen et al. (1982), Morell et al. (1985), Nguyen et al. (1985), Eckardt et al. (1986), Hirabayashi et al. (1988), Lucchi et al. (1989), Roccatello et al. (1989)
Burns	Decreased H_2O_2 formation (neutrophils).	Bjerknes et al. (1990)
Malignant infantile osteopetrosis	Decreased NBT reduction (neutrophils, monocytes).	Beard et al. (1986)
Disorders in other species		
Systemic Lupus erythematosus-like syndrome (MRL/1 mice)	Enhanced $O_2^{\cdot-}$ formation, decreased K_m for NADPH of NADPH oxidase (peritoneal macrophages).	Rokutan et al. (1988)
Osteopetrosis	Decreased NBT reduction (rat peritoneal macrophages).	Schneider (1982)

Hamon 1972; Rosen and Klebanoff 1976; Nauseef et al. 1983b; Stendahl et al. 1984). These data suggest that myeloperoxidase or some of its products plays a role in the termination of the respiratory burst (see also Sect. 3.3.1.1.3).

In patients with glucose-6-phosphate dehydrogenase deficiency, NADPH cannot be generated in the hexose monophosphate shunt, resulting in substrate depletion for NADPH oxidase (Roos 1980). Neutrophils of patients with glucose-6-phosphate dehydrogenase deficiency do not undergo a respiratory burst upon stimulation, and the functional defect is similar to that in CGD (Baehner et al. 1972; Cooper et al. 1972; Gray et al. 1973; Roos 1980; see also Sect. 6.1).

The neutrophil glutathione redox system is involved in the protection of the cell against oxidative damage (Roos 1980; see Sect. 1). Glutathione synthetase deficiency has been reported to be associated with a shortened respiratory burst (Roos et al. 1979; Roos 1980). In addition, glutathione peroxidase deficiency has been reported to be accompanied by a decreased ability to generate reactive oxygen intermediates (Matsuda et al. 1976; Roos 1980). Selenium is a cofactor for glutathione peroxidase, and O_2^- formation is decreased in selenium-deficient rat neutrophils (Baker and Cohen 1983). These data suggest that a decrease in glutathione peroxidase activity leads to H_2O_2 accumulation and results in inhibition of NADPH oxidase (see also Sect. 3.3.1.1.3).

Chediak-Higashi syndrome is a rare disorder of human neutrophils and is characterized by giant lysosomes, delayed fusion of these granules with phagosomes and increased susceptibility to infections (Tauber 1981; Newburger et al. 1983; Volkman et al. 1984). Neutrophils and EBV-transformed B-lymphocyte cell lines of patients with Chediak-Higashi syndrome have been reported to show an increased respiratory burst in comparison to control subjects (Volkman et al. 1984; see also Sect. 3.4.4.2.1). Neutrophils from a patient with a syndrome showing morphological similarities and biochemical dissimilarities to Chediak-Higashi syndrome generated O_2^- at substantially reduced rates upon stimulation with PMA, whereas O_2^- formation induced by zymosan was not substantially affected (Newburger et al. 1980b; see also Sect. 3.2.2.1, 3.3.1.5.5).

Neutrophils from patients with paroxysmal nocturnal hemoglobinuria show a normal respiratory burst upon stimulation with IgG complexes and opsonized zymosan, whereas O_2^- formation upon stimulation with PMA is impaired (Tauber et al. 1983; Huizinga et al. 1989). The binding of phorbol esters to their binding sites is not impaired, suggesting that the defect of the PMA-induced signal transduction pathway is localized distally to protein kinase C but proximally to NADPH oxidase (Tauber et al. 1983; see also Sects. 3.2.2.1, 3.3.1.5.5).

CR3 receptor deficiency is a very rare inherited condition which is characterized by life-threatening infections (Arnaout 1990). Phagocytes of these patients show various defects of adhesion-dependent functions such as aggregation, phagocytosis and binding of C3bi (Arnaout 1990). In addition, neutrophils of patients with a defect of the CR3 receptor show impaired respiratory burst upon exposure to C3bi-coated latex particles and a defect of adherence-dependent H_2O_2 formation (Hoogerwerf et al. 1990; Shappell et al. 1990; see also Sects. 3.3.1.5.2, 3.4.3).

6.2.2 Infections

The activity of the respiratory burst has been reported to be altered in bacterial, viral, fungal, and protozoal infections (see also Sect. 4.3.1). Neutrophils from patients with bacterial infections have been reported to be hyperresponsive to PMA or fMet-Leu-Phe in comparison to control neutrophils, i.e., the phagocytes are primed (Bass et al. 1986; Briheim et al. 1989; see also Sects. 3.2.2.2, 3.3.1.1.4). Priming of neutrophils in bacterial infections affects about 40% of the cell population, but the size of the population primed shows considerable interindividual variation (Bass et al. 1986; see also Sect. 1). It has been suggested that the increase in responsiveness of the respiratory burst during infection may contribute to host defense and to the pathogenesis of tissue damage (Bass et al. 1986).

Vaccinia virus has been shown to stimulate oxygen consumption and chemiluminescence in human neutrophils, and opsonization of the virus substantially enhances the respiratory burst (Jones 1982). In contrast, influenza virus has been reported to depress various neutrophil functions including the respiratory burst (Abramson et al. 1984; Cassidy et al. 1989). The decrease in neutrophil bactericidal activity may contribute to the enhanced susceptibility to bacterial infections subsequent to infection with influenza virus. Influenza virus may bind to sialic acid-containing receptors on neutrophils through its hemagglutinin glycoprotein, and this process leads to a rapid and long-lasting inhibition of the respiratory burst (Cassidy et al. 1989).

Stimulatory effects of influenza virus on the respiratory burst have also been observed. Recently, Hartshorn et al. (1990a,b) reported that influenza A virus induces H_2O_2 formation but not O_2^- formation in human neutrophils. The respiratory burst induced by the virus is anteceded by an increase in cytoplasmic Ca^{2+} which does not depend on the presence of extracellular Ca^{2+}. In addition, the virus induces membrane depolarization and formation of inositol phosphates. Furthermore, pertussis toxin does not inhibit the influenza A virus-induced responses (Hartshorn et al. 1990a,b).

Recently, plasma of patients with the arenavirus infection Lassa fever has been shown to inhibit the respiratory burst induced by fMet-Leu-Phe in neutrophils of healthy subjects (Roberts et al. 1989). The mechanism by which plasma of patients with Lassa fever inhibits the respiratory burst, may involve interaction with yet unknown steps of the signal transduction process (Roberts et al. 1989). Finally, infection of porcine alveolar macrophages with pseudorabies virus is associated with a reduction in zymosan-induced $O_2^{\cdot-}$ formation (Iglesias et al. 1989), and intraperitoneal cytomegalovirus infection in the rat causes a significant decrease in zymosan-induced $O_2^{\cdot-}$ formation in macrophages (Engels et al. 1989).

6.2.3 Pulmonary and Allergic Disorders

Neutrophils of patients with atopic dermatitis have been reported to show no alterations of $O_2^{\cdot-}$ formation, but macrophages of these patients show a moderate enhancement of $O_2^{\cdot-}$ formation upon exposure to opsonized zymosan (Mrowietz et al. 1988). Concomitant infections in these patients are associated with increased $O_2^{\cdot-}$ formation in macrophages and decreased $O_2^{\cdot-}$ formation in neutrophils (see also Sect 6.2.2). Eosinophils form atopic individuals show an increased respiratory burst upon stimulation with fMet-Leu-Phe or PAF (Koenderman and Bruijnzeel 1989), and alveolar macrophages from patients with allergic asthma have been reported to be permanently activated, i.e., they continuously generate inositolphosphates and $O_2^{\cdot-}$ (Damon et al. 1988; see also Sect. 3.4.4.1.1). Upon exposure to chemotactic peptides, these cells show only a slight increase in phosphoinositide degradation but enhanced $O_2^{\cdot-}$ formation (Damon et al. 1988). The enhanced $O_2^{\cdot-}$ formation may be due to in vivo priming by PAF, LTB_4, or other intercellular signal molecules (Damon et al. 1988; see also Sects. 3.3.1.6, 3.3.1.7). Moreover, the activity of the respiratory burst in neutrophils of patients with chronic obstructive airway disease has been reported to be increased (Renkema et al. 1989).

Coal workers' pneumoconiosis is associated with increased basal and PMA-stimulated $O_2^{\cdot-}$ formation in their alveolar macrophages in comparison to control subjects (Wallaert et al. 1990). These data suggest that alveolar macrophages from pneumoconiotic patients are primed, and that enhanced $O_2^{\cdot-}$ formation plays a part in the pathogenesis of lung injury in this condition (Wallaert et al. 1990). Idiopathic pulmonary fibrosis is also associated with an augmentation of basal and stimulated formation of reactive oxygen intermediates in the corresponding alveolar macrophages (Strausz et al. (1990).

Sarcoidosis is characterized by the accumulation of T-lymphocytes and macrophages in the alveoli, resulting in chronic inflammatory injury of the lung (Cassatella et al. 1989a). The activity of the respiratory burst in alveolar macrophages from patients with active sarcoidosis has been shown to be significantly higher than that in the corresponding cells from patients with inactive sarcoidosis patients or from healthy subjects (Fels et al. 1987; Cassatella et al. 1989a). Interestingly, IFN-γ primes macrophages of patients with inactive sarcoidosis but not those of patients with active sarcoidosis for enhanced $O_2^{\bullet-}$ formation (Cassatella et al. 1989a). In contrast, blood monocytes of patients with active sarcoidosis show no increased respiratory burst activity, indicating that priming of macrophages in sarcoidosis is a local process. IFN-γ has been suggested to prime the respiratory burst in vivo, and this process may be involved in the pathogenesis of sarcoidosis (Cassatella et al. 1989a; see also Sect. 3.3.1.3.2).

Cigarette smoking alters the morphology of alveolar marcophages and impairs their phagocytic activity (Finch et al. 1982; Fisher et al. 1982). Nicotine has been suggested to bind to noncholinergic nicotine binding sites on human phagocytes, suggesting that the effects of nicotine are mediated via specific receptors (Davies et al. 1982). Alveolar macrophages from smokers have been reported to generate substantially greater amounts of H_2O_2 than those of nonsmokers, and the increased release of reactive oxygen intermediates may contribute to the development of emphysema (Greening and Lowrie 1983). In contrast, Thomassen et al. (1988) reported that alveolar macrophages from smokers generate less $O_2^{\bullet-}$ than control macrophages. Totti et al. (1984) reported that nicotine is chemotactic to neutrophils and does not affect $O_2^{\bullet-}$ formation. In contrast, Sasagawa et al. (1985) reported that nicotine is not chemotactic for human neutrophils, whereas nicotine inhibits fMet-Leu-Phe-induced $O_2^{\bullet-}$ formation.

6.2.4 Essential Hypertension

There is a current discussion concerning the regulation of NADPH oxidase in neutrophils of hypertensive patients. Hypertensive subjects receiving no antihypertensive medication have been suggested to generate $O_2^{\bullet-}$ at rates about three- to fourfold higher than those of healthy subjects upon exposure to fMet-Leu-Phe at a maximally effective concentration (Pontremoli et al. 1989). In contrast, neutrophils of hypertensive patients have shown to be less sensitive to homologous priming by fMet-Leu-Phe (see also Sect. 3.3.1.1.4). No differences between hypertensive and normotensive subjects are apparent with respect to the activity of the membrane components and of the cytosolic activation factors of NADPH oxidase in the cell-free system (see also Sect. 5.1.1). Pontremoli et al. (1989) suggested that the functional organization of NADPH oxidase in the plasma membrane of neutrophils is altered in essential hypertension.

Seifert et al. (1991) reexamined the regulation of O_2^- formation in untreated patients with essential hypertension and in age- and sex-matched normotensive subjects. In this study, neutrophils were stimulated with various intercellular signal molecules including fMet-Leu-Phe, PAF, and LTB_4 and stimuli which circumvent receptor activation, i.e., PMA, dioctanoylglycerol, γ-hexachlorocyclohexane and arachiclonic acid. In addition, the inhibitory effects of isoproterenol, PGE_1 and histamine on fMet-Leu-Phe-induced O_2^- formation were assessed. With respect to none of the parameters studied were significant differences evident between the hypertensive and the normotensive subjects, suggesting that regulation of NADPH oxidase is not altered in essential hypertension (see also Sects. 1, 3.2.6.2, 3.3.1.2.4). With respect to cytoplasmic Ca^{2+}, there are also no differences between neutrophils of normotensive and hypertensive subjects (Lew et al. 1985).

6.2.5 Myotonic Dystrophy

Myotonic dystrophy is an autosomal dominant disease which is characterized by progressive myotonia and muscle weakness (Friedenberg et al. 1986). Patients with myotonic dystrophy show a variety of abnormalities of plasma membrane functions, suggesting that a defect in membrane structure is underlying this disease. Several neutrophil functions are abnormal in patients with myotonic dystrophy (Friedenberg et al. 1986). Neutrophils of patients with myotonic dystrophy have been reported to generate less O_2^- than those of healthy subjects upon stimulation with fMet-Leu-Phe or PMA (Mege et al. 1988b). This defect of O_2^- formation is apparently not due alterations in the K_m values of NADPH oxidase, alterations in membrane potential, or alterations in the regulation of protein kinase C (Mege et al. 1988b).

6.2.6 Diabetes Mellitus

Bacterial infections in patients with diabetes mellitus are more severe and prolonged than in healthy subjects, and diabetes mellitus is associated with blood vessel and kidney damage (Brownlee et al. 1988; Taylor and Agius

1988). Human monocytes and U-937 cells have been reported to possess insulin receptors, but their role in the regulation of the respiratory burst is not known (Schwartz et al. 1975; Grunberger et al. 1983; Carpentier et al. 1984; Taylor 1986). Chemiluminescence and $O_2^{\cdot-}$ formation in monocytes of patients with poorly controlled diabetes have been found to be significantly enhanced in comparison to control subjects (Kitahara et al. 1980). In addition, neutrophils from diabetic patients generate larger amounts of $O_2^{\cdot-}$ than those from control subjects, suggesting that diabetic neutrophils are primed, and that increased activity of the respiratory burst may contribute to cell damage in this disease (Wierusz-Wysocka et al. 1987).

In contrast, Shah et al. (1983) reported that diabetic neutrophils generate $O_2^{\cdot-}$ at lower rates than control cells upon stimulation with PMA or opsonized zymosan. The addition of insulin to diabetic neutrophils in vitro has been reported to be without effect on $O_2^{\cdot-}$ formation (Shah et al. 1983). The impaired activation of the respiratory burst in diabetes mellitus has been suggested to contribute to the increased morbidity and mortality in these patients (Shah et al. 1983).

6.2.7 Renal Disorders

Patients with chronic renal failure and patients undergoing chronic hemodialysis are susceptible to bacterial and fungal infections (Lewis and van Epps 1987). A number of studies have been performed addressing the question whether the activity of the respiratory burst is altered in phagocytes of these patients. Both increased and decreased activity of the respiratory burst has been observed in chronic renal failure, but it is not yet possible to explain all the reasons for the controversial experimental data (see also Sect. 1).

On one hand, the PMA- or opsonized zymosan-induced chemiluminescence in neutrophils from patients with chronic renal failure has been reported to be significantly higher than in healthy subjects (Eckardt et al. 1986), and neutrophils of patients with cystinosis show increased chemiluminescence upon exposure to soluble stimuli but not upon exposure to particulate stimuli (Morell et al. 1985). Serum and dialysis fluid from patients with chronic renal failure have been suggested to contain a yet unidentified low molecular mass factor which stimulates the respiratory burst, and restoration of renal function by kidney transplantation may be associated with the disappearance of this factor (Rhee et al. 1986). Neutrophils of hemodialysis patients may be primed for an enhanced respiratory burst prior to dialysis (Jacobs et al. 1989). In addition, chemiluminescence may be increased after dialysis, and this effect is explained, at least in part, by priming of neutrophils by certain dialysis membranes (Nguyen et al. 1985).

On the other hand, the ability of neutrophils of patients with chronic renal failure to undergo a respiratory burst has been reported to be impaired prior to hemodialysis, and the defect may be restored subsequently to dialysis (Hirabayashi et al. 1988). In addition, the respiratory burst has been reported to be impaired in patients with chronic renal failure not undergoing dialysis (Hirabayashi et al. 1988). Whereas basal H_2O_2 production in neutrophils from patients undergoing continuous ambulatory peritoneal dialysis is reduced in comparison to healthy subjects, no differences are apparent in PMA-stimulated neutrophils (Hirabayashi et al. 1988). In contrast, Lucchi et al. (1989) reported that phagocytes of patients with chronic renal failure undergoing or not undergoing dialysis show an enhanced basal chemiluminescence, but neutrophils of these patients show decreased chemiluminescence upon exposure to opsonized zymosan. Dialysis per se may result in inhibition of the respiratory burst stimulated by PMA and receptor agonists (M.S. Cohen et al. 1982). With respect to chemotactic peptides, a decrease in the number of formyl peptide receptors may contribute to this inhibition (M.S. Cohen et al. 1982). Moreover, the PMA- but not the fMet-Leu-Phe-induced chemiluminescence in diluted whole blood has been shown to be decreased prior to dialysis (Nguyen et al. 1985). Finally, exposure of human monocytes to various dialysis membranes may be associated with a decreased activity of the respiratory burst (Roccatello et al. 1989).

Recently, Hörl et al. (1990) reported on the purification of a protein with an apparent molecular mass of 28 kDa and a pI of 4.0–4.5 from uremic serum. This protein shows no similarity to serum proteins associated with inflammatory states, inhibits the respiratory burst in neutrophils, and may be responsible, at least in part, for impaired activation of NADPH oxidase in uremia.

6.2.8 Osteopetrosis

Osteopetrosis is a hereditary disease which is characterized by a failure of normal bone remodeling, resulting in excessive bone formation. Neutrophils of patients with malignant infantile osteopetrosis show an impaired plasma membrane depolarization response upon exposure to PMA or fMet-Leu-Phe (Beard et al. 1986). In addition, the patients' neutrophils and blood monocytes show a severely impaired respiratory burst as assessed by NBT reduction (Beard et al. 1986). In the rat, a defect of osteoclasts is responsible for skeletal sclerosis and reduced bone resorption, and the percentage of NBT-positive resident peritoneal macrophages in osteopetrotic rats is reduced in comparison to control animals (Schneider 1982). In addition, the intensity of NBT reduction is reduced in macrophages of osteopetrotic rats. These data suggest that the respiratory burst is defect in macrophages of osteopetrotic rats (Schneider 1982).

6.2.9 Glycogen Storage Disease

Glycogen storage disease type 1b is associated with susceptibility to infection (Kilpatrick et al. 1990). In comparison to patients with glycogen storage disease type 1a, who are not prone to infections, or to healthy subjects, neutrophils and monocytes of patients with type 1b disease show decreased respiratory burst activity. Their phagocytes show also decreased abilities of fMet-Leu-Phe and ionomycin to increase cytoplasmic Ca^{2+}. These data suggest that the defect of NADPH oxidase activation in glycogen storage disease type 1b may be associated with a defect in the regulation of the cytoplasmic Ca^{2+} concentration.

7   Age- and Sex-Related Alterations
of the Activity of NADPH Oxidase

In comparison to alveolar macrophages from adult rabbits, these of neonatal animals show a substantially decreased respiratory burst (Sugimoto et al. 1980). This finding may explain, at least in part, the susceptibility of neonatal animals to bacterial infections (Sugimoto et al. 1980). Newborn calf neutrophils generate less $O_2^{\cdot-}$ than neutrophils from fetal and adult animals upon stimulation with PMA (Clifford et al. 1989). The decreased activity of the respiratory burst in neutrophils from newborn calves has been reported to persist for about 7–10 days (Clifford et al. 1989). In neutrophils of human neonates and adults, PMA-induced $O_2^{\cdot-}$ formation is similar in magnitude, and there are no substantial differences in the respiratory burst among healthy adults and children with ages ranging from 11–18 months (Curnutte et al. 1974; Shigeoka et al. 1981). In contrast, neonates stressed by various factors such as premature delivery, respiratory distress syndrome, hypocalcemia, or sepsis show enhanced PMA-induced $O_2^{\cdot-}$ formation in comparison to control subjects (Shigeoka et al. 1981). Moreover, NADPH oxidase of neutrophils from vaginally delivered children has been reported to show a higher V_{max} than that of children delivered by caesarean section or that of adults, suggesting that parturition is associated with priming of the respiratory burst (Ambruso et al. 1987).

Neutrophil functions of elderly individuals are discussed to be impaired. For example, the chemoattractant-induced phosphoinositide degradation in neutrophils of persons older than 65 years has been reported to be decreased in comparison to younger subjects (Fülöp et al. 1989). However, with respect to the respiratory burst, Niwa et al. (1989) did not obtain positive evidence for a defect in neutrophils of aged humans. Studying the regulation of NADPH oxidase in normotensive and hypertensive subjects with ages ranging from 17 to 64 years, Seifert et al. (1991) did not find a correlation between the age of the subjects and fMet-Leu-Phe-induced $O_2^{\cdot-}$ formation in the neutrophils.

In the rat, aging is associated with a decreased ability of peritoneal macrophages to undergo a respiratory burst (Davila et al. 1990). This defect is restored by implantation of syngeneic pituitary grafts from young rats.

Regulation of the respiratory burst may be sex related. Neutrophils of women have been reported to show a relatively higher ability to generate $O_2^{\cdot-}$ than those of men, whereas neutrophils from women may generate less prostaglandins than men (Mallery et al. 1986). The ability of neutrophils from women to release prostaglandins correlates with the menstrual cycle, and $O_2^{\cdot-}$ formation and prostaglandin formation may be inversely related functions (see also Sect. 4.1). These sex-related differences in the respiratory burst may be attributable, at least in part, to variations in the concentration of circulating sex steroids (Mallery et al. 1986). We did not find significant differences between male and female subjects with regard to fMet-Leu-Phe-induced $O_2^{\cdot-}$ formation in neutrophils (Seifert et al. 1991).

8 Concluding Remarks

During the past few years, our knowledge of the regulation of the respiratory burst has increased tremendously. The reader of this review may be confused by the conflicting data and the multitude of mechanisms involved and may ask the crucial question of what, as a condensed scheme, NADPH oxidase regulation actually is. The reader may be disappointed by the fact that we cannot yet give a conclusive answer to this question.

We did not include schematic presentations depicting *the* regulation of NADPH oxidase for several reasons. It is evident from the data discussed in this review that more questions have been raised in recent years than have been answered. Besides Ca^{2+} and protein kinase C, additional signal transduction mechanisms are presently discussed to be involved in the regulation of NADPH oxidase, e.g., phospholipase D activation, protein tyrosine phosphorylation, and direct control by G-proteins and/or low molecular mass GTP-binding proteins. The number of intercellular signal molecules which activate NADPH oxidase through pertussis toxin-sensitive or -insensitive mechanisms has increased substantially, and several studies suggest that Ca^{2+} and protein kinase C play less crucial roles in receptor agonist-induced activation of $O_2^{\cdot-}$ formation than was previously assumed. In addition, the relative importance of the biochemical changes which precede or accompany $O_2^{\cdot-}$ formation cannot yet be exactly estimated. Moreover, a recent study suggests that chemotactic peptides, which have been shown to activate phagocytes through pertussis toxin-sensitive G-proteins in all studies published so far, primes the respiratory burst through pertussis toxin-insensitive mechanisms (Karnad et al. 1989). Furthermore, a "classical" activator of protein kinase C, i.e., dioctanoylglycerol, has recently been suggested to activate NADPH oxidase through protein kinase C-independent mechanisms (Badwey et al. 1989c). Finally, there is substantial evidence for the assumption that numerous cytosolic activation factors are involved in the regulation of NADPH oxidase, but it is still unknown how many there are. *In pars pro toto*, the above mentioned problems clearly show that it is yet premature to present a generally acceptable model of *the* regulation of NADPH oxidase.

The question arises of what may be important future lines of investigation concerning the physiology, biochemistry, and pharmacology of NADPH oxidase regulation.

With respect to physiology, one pertinent question regards how NADPH oxidase is regulated in vivo. In vivo, phagocytes are likely to interact simultaneously with a multitude of cell types and stimulatory and inhibitory signal molecules. In what manner different cell types and signal molecules interact to regulate NADPH oxidase is still very incompletely understood. In particular, the interaction of phagocytes with platelets, endothelium, lymphocytes, fibroblasts, and extracellular matrix proteins remains to be studied. In addition, the interaction of various types of cytokines and other intercellular signal molecules must be analyzed in much more detail. Moreover, recognition of the fact that NADPH oxidase-related enzyme systems apparently occur in many cellular systems, such as lymphocytes, fibroblasts, glia cells, and carotid body, raises the question of how these enzymes are regulated, and what their physiological function may be.

With regard to the biochemistry of NADPH oxidase regulation, there are also many interesting routes to pursue. For example, it is to yet be clarified which type of plasma membrane receptor interacts with which type of G-proteins and/or low molecular mass GTP-binding proteins. In addition, the relative importance of the putative intracellular signals for the activation of NADPH oxidase deserves clarification. Another line of investigation must focus on the pathobiochemistry of signal transduction pathways for NADPH oxidase in various disease states, as this type of research is still in its infancy (see Sect. 6.2). Most importantly, we anticpate that the cloning and expression of additional cytosolic activation factors (see Sect. 5.1.5.2.3) and their manipulation by site-directed mutagenesis will greatly help to understand the complex interaction of the regulatory and structural components of NADPH oxidase. Moreover, it is an ambitious undertaking to reconstitute purified and/or recombinant components of NADPH oxidase to a functionally intact enzyme system. In particular, the question must be answered as to why chemoattractants do not activate NADPH oxidase in cell-free systems derived from phagocytes.

In comparison to the physiology and biochemistry of NADPH oxidase, its pharmacology is perhaps the least elaborated part. With respect to stimulation of the respiratory burst, certain cytokines may be of therapeutic value as activators and/or primers of NADPH oxidase, resulting in improved host defense (see Sect. 3.3.1.3.8). It is probable that substantial progress will be achieved in this area during the next few years.

With respect to inhibitors of NADPH oxidase, the situation is somewhat unsatisfying. Many inhibitors known so far are rather nonspecific, and inhibitory effects of certain drugs on the respiratory burst in vitro, e.g., glucocorticoids, cyclosporin A, and nonsteroidal anti-inflammatory drugs, are of questionable clinical relevance. Thus, there is a need for potent and selective inhibitors of NADPH oxidase, and in this area the diphenylene iodonium compounds are certainly a promising class of substances. The search for inhibitors of $O_2^{\cdot-}$ formation of microbial origin may be another promising approach. Unexpectedly, we found very recently that cyclosporin H, which is generally assumed to be immunologically inactive, potently and effectively inhibits fMet-Leu-Phe-induced $O_2^{\cdot-}$ formation in human neutrophils (unpublished results).

Finally, a substantial portion of the research on NADPH oxidase regulation relies on the use of drugs which are assumed to interfere with various signal transduction processes, among others phospholipid degradation, protein kinase C activation, Ca^{2+} mobilization, and organization of the cytoskeleton. Unfortunately, many drugs used for these purposes are nonspecific. For example, there is much confusion and controversy in the field of protein kinase C inhibitors. Therefore, it is very important to develop potent and selective pharmacological tools to interfere with the above mentioned processes.

Acknowledgements. The authors are most grateful to Dr. H.H.H.W. Schmidt for many stimulating discussions and to Mrs. R. Krüger for expert help in preparation of the manuscript. The work of the authors cited herein was supported by the Deutsche Forschungsgemeinschaft and the Fonds der Chemischen Industrie.

References

Aaku E, Sorsa T, Wikström M (1990) Human immunoglobulin G potentiates superoxide production induced by chemotactic peptides and causes degranulation in isolated human neutrophils. Biochim Biophys Acta 1052:243–247

Abdel-Latif AA (1986) Calcium-mobilizing receptors, polyphosphoinositides, and the generation of second messengers. Pharmacol Rev 38:227–272

Abramovitz AS, Hong JY, Randolph V (1983a) Pseudo-inhibitors of neutrophil superoxide production: evidence that soybean-derived polypeptides are superoxide dismutases. Biochem Biophys Res Commun 117:22–29

Abramovitz AS, Yavelow J, Randolph V, Troll W (1983b) Inhibition of superoxide production in human neutrophils by purified soybean polypeptides. Re-evaluation of the involvement of proteases. J Biol Chem 258:15153–15157

Abramson JS, Parce JW, Lewis JC, Lyles DS, Mills EL, Nelson RD, Bass DA (1984) Characterization of the effect of influenza virus on polymorphonuclear leukocyte membrane responses. Blood 64:131–138

Abramson S, Hoffstein ST, Weissmann G (1982) Superoxide anion generation by human neutrophils exposed to monosodium urate. Effect of protein adsorption and complement activation. Arthritis Rheum 25:174–180

Abramson S, Korchak H, Ludewig R, Edelson H, Haines K, Levin RI, Herman R, Rider L, Kimmel S, Weissmann G (1985) Modes of action of aspirin-like drugs. Proc Natl Acad Sci USA 82:7227–7231

Abramson SL, Gallin JI (1990) IL-4 inhibits superoxide production by human mononuclear phagocytes. J Immunol 144:625–630

Absolom DR (1986) Basic methods for the study of phagocytosis. Methods Enzymol 132:95–180

Acker H, Dufau E, Huber J, Sylvester D (1989) Indications to an NADPH oxidase as a possible pO_2 sensor in the rat carotid body. FEBS Lett 256:75–78

Adam A, Petit JF, Lefrancier P, Lederer E (1981) Muramyl peptides. Chemical structure, biological activity and mechanism of action. Mol Cell Biochem 41:27–47

Adams DO (1989) Molecular interactions in macrophage activation. Immunology Today 10:33–35

Aderem AA, Keum MM, Pure E, Cohn ZA (1986a) Bacterial lipopolysaccharides, phorbol myristate acetate, and zymosan induce the myristoylation of specific macrophage proteins. Proc Natl Acad Sci USA 83:5817–5821

Aderem AA, Cohen DS, Wright SD, Cohn ZA (1986b) Bacterial lipopolysaccharides prime macrophages for enhanced release of arachidonic acid metabolites. J Exp Med 164:165–179

Adler S, Baker PJ, Johnson RJ, Ochi RF, Pritzl P, Couser WG (1986) Complement membrane attack complex stimulates production of reactive oxygen metabolites by cultured rat mesangial cells. J Clin Invest 77:762–767

Aerts C, Wallaert B, Grosbois JM, Voisin C (1986) Release of superoxide anion by alveolar macrophages in pulmonary sarcoidosis. Ann NY Acad Sci 465:193–200

Aggarwal BB, Kohr WJ, Hass PE, Moffat B, Spencer SA, Henzel WJ, Bringman TS, Nedwin GE, Goeddel DV, Harkins RN (1985) Human tumor necrosis factor. Production, purification, and characterization. J Biol Chem 260:2345–2354

Aharoni I, Pick E (1990) Activation of the superoxide-generating NADPH oxidase of macrophages by sodium dodecyl sulfate. Evidence for involvement of a G-protein. J Leukocyte Biol 48:107–115

Aida Y, Pabst MJ (1990) Priming of neutrophis by lipopolysaccharide for enhanced release of superoxide. Requirement for plasma but not for tumor necrosis factor-α. J Immunol 145:3017–3025

Aida Y, Pabst MJ, Rademacher JM, Hatakeyama T, Aono M (1990) Effects of polymyxin B on superoxide anion release and priming in human polymorphonuclear leukocytes. J Leukocyte Biol 47:283–291

Akard LP, English D, Gabig TG (1988) Rapid deactivation of NADPH oxidase in neutrophils: continuous replacement by newly activated enzyme sustains the respiratory burst. Blood 72:322–327

Aktories K, Ankenbauer T, Schering B, Jakobs KH (1986a) ADP-ribosylation of platelet actin by botulinum C2 toxin. Eur J Biochem 161:155–162

Aktories K, Bärmann M, Ohishi I, Tsuyama S, Jakobs KH, Habermann E (1986b) Botulinum C2 toxin ADP-ribosylates actin. Nature 322:390–392

Aktories K, Bärmann M, Chhatwai GS, Presek P (1987) New class of microbial toxins ADP-ribosylates actin. Trends Pharmacol Sci 8:158–160

Aktories K, Rösener S, Blaschke U, Chatwal GS (1988) Botulinum ADP-ribosyltransferase C3. Purification of the enzyme and characterization of the ADP-ribosylation reaction in platelet membranes. Eur J Biochem 172:445–450

Al-Mohanna FA, Hallett MB (1987) Actin polymerization modifies stimulus—oxidase coupling in rat neutrophils. Biochim Biophys Acta 927:366–371

Al-Mohanna FA, Hallett MB (1990) "Clamping" actin in polymerized form in electropermeabilized neutrophils inhibits oxidase activation. Biochem Biophys Res Commun 169:1222–1228

Al-Mohanna FA, Ohishi I, Hallett MB (1987) Botulinum C_2 toxin potentiates activation of the neutrophil oxidase. Further evidence of a role for actin polymerization. FEBS Lett 219:40–44

Albina JE, Caldwell MD, Henry WL Jr, Mills CD (1989a) Regulation of macrophage functions by L-arginine. J Exp Med 169:1021–1029

Albina JE, Mills CD, Henry WL Jr, Caldwell MD (1989b) Regulation of macrophage physiology by L-arginine: role of the oxidative deiminase pathway. J Immunol 143:3641–3646

Albrechtsen M, Yeaman GR, Kerr MA (1988) Characterization of the IgA receptor from human polymorphonuclear leucocytes. Immunology 64:201–205

Allen BR (1983) Benoxaprofen and the skin. Br J Dermatol 109:361–364

Allen RA, Tolley JO, Jesaitis AJ (1986a) Preparation and properties of an improved photoaffinity ligand for the N-formyl peptide receptor. Biochim Biophys Acta 882:271–280

Allen RA, Jesaitis AJ, Sklar LA, Cochrane CG, Painter RG (1986b) Physicochemical properties of the N-formyl peptide receptor on human neutrophils. J Biol Chem 261:1854–1857

Allen RA, Erickson RW, Jesaitis AJ (1989) Identification of a human neutrophil protein of M_r 24 000 that binds N-formyl peptides: co-sedimentation with specific granules. Biochim Biophys Acta 991:123–133

Allen RC, Stjernholm RL, Steele RH (1972) Evidence for the generation of an electronic excitation state(s) in human polymorphonuclear leukocytes and its participation in bactericidal activity. Biochem Biophys Res Commun 47:679–684

Alobaidi T, Naccache PH, Sha'afi RI (1981) Calmodulin antagonists modulate rabbit neutrophil degranulation, aggregation and stimulated oxygen consumption. Biochim Biophys Acta 675:316–321

Amar M, Amit N, Pham Huu T, Chollet-Martin S, Labro MT, Gougerot-Pocidalo MA, Hakim J (1990) Production by K 562 cells of an inhibitor of adherence-related functions of human neutrophils. J Immunol 144:4749–4756

Ambruso DR, Stork LC, Gibson BE, Thurman GW (1987) Increased activity of the respiratory burst in cord blood neutrophils: kinetics of the NADPH oxidase enzyme system in subcellular fractions. Pediatr Res 21:205–210

Anand-Srivastava MB (1989) Amiloride interacts with guanine nucleotide regulatory proteins and attenuates the hormonal inhibition of adenylate cyclase. J Biol Chem 264:9491–9496

Anderson CL, Looney RG (1986) Human leukocyte IgG Fc receptors. Immunology Today 7:264–266

Anderson R, Eftychis HA (1986) Potentiation of the generation of reactive oxidants by human phagocytes during exposure to benoxaprofen and ultraviolet radiation in vitro. Br J Dermatol 115:285–295

Anderson R, Niedel J (1984) Processing of the formyl peptide receptor by HL-60 cells. J Biol Chem 259:13309–13315

Anderson R, Beyers AD, Savage JE, Nel AE (1988) Apparent involvement of phospholipase A_2, but not protein kinase C, in the pro-oxidative interactions of clofazimine with human phagocytes. Biochem Pharmacol 37:4635–4641

Andersson T, Dahlgren C, Pozzan T, Stendahl O, Lew DP (1986a) Characterization of fMet-Leu-Phe receptor-mediated Ca^{2+} influx across the plasma membrane of human neutrophils. Mol Pharmacol 30:437–443

Andersson T, Schlegel W, Monod A, Krause KH, Stendahl O, Lew DP (1986b) Leukotriene B_4 stimulation of phagocytes results in the formation of inositol 1,4,5-triphosphate. A second messenger for Ca^{2+} mobilization. Biochem J 240:333–340

Andersson T, Fällman M, Lew DP, Stendahl O (1988) Does protein kinase C control receptor-mediated phagocytosis in human neutrophils? FEBS Lett 239:371–375

Andre P, Capo C, Mege JL, Benoliel AM, Bongrand P (1988) Zymosan but not phorbol myristate acetate induces an oxidative burst in rat bone marrow-derived macrophages. Biochem Biophys Res Commun 151:641–648

Andrew PW, Robertson AK, Lowrie DB, Cross AR, Jones OTG (1987) Induction of synthesis of components of the hydrogen peroxide-generating oxidase during activation of the human monocytic cell line U937 by interferon-γ. Biochem J 248:281–283

Andrews PC, Babior BM (1983) Endogenous protein phosphorylation by resting and activated human neutrophils. Blood 61:333–340

Andrews PC, Babior BM (1984) Phosphorylation of cytosolic proteins by resting and activated human neutrophils. Blood 64:883–890

Anthes JC, Billah MM, Egan RW, Siegel MI (1987) Chemotactic peptide, calcium and guanine nucleotide regulation of phospholipase C activity in membranes from DMSO-differentiated HL60 cells. Biochem Biophys Res Commun 145:825–833

Anthes JC, Eckel S, Siegel MI, Egan RW, Billah MM (1989) Phospholipase D in homogenates from HL-60 granulocytes: implications of calcium and G protein control. Biochem Biophys Res Commun 163:657–664

Apfeldorf WJ, Melnick DA, Meshulam T, Rasmussen H, Malech HL (1985) A transient rise in intracellular free calcium is not a sufficient stimulus for respiratory burst activation in human polymorphonuclear leukocytes. Biochem Biophys Res Commun 132:674–680

Arnaout MA (1990) Structure and function of the leukocyte adhesion molecules CD11/CD18. Blood 75:1037–1050

Arnaout MA, Wang EA, Clark SC, Sieff CA (1986) Human recombinant granulocyte-macrophage colony-stimulating factor increases cell-to-cell adhesion and surface expression of adhesion-promoting surface glycoproteins on mature granulocytes. J Clin Invest 78:597–601

Arthur MJP, Kowalski-Saunders P, Wright R (1986) *Corynebacterium parvum*-elicited hepatic macrophages demonstrate enhanced respiratory burst activity compared with resident Kupffer cells in the rat. Gastroenterology 91:174–181

Ashendel CL (1985) The phorbol ester receptor: a phospholipid-regulated protein kinase. Biochim Biophys Acta 822:219–242

Ashkenazi A, Peralta EG, Winslow JW, Ramachandran J, Capon DJ (1989) Functionally distinct G proteins selectively couple different receptors to PI hydrolysis in the same cell. Cell 56:487–493

Atkinson YH, Marasco WA, Lopez AF, Vadas MA (1988) Recombinant human tumor necrosis factors-α. Regulation of N-formylmethionylleucylphenylalanine receptor affinity and function on human neutrophils. J Clin Invest 81:759–765

Auch-Schwelk W, Katusic ZS, Vanhoutte PM (1989) Contractions to oxygen-derived free radicals are augmented in aorta of the spontaneously hypertensive rat. Hypertension 13:859–864

Aviram I, Faber A (1990) A protein of neutrophil granules interferes with activation of NADPH oxidase in a cell-free system. Biochem Biophys Res Commun 169:198–202

Aviram I, Henis YI (1984) Activation of human neutrophil NADPH oxidase and lateral mobility of membrane proteins. A study with crosslinkers. Biochim Biophys Acta 805:227–231

Aviram I, Sharabani M (1986) Kinetic studies of the reduction of neutrophil cytochrome b-558 by dithionite. Biochem J 237:567–572

Aviram I, Sharabani M (1989a) Inositol lipids and phosphatidic acid inhibit cell-free activation of neutrophil NADPH oxidase. Biochem Biophys Res Commun 161:712–719

Aviram I, Sharabani M (1989b) Kinetics of cell-free activation of neutrophil NADPH oxidase. Effects of neomycin and guanine nucleotides. Biochem J 261:477–482

Aviram I, Simons ER, Babior BM (1984) Reversible blockade of the respiratory burst in human neutrophils by a cleavable corss-linking agent. J Biol Chem 259:306–311

Axtell RA, Sandborg RR, Smolen JE, Ward PA, Boxer LA (1990) Exposure of human neutrophils to exogenous nucleotides causes elevation in intracellular calcium, transmembrane calcium fluxes, and an alteration of a cytosolic factor resulting in enhanced superoxide production in response to FMLP and arachidonic acid. Blood 75:1324–1332

Babior GM (1978a) Oxygen-dependent microbial killing by phagocytes. I. N Engl J Med 298:659–668

Babior GM (1978b) Oxygen-dependent microbial killing by phagocytes. II. N Engl J Med 298:721–725

Babior BM (1984) Oxidants from phagocytes. Agents of defense and destruction. Blood 64:959–966

Babior BM, Kipnes RS (1977) Superoxide-forming enzyme from human neutrophils: evidence for a flavin requirement. Blood 50:517–524

Babior BM, Peters WA (1981) The O_2^--producing enzyme of human neutrophils. Further properties. J Biol Chem 256:2321–2323

Babior BM, Kipnes RS, Curnutte JT (1973) Pathological defense mechanisms. The production by leukocytes of superoxide, a potential bactericidal agent. J Clin Invest 52:741–744

Babior BM, Curnutte JT, Kipnes RS (1975) Pyridine nucleotide-dependent superoxide production by a cell-free system from human granulocytes. J Clin Invest 56:1035–1042

Babior BM, Curnutte JT, McMurrich BJ (1976) The particulate superoxide-forming system from human neutrophils. Properties of the system and further evidence supporting its participation in the respiratory burst. J Clin Invest 58:989–996

Babior BM, Kuver R, Curnutte JT (1988) Kinetics of activation of the respiratory burst oxidase in a fully soluble system from human neutrophils. J Biol Chem 263:1713–1718

Babior GL, Rosin RE, McMurrich BJ, Peters WA, Babior BM (1981) Arrangement of the respiratory burst oxidase in the plasma membrane of the neutrophil. J Clin Invest 67:1724–1728

Badwey JA, Karnovsky ML (1979) Production of superoxide and hydrogen peroxide by an NADH oxidase in guinea pig polymorphonuclear leukocytes. Modulation by nucleotides and divalent cations. J Biol Chem 254:11530–11537

Badwey JA, Karnovsky ML (1980) Active oxygen species and the functions of phagocytic leukocytes. Annu Rev Biochem 49:695–726

Badwey JA, Karnovsky ML (1986) Production of superoxide by phagocytic leukocytes: a paradigm for stimulus-response phenomena. Current Top Cell Regul 28:183–208

Badwey JA, Curnutte JT, Karnovsky ML (1979) The enzyme of granulocytes that produces superoxide and peroxide. An elusive pimpernel. N Engl J Med 300:1157–1160

Badwey JA, Curnutte JT, Robinson JM, Lazdins JK, Briggs RT, Karnovsky MJ, Karnovsky ML (1980) Comparative aspects of oxidative metabolism of neutrophils from human blood and guinea pig peritonea: magnitude of the respiratory burst, dependence upon stimulating agents, and localization of the oxidases. J Cell Physiol 105:541–551

Badwey JA, Curnutte JT, Karnovsky ML (1981) Cis-polyunsaturated fatty acids induce high levels of superoxide production by human neutrophils. J Biol Chem 256:12640–12643

Badwey JA, Curnutte JT, Berde CB, Karnovsky ML (1982) Cytochalasin E diminishes the lag phase in the release of superoxide by human neutrophils. Biochem Biophys Res Commun 106:170–174

Badwey JA, Robinson JM, Lazdins JK, Briggs RT, Karnovsky MJ, Karnovsky ML (1983) Comparative biochemical and cytochemical studies on superoxide and peroxide in mouse macrophages. J Cell Physiol 115:208–216

Badwey JA, Curnutte JT, Robinson JA, Berde CB, Karnovsky MJ, Karnovsky ML (1984) Effects of free fatty acids on release of superoxide and on change of shape by human neutrophils. Reversibility by albumin. J Biol Chem 259:7870–7877

Badwey JA, Robinson JM, Curnutte JT, Karnovsky MJ, Karnovsky ML (1986) Retinoids stimulate the release of superoxide by neutrophils and change their morphology. J Cell Physiol 127:223–228

Badwey JA, Robinson JM, Horn W, Soberman RJ, Karnovsky MJ, Karnovsky ML (1988) Synergistic stimulation of neutrophils. Possible involvement of 5–hydroxy-6,8,11,14-eicosatetraenoate in superoxide release. J Biol Chem 263:2779–2786

Badwey JA, Heyworth PG, Karnovsky ML (1989a) Phosphorylation of both 47 and 49 kDa proteins accompanies superoxide release by neutrophils. Biochem Biophys Res Commun 158:1029–1035

Badwey JA, Horn W, Heyworth PG, Robinson JM, Karnovsky ML (1989b) Paradoxical effects of retinal in neutrophil stimulation. J Biol Chem 264:14947–14953

Badwey JA, Robinson JM, Heyworth PG, Curnutte JT (1989c) 1,2-Dioctanoyl-sn-glycerol can stimulate neutrophils by different mechanisms. Evidence for a pathway that does not involve phosphorylation of the 47 kDa protein. J Biol Chem 264:20676–20682

Baehner RL, Karnovsky ML (1968) Deficiency of reduced nicotinamide-adenine dinucleotide oxidase in chronic granulomatous disease. Science 162:1277–1279

Baehner RL, Nathan DG (1967) Leukocyte oxidase: defective activity in chronic granulomatous disease. Science 155:835–836

Baehner RL, Johnston RB Jr, Nathan DG (1972) Comparative study of the metabolic and bactericidal characteristics of severely glucose-6-phosphate dehydrogenase-deficient polymorphonuclear leukocytes and leukocytes from children with chronic granulomatous disease. J Reticuloendothel Soc 12:150–169

Baehner RL, Murrmann SK, Davis J, Johnston RB Jr (1975) The role of superoxide anion and hydrogen peroxide in phagocytosis-associated oxidative metabolic reactions. J Clin Invest 56:571–576

Baehner RL, Boxer LA, Ingraham LM, Butterick C, Haak RA (1982) The influence of vitamin E on human polymorphonuclear cell metabolism and function. Ann NY Acad Sci 393:237–250

Baggiolini M (1984) Phagocytes use oxygen to kill bacteria. Experientia 40:906–909

Baggiolini M, Walz A, Kunkel SL (1989) Neutrophil-activating peptide-1/interleukin 8, a novel cytokine that activates neutrophils. J Clin Invest 84:1045–1049

Bahr GM, Chedid L (1986) Immunological activities of muramyl peptides. Fed Proc 45:2541–2544

Baker SS, Cohen HJ (1983) Altered oxidative metabolism in selenium-deficient rat granulocytes. J Immunol 130:2856–2860

Balazovich KJ, Boxer LA (1990) Extracellular adenosine nucleotides stimulate protein kinase C activity and human neutrophil activation. J Immunol 144:631–637

Balazovich KJ, Smolen JE, Boxer LA (1986a) Ca^{2+} and phospholipid-dependent protein kinase (protein kinase C) activity is not necessarily required for secretion by human neutrophils. Blood 68:810–817

Balazovich KJ, Smolen JE, Boxer LA (1986b) Endogenous inhibitor of protein kinase C: association with human peripheral blood neutrophils but not with specific granule-deficient neutrophils or cytoplasts. J Immunol 137:1665–1673

Baldridge GW, Gerard RW (1933) The extra respiration of phagocytosis. Am J Physiol 103:235–236

Baldwin GC, Gasson JC, Quan SG, Fleischmann J, Weisbart R, Oette D, Mitsuyasu RT, Golde DW (1988) Granulocyte-macrophage colony-stimulating factor enhances neutrophil function in acquired immunodeficiency syndrome patients. Proc Natl Acad Sci USA 85:2763–2766

Baldwin JM, Bennett JP, Gomperts BD (1983) Detergent solubilisation of the rabbit neutrophil receptor for chemotactic formyl peptides. Eur J Biochem 135:515–518

Balsinde J, Mollinedo F (1988) Specific activation by concanavalin A of the superoxide anion generation capacity during U937 differentiation. Biochem Biophys Res Commun 151:802–808

Balsinde J, Mollinedo F (1990) Induction of the oxidative response and of concanavalin A-binding capacity in maturing human U937 cells. Biochim Biophys Acta 1052:90–95

Bamberg E, Noda K, Gross E, Läuger P (1976) Single-channel parameters of gramicidin A, B and C. Biochim Biophys Acta 419:223–228

Banga HS, Gupta SK, Feinstein MB (1988) Botulinum toxin D ADP-ribosylates a 22-24 kDa membrane protein in platelets and HL-60 cells that is distinct from $p21^{N-ras}$. Biochem Biophys Res Commun 155:261–269

Banks P, Barker MD, Burton DR (1988) Recruitment of actin to the cytoskeletons of human monocyte-like cells activated by complement fragment C5a. Is protein kinase C involved? Biochem J 252:765–769

Barnes NC, Costello JF (1987) Airway hyperresponsiveness and inflammation. Br Med Bull 43:445–459

Barnes PJ, Chung KF, Page CP (1988) Inflammatory mediators and asthma. Pharmacol Rev 40:49–84

Barondes SH (1981) Lectins: their multiple endogenous cellular functions. Annu Rev Biochem 50:207–231

Bar-Shavit Z, Goldman R, Stabinsky Y, Gottlieb P, Fridkin M, Teichberg VI, Blumberg S (1980) Enhancement of phagocytosis—a newly found activity of substance P residing in its N-terminal tetrapeptide sequence. Biochem Biophys Res Commun 94:1445–1451

Barzaghi G, Sarau HM, Mong S (1989) Platelet-activating factor-induced phosphoinositide metabolism in differentiated U-937 cells in culture. J Pharmacol Exp Ther 248:559–566

Basci A, Shah SV (1987) Trypsin- and chymotrypsin-induced chemiluminescence by isolated rat glomeruli. Am J Physiol 21:C611–C617

Bass DA, Dechatelet LR, McCall CE (1978) Independent stimulation of motility and the oxidative metabolic burst of human polymorphonuclear leukocytes. J Immunol 121:172–178

Bass DA, Grover WH, Lewis JC, Szejda P, DeChatelet LR, McCall CE (1980) Comparison of human eosinophils from normals and patients with eosinophilia. J Clin Invest 66:1265–1273

Bass DA, Parce JW, DeChatelet LR, Szejda P, Seeds MC, Thomas M (1983) Flow cytometric studies of oxidative product formation by neutrophils: a graded response to membrane stimulation. J Immunol 130:1910–1917

Bass DA, Olbrantz P, Szejda P, Seeds MC, McCall CE (1986) Subpopulations of neutrophils with increased oxidative product formation in blood of patients with infection. J Immunol 136:860–866

Bass DA, Gerard C, Olbrantz P, Wilson J, McCall CE, McPhail LC (1987) Priming of the respiratory burst of neutrophils by diacylglycerol. Independence from activation or translocation of protein kinase C. J Biol Chem 262:6642–6649

Bass DA, McPhail LC, Schmitt JD, Morris-Natschke S, McCall CE, Wykle RL (1988) Selective priming of rate and duration of the respiratory burst of neutrophils by 1,2-diacyl and 1-O-alkyl-2-acyl diglycerides. Possible relation to effects on proteins kinase C. J Biol Chem 264:19610–19617

Baud L, Hagege J, Sraer J, Rondeau E, Perez J, Ardaillou R (1983) Reactive oxygen production by cultured rat glomerular mesangial cells during phagocytosis is associated with stimulation of lipoxygenase activity. J Exp Med 158:1836–1852

Baud L, Perez J, Ardaillou R (1986) Dexamethasone and hydrogen peroxide production by mesangial cells during phagocytosis. Am J Physiol 19:F596–F604

Bauldry SA, Wykle RL, Bass DA (1988) Phospholipase A_2 activation in human neutrophils. Differential actions of diacylglycerols and alkylacylglycerols in priming cells for stimulation by N-formyl-Met-Leu-Phe. J Biol Chem 263:16787–16795

Baumgartner MK, Dennison RL, Narayanan TK, Aronstam RS (1990) Halothane disruption of α_2-adrenergic receptor-mediated inhibition of adenylate cyclase and receptor G-protein coupling in rat brain. Biochem Pharmacol 39:223–225

Bazzi MD, Nelsestuen GL (1987) Mechanisms of protein kinase C inhibition by sphingosine. Biochem Biophys Res Commun 146:203–207

Beard CJ, Key L, Newburger PE, Ezekowitz AB, Arceci R, Miller B, Proto P, Ryan T, Anast C, Simons ER (1986) Neutrophil defect associated with malignant infantile osteopetrosis. J Lab Clin Med 108:498–505

Beaumier L, Faucher N, Naccache PH (1987) Arachidonic acid-induced release of calcium in permeabilized human neutrophils. FEBS Lett 221:289–292

Becker EL (1990) The short and happy life of neutrophil activation. J Leukocyte Biol 47:378–389

Becker JL, Grasso RJ, Davis JS (1988) Dexamethasone action inhibits the release of arachidonic acid from phosphatidylcholine during the suppression of yeast phagocytosis in macrophage cultures. Biochem Biophys Res Commun 153:583–590

Beckman JK, Gay JC, Brash AB, Lukens JN, Oates JA (1985) Investigation of the biological activity of leukotriene A_4 in human polymorphonuclear leukocytes. Biochem Biophys Res Commun 133:23–29

Beckman JS, Beckman TW, Chen J, Marshall PA, Freeman BA (1990) Apparent hydroxyl radical production by peroxynitrite: implications for endothelial injury from nitric oxide and superoxide. Proc Natl Acad Sci USA 87:1620–1624

Bellavite P, Berton G, Dri P, Soranzo MR (1981) Enzymatic basis of the respiratory burst of guinea pig resident peritoneal macrophages. J Reticuloendothel Soc 29:47–60

Bellavite P, Dri P, Della Bianca V, Serra MC (1983) The measurement of superoxide anion production by granulocytes in whole blood. A clinical test for the evaluation of phagocyte function and serum opsonic capacity. Eur J Clin Invest 13:363–368

Bellavite P, Jones OTG, Cross AR, Papini E, Rossi F (1984) Composition of partially purified NADPH oxidase from pig neutrophils. Biochem J 223:639–648

Bellavite P, Cassatella MA, Papini E, Megyeri P, Rossi F (1986) Presence of cytochrome b_{-245} in NADPH oxidase preparations from human neutrophils. FEBS Lett 199:159–163

Bellavite P, Corso F, Dusi S, Grzeskowiak M, Della Bianca V, Rossi F (1988) Activation of NADPH-dependent superoxide production in plasma membrane extracts of pig neutrophils by phosphatidic acid. J Biol Chem 263:8210–8214

Bender JG, McPhail LC, van Epps DE (1983) Exposure of human neutrophils to chemotactic factors potentiates activation of the respiratory burst enzyme. J Immunol 130:2316–2323

Bender JG, van Epps DE, Chenoweth DE (1987) Independent regulation of human neutrophil chemotactic receptors after activation. J Immunol 139:3028–3033

Bengtsson T, Rundquist I, Stendahl O, Wymann MP, Andersson T (1988) Increased breakdown of phosphatidylinositol 4,5-biphosphate is not an initiating factor for actin assembly in human neutrophils. J Biol Chem 263:17385–17389

Bennett JP, Cockcroft S, Gomperts BD (1980a) Use of cytochalasin B to distinguish between early and late events in neutrophil activation. Biochim Biophys Acta 601:584–591

Bennett JP, Hirth KP, Fuchs E, Sarvas M, Warren GB (1980b) The bacterial factors which stimulate neutrophils may be derived from procaryote signal peptides. FEBS Lett 116:57–61

Bennett WM, Norman DJ (1986) Action and toxicity of cyclosporine. Annu Rev Med 37:215–224

Benninghoff B, Dröge W, Lehmann V (1989) The lipopolysaccharide-induced stimulation of peritoneal macrophages involves at least two signal pathways. Partial stimulation by lipid A precursors. Eur J Biochem 179:589–594

Bentley JK, Reed PW (1981) Activation of superoxide production and differential exocytosis in polymorphonuclear leukocytes by cytochalasins A, B, C, D and E. Effects of various ions. Biochim Biophys Acta 678:238–244

Berger M, O'Shea J, Cross AS, Folks TM, Chused TM, Brown EJ, Frank MM (1984) Human neutrophils increase expression of C3bi as well as C3b receptors upon activation. J Clin Invest 74:1566–1571

Berkow RL, Dodson MR (1988) Biochemical mechanisms involved in the priming of neutrophils by tumor necrosis factor. J Leukocyte Biol 44:345–352

Berkow RL, Kraft AS (1985) Bryostatin, a non-phorbol macrocyclic lactone, activates intact human polymorphonuclear leukocytes and binds to the phorbol ester receptor. Biochem Biophys Res Commun 131:1109–1116

Berkow RL, Weisman SJ, Tzeng D, Haak RA, Kleinhans FW, Barefoot S, Baehner RL (1984) Comparative responses of human polymorphonuclear leukocytes obtained by counterflow centrifugal elutriation and Ficoll-Hypaque density centrifugation. II. Membrane potential changes, membrane receptor analysis, membrane fluidity, and analysis of the effects of the preparative techniques. J Lab Clin Med 104:698–710

Berkow RL, Dodson RW, Kraft AS (1987a) Dissociation of human neutrophil activation events by prolonged treatment with amiloride. J Lab Clin Med 110:97–105

Berkow RL, Dodson RW, Kraft AS (1987b) The effect of a protein kinase C inhibitor, H-7, on human neutrophil oxidative burst and degranulation. J Leukocyte Biol 41:441–446

Berkow RL, Wang D, Larrick JW, Dodson RW, Howard TH (1987c) Enhancement of neutrophil superoxide production by preincubation with recombinant human tumor necrosis factor. J Immunol 139:3783–3791

Berkow RL, Dodson RW, Kraft AS (1989) Human neutrophils contain distinct cytosolic and particulate tyrosine kinase activities: possible role in neutrophil activation. Biochim Biophys Acta 997:292–301

Bernardo J, Brink HF, Simons ER (1988) Time dependence of transmembrane potential changes and intracellular calcium flux in stimulated human monocytes. J Cell Physiol 134:131–136

Berridge MJ (1984) Inositol trisphosphate and diacylglycerol as second messengers. Biochem J 220:345–360

Berridge MJ (1987) Inositol trisphosphate and diacylglycerol: two interacting second messengers. Annu Rev Biochem 56:159–193

Berridge MJ (1989) Inositol trisphosphate, calcium, lithium, and cell signaling. JAMA 262:1834–1841

Berridge MJ, Irvine RF (1984) Inositol trisphosphate, a novel second messenger in cellular signal transduction. Nature 312:315–321

Bersani L, Colotta F, Peri G, Mantovani A (1987) Cytotoxic effector function of B lymphoblasts. J Immunol 139:645–648

Berton G, Gordon S (1983a) Superoxide release by peritoneal and bone marrow-derived mouse macrophages. Modulation by adherence and cell activation. Immunology 49:693–704

Berton G, Gordon S (1983b) Modulation of macrophage mannosyl-specific receptors by cultivation on immobilized zymosan. Effects on superoxide-anion release and phagocytosis. Immunology 49:705–715

Berton G, Gordon S (1983c) Desensitization of macrophages to stimuli which induce secretion of superoxide anion. Down-regulation of receptors for phorbol myristate acetate. Eur J Immunol 13:620–627

Berton G, Cassatella M, Cabrini G, Rossi F (1985a) Activation of mouse macrophages causes no change in expression and function of phorbol diesters' receptors, but is accompanied by alterations in the activity and kinetic parameters of NADPH oxidase. Immunology 54:371–379

Berton G, Papini E, Cassatella MA, Bellavite P, Rossi F (1985b) Partial purification of the superoxide-generating system of macrophages. Possible association of the NADPH oxidase activity with a low-potential (- 247 mV) cytochrome b. Biochim Biophys Acta 810:164–173

Berton G, Cassatella MA, Bellavite P, Rossi F (1986a) Molecular basis of macrophage activation. Expression of the low potential cytochrome b and its reduction upon cell stimulation in activated macrophages. J Immunol 136:1393–1399

Berton G, Rosen H, Ezekowitz RAB, Bellavite P, Serra MC, Rossi F, Gordon S (1986b) Monoclonal antibodies to a particulate superoxide-forming system stimulate a respiratory burst in intact guinea pig neutrophils. Proc Natl Acad Sci USA 83:4002–4006

Berton G, Zeni L, Cassatella MA, Rossi F (1986c) Gamma interferon is able to enhance the oxidative metabolism of human neutrophils. Biochem Biophys Res Commun 138:1276–1282

Berton G, Dusi S, Serra MC, Bellavite P, Rossi F (1989) Studies on the NADPH oxidase of phagocytes. Production of a monoclonal antibody which blocks the enzymatic activity of pig neutrophil NADPH oxidase. J Biol Chem 264:5564–5568

Besemer J, Hujber A, Kuhn B (1989) Specific binding, internalization, and degradation of human neutrophil activating factor by human polymorphonuclear leukocytes. J Biol Chem 264:17409–17415

Bessler H, Gilgal R, Djaldetti M, Zahavi I (1986) Effect of pentoxifylline on the phagocytic activity, cAMP levels, and superoxide anion production by monocytes and polymorphonuclear cells. J Leukocyte Biol 40:747–754

Bessler WG, Ottenbreit BP (1977) Studies on the mitogenic principle of the lipoprotein from the outer membrane of Escherichia coli. Biochem Biophys Res Commun 76:239–246

Besterman JM, Cuatrecasas P (1984) Phorbol esters rapidly stimulate amiloride-sensitive Na^+/H^+ exchange in a human leukemic cell line. J Cell Biol 99:340–343

Besterman JM, May WS Jr, LeVine H III, Cragoe EJ, Cuatrecasas P (1985) Amiloride inhibits phorbol ester-stimulated Na^+/H^+ exchange and protein kinase C. An amiloride analog selectively inhibits Na^+/H^+ exchange. J Biol Chem 260:1155–1159

Beswick PH, Kay AB (1981) The effects of an ECF-A and formyl methionyl chemotactic peptides on oxidative metabolism of human eosinophils and neutrophils. Clin Exp Immunol 43:399–407

Beutler B, Cerami A (1986) Cachektin and tumour necrosis factor as two sides of the same biological coin. Nature 320:584–588

Beutler B, Cerami A (1989) The biology of cachektin/TNF—a primary mediator of the host response. Annu Rev Immunol 7:625–655

Bhatnagar R, Schirmer R, Ernst M, Decker K (1981) Superoxide release by zymosan-stimulated rat Kupffer cells in vitro. Eur J Biochem 119:171–175

Bigay J, Deterre P, Pfister C, Chabre M (1985) Fluoroaluminates activate transducin-GDP by mimicking the γ-phosphate of GTP in its binding site. FEBS Lett 191:181–185

Billah MM, Siegel MI (1987) Phospholipase A_2 activation in chemotactic peptide-stimulated HL60 granulocytes: synergism between diacylglycerol and Ca^{2+} in a protein kinase C-independent mechanism. Biochem Biophys Res Commun 144:683–691

Billah MM, Eckel S, Mullmann TJ, Egan RW, Siegel MI (1989) Phosphatidylcholine hydrolysis by phospholipase D determines phosphatidate and diglyceride levels in chemotactic peptide-stimulated human neutrophils. Involvement of phosphatidate phosphohydrolase in signal transduction. J Biol Chem 264:17069–17077

Billingham MEJ (1987) Cytokines as inflammatory mediators. Br Med Bull 43:350–370

Bird J, Pelletier M, Tissot M, Giroud JP (1985) The modification of the oxidative metabolism of cells derived both locally and at distance from the site of an acute inflammatory reaction. J Leukocyte Biol 37:109–120

Birnbaumer L, Yatani A, Codina J, van Dongen A, Graf R, Mattera R, Sanford J, Brown AM (1989) Multiple roles of G proteins in coupling of receptors to ionic channels and other effectors. In: Gehring U, Helmreich E, Schultz G (eds) Molecular mechanisms of hormone action. Springer, Berlin Heidelberg New York, pp 147–177

Bjerknes R, Vindenes H, Laerum OD (1990) Altered neutrophil functions in patients with large burns. Blood Cells 16:127–143

Bjornson AB, Knippenberg RW, Bjornson HS (1989) Bactericidal defect of neutrophils in a guinea pig model of thermal injury is related to elevation of intracellular cyclic 3′,5′-adenosine monophosphate. J Immunol 143:2609–2616

Blackburn WD Jr, Heck LW (1988) Neutrophil activation by surface bound IgG: pertussis toxin insensitive activation. Biochem Biophys Res Commun 152:136–142

Blackburn WD Jr, Heck LW (1989) Neutrophil activation by surface-bound IgG is via a pertussis toxin insensitive G protein. Biochem Biophys Res Commun 164:983–989

Blackburn WD Jr, Heck LW, Wallace RW (1987) The bioflavonoid quercetin inhibits neutrophil degranulation, superoxide production, and the phosphorylation of specific neutrophil proteins. Biochem Biophys Res Commun 144:1229–1236

Blake DR, Allen RE, Lunec J (1987) Free radicals in biological systems—a review orientated to inflammatory processes. Br Med Bull 43:371–385

Blaustein RO, Koehler TM, Collier RJ, Finkelstein A (1989) Anthrax toxin: channel-forming activity of protective antigen in planar phospholipid bilayers. Proc Natl Acad Sci USA 86:2209–2213

Bokoch GM (1987) The presence of free G protein β/γ subunits in human neutrophils results in suppression of adenylate cyclase activity. J Biol Chem 262:589–594

Bokoch GM, Gilman AG (1984) Inhibition of receptor-mediated release of arachidonic acid by pertussis toxin. Cell 39:301–308

Bokoch GM, Parkos CA (1988) Identification of novel GTP-binding proteins in the human neutrophil. FEBS Lett 227:66–70

Bokoch GM, Quilliam LA (1990) Guanine nucleotide binding properties of rap1 purified from human neutrophils. Biochem J 267:407–411

Bokoch GM, Reed PW (1979) Inhibition of the neutrophil oxidative response to a chemotactic peptide by inhibitors of arachidonic acid oxygenation. Biochem Biophys Res Commun 90:481–487

Bolscher BGJM, van Zwieten R, Kramer IM, Weening RS, Verhoeven AJ, Roos D (1989) A phosphoprotein of M_r 47,000, defective in autosomal chronic granulomatous disease, copurifies with one of two soluble components required for NADPH:O_2 oxidoreductase activity in human neutrophils. J Clin Invest 83:757–763

Bolscher BGJM, Denis SW, Verhoeven AJ, Roos D (1990) The activity of one soluble component of the cell-free NADPH: O_2 oxidoreductase of human neutrophils depends on guanosine 5'-O-(3-thio)triphosphate. J Biol Chem 265:15782–15787

Bomalaski JS, Hirata F, Clark MA (1986) Aspirin inhibits phospholipase C. Biochem Biophys Res Commun 139:115–121

Bomalaski JS, Baker DG, Brophy L, Resurreccion NV, Spilberg I, Muniain M, Clark MA (1989) A phospholipase A_2-activating protein (PLAP) stimulates human neutrophil aggregation and release of lysosomal enzymes, superoxide, and eicosanoids. J Immunol 142:3957–3962

Bonora GM, Toniolo C, Freer RJ, Becker EL (1986) Retro-all-D and retro isomers of a formyl-methionyl peptide chemoattractant: an insight into the mode of binding at the receptor on rabbit neutrophils. Biochim Biophys Acta 884:545–549

Bonser RW, Dawson J, Thompson NT, Hodson HF, Garland LG (1986) Inhibition of phorbol ester stimulated superoxide production by 1-oleoyl-2-acetyl-sn-glycerol (OAG); fact or artefact? FEBS Lett 209:134–138

Bonser RW, Thompson NT, Randall RW, Garland LG (1989) Phospholipase D activation is functionally linked to superoxide generation in the human neutrophil. Biochem J 264:617–620

Borish LC, Rocklin RE (1985) Physiological studies with human leukocyte inhibitory factory. Surv Immunol Res 4:230–237

Borish LC, Rocklin RE (1987) Effects of leukocyte inhibitory factor (LIF) on neutrophil phagocytosis and bactericidal activity. J Immunol 138:1475–1479

Borish L, O'Reilly D, Klempner MS, Rocklin RE (1986) Leukocyte inhibitory factor (LIF) potentiates neutrophil responses to formyl-methionyl-leucyl-phenyl-alanine. J Immunol 137:1897–1903

Borish L, Rosenbaum R, Albury L, Clark S (1989) Activation of neutrophils by recombinant interleukin 6. Cell Immunol 121:280–289

Bormann BJ, Huang C-K, Mackin WM, Becker EL (1984) Receptor-mediated activation of a phospholipase A_2 in rabbit neutrophil plasma membrane. Proc Natl Acad Sci USA 81:767–770

Bornstein P, Sage H (1980) Structurally distinct collagen types. Annu Rev Biochem 49:957–1003

Borregaard N, Tauber AI (1984) Subcellular localization of the human neutrophil NADPH oxidase. b-Cytochrome and associated flavoprotein. J Biol Chem 259:47–52

Borregaard N, Heiple JM, Simons ER, Clark RA (1983) Subcellular localization of the b-cytochrome component of the human neutrophil microbicidal oxidase: translocation during activation. J Cell Biol 97:52–61

Borregaard N, Schwartz JH, Tauber AI (1984) Proton secretion by stimulated neutrophils. Significance of hexose monophosphate shunt activity as source of electrons and protons for the respiratory burst. J Clin Invest 74:455–459

Boukili MA, Bureau M, Lagente V, Lefort J, Lellouch-Tubiana A, Malanchère E, Vargaftig BB (1986) Pharmacological modulation of the effects of N-formyl-L-methionyl-L-leucyl-L-phenylalanine in guinea pigs: involvement of the arachidonic acid cascade. Br J Pharmacol 89:349–359

Boulay F, Tardif M, Brouchon L, Vignais P (1990) Synthesis and use of a novel N-formyl peptide derivative to isolate a human N-formyl peptide receptor cDNA. Biochem Biophys Res Commun 168:1103–1109

Bourgcin S, Plante E, Gaudry M, Naccache PH, Borgeat P, Poubelle PE (1990) Involvement of a phospholipase D in the mechanism of action of granulocyte-macrophage colony-stimulating factor (GM-CSF): Priming of human neutrophils in vitro with GM-CSF is associated with accumulation of phosphatidic acid and diradylglycerol. J Exp Med 172:767–777

Bourne HR, Melmon KL (1971) Adenyl cyclase in human leukocytes: Evidence for activation by separate beta adrenergic an prostaglandin receptors. J Pharmacol Exp Ther 178:1–7

Bourne HR, Lehrer RI, Lichtenstein LM, Weissmann G, Zurier R (1973) Effects of cholera enterotoxin on adenosine 3′,5′-monophosphate and neutrophil function. Comparison with other compounds which stimulate leukocyte adenyl cyclase. J Clin Invest 52:698–708

Boutin JA, Ernould AP, Genton A, Cudennec CA (1989) Partial purification and characterization of a new p36/40 tyrosine protein kinase from HL-60. Biochem Biophys Res Commun 160:1203–1211

Boxer LA, Coates TD, Haak RA, Wolach JB, Hoffstein S, Baehner RL (1982) Lactoferrin deficiency associated with altered neutrophil function. N Engl J Med 307:404–410

Bradford PG, Rubin RP (1985) Characterization of formylmethionyl-leucyl-phenylalanine stimulation of inositol trisphosphate accumulation in rabbit neutrophils. Mol Pharmacol 27:74–78

Bradford PG, Rubin RP (1986) Guanine nucleotide regulation of phospholipase C activity in permeabilized rabbit neutrophils. Inhibition by pertussis toxin and sensitization to submicromolar calcium concentrations. Biochem J 239:97–102

Brandt SJ, Dougherty RW, Lapetina EG, Niedel EJ (1985) Pertussis toxin inhibits chemotactic peptide-stimulated generation of inositol phosphates and lysosomal enzyme secretion in human leukemic (HL-60) cells. Proc Natl Acad Sci USA 82:3277–3280

Braquet P, Touqui L, Shen Y, Vargaftig BB (1987) Perspectives in platelet-activating factor research. Pharmacol Rev 39:97–145

Braun U, Habermann B, Just I, Aktories K, Vandekerckhove J (1989) Purification of the 22 kDa protein substrate of botulinum ADP-ribosyltransferase C3 from porcine brain cytosol and its characterization as a GTP-binding protein highly homologous to the *rho* gene product. FEBS Lett 243:70–76

Braun V (1975) Covalent lipoprotein from the outer membrane of *Escherichia coli*. Biochim Biophys Acta 415:335–377

Breitman TR, Selonick SE, Collins SJ (1980) Induction of differentiation of the human promyelocytic leukemia cell line (HL-60) by retinoic acid. Proc Natl Acad Sci USA 77:2936–2940

Briggs RT, Drath DB, Karnovsky ML, Karnovsky MJ (1975) Localization of NADH oxidase on the surface of human polymorphonuclear leukocytes by a new cytochemical method. J Cell Biol 67:566–586

Briheim G, Follin P, Sandstedt S, Dahlgren C (1989) Relationship between intracellularly and extracellulary generated oxygen metabolites from primed polymorphonuclear leukocytes differs from that obtained from nonprimed cells. Inflammation 13:455–464

Britigan BE, Cohen MS, Rosen GM (1987) Detection of the production of oxygen-centered free radicals by human neutrophils using spin trapping techniques: a critical perspective. J Leukocyte Biol 41:349–362

Bromberg Y, Pick E (1983) Unsaturated fatty acids as second messengers of superoxide generation by macrophages. Cell Immunol 79:240–252

Bromberg Y, Pick E (1984) Unsaturated fatty acids stimulate NADPH-dependent superoxide production by cell-free system derived from macrophages. Cell Immunol 88:213–221

Bromberg Y, Pick E (1985) Activation of NADPH-dependent superoxide production in a cell-free system by sodium dodecyl sulfate. J Biol Chem 260:13539–13545

Brown SS, Spudich JA (1981) Mechanism of action of cytochalasin: evidence that it binds to actin filament ends. J Cell Biol 88:487–491

Brownlee M, Cerami A, Vlassara H (1988) Advanced glycosylation end products in tissue and the biochemical basis of diabetic complications. N Engl J Med 318:1315–1321

Brumley LM, Wallace RW (1989) Calmodulin and protein kinase C antagonists also inhibit the Ca^{2+}-dependent protein protease, calpain I. Biochem Biophys Res Commun 159:1297–1303

Bryant SM, Lynch RE, Hill HR (1982) Kinetic analysis of superoxide anion production by activated resident murine peritoneal macrophages. Cell Immunol 69:46–58

Buchmüller-Rouiller Y, Ransijin A, Mauel J (1989) Lead inhibits oxidative metabolism of macrophages exposed to macrophage-activating factor. Biochem J 260:325–332

Buchta R, Fridkin M, Pontet M, Contessi E, Scaggiante B, Romeo D (1987a) Modulation of human neutrophil function by C-reactive protein. Eur J Biochem 163:141–146

Buchta R, Pontet M, Fridkin M (1987b) Binding of C-reactive protein to human neutrophils. FEBS Lett 211:165–168

Bump NJ, Najjar VA (1988) Tuftsin stimulates growth of HL60 cells. FEBS Lett 226:303–306

Burch RM, Luini A, Axelrod J (1986) Phospholipase A_2 and phospholipase C are activated by distinct GTP-binding proteins in response to α_1-adrenergic stimulation in FRTL5 thyroid cells. Proc Natl Acad Sci USA 83:7201–7205

Burde R, Seifert R, Buschauer A, Schultz G (1989) Histamine inhibits activation of human neutrophils and HL-60 leukemic cells via H_2-receptors. Naunyn Schmiedebergs Arch Pharmacol 340:671–678

Burde R, Buschauer A, Seifert R (1990) Characterization of histamine H_2-receptors in human neutrophils with a series of guanidine analogues of impromidine: are cell type-specific H_2-receptors involved in the regulation of NADPH oxidase? Naunyn Schmiedebergs Arch Pharmacol 341:455–461

Burnham DN, Uhlinger DJ, Lambath JD (1990) Diradylglycerol synergizes with an anionic amphiphile to activate superoxide generation and phosphorylation of p47$_{phox}$ in a cell-free system from human neutrophils. J Biol Chem 265:17550–17559

Burnstock G, Kennedy C (1985) Is there a basis for distinguishing two types of P$_2$-purinoceptor? Gen Pharmacol 16:433-440

Buschauer A (1989) Synthesis and in vitro pharmacology of arpromidine and related phenyl(pyridylalkyl)guanidines, a potential new class of positive inotropic drugs. J Med Chem 32:1963–1970

Busse WW, Sosman J (1976) Histamine inhibition of neutrophil lysosomal enzyme release: an H$_2$ histamine receptor response. Science 194:737–738

Butcher EC, Lewinsohn D, Duijvestijn A, Bargatze R, Wu N, Jalkanen S (1986) Interactions between endothelial cells and leukocytes. J Cell Biochem 30:121–131

Butler J, Koppenol WH, Margoliash E (1982) Kinetics and mechanism of the reduction of ferricytochrome c by the superoxide anion. J Biol Chem 257:10747–10750

Butterick CJ, Baehner RL, Boxer LA, Jersild RA (1983) Vitamin E—a selective inhibitor of the NADPH oxidoreductase enzyme system in human granulocytes. Am J Pathol 112:287–293

Butterworth AE, David JR (1981) Eosinophil function. N Engl J Med 304:154–156

Cadenas E (1989) Biochemistry of oxygen toxicity. Annu Rev Biochem 58:79–110

Cairo MS, Toy C, Sender L, van de Ven C (1990) Effect of idarubicin and epirubicin on in vitro polymorphonuclear function: diminished superoxide radical formation compared to their parent compounds daunorubicin and doxorubicin. J Leukocyte Biol 47:224–233

Caldwell SE, McCall CE, Hendricks CL, Leone PA, Bass DA, McPhail LC (1988) Coregulation of NADPH oxidase and phosphorylation of a 48-kD protein(s) by a cytosolic factor defective in autosomal recessive chronic granulomatous disease. J Clin Invest 81:1485–1496

Cameron AR, Nelson J, Forman HJ (1983) Depolarization and increased conductance precede superoxide release by concanavalin A-stimulated rat alveolar macrophages. Proc Natl Acad Sci USA 80:3726–3728

Cameron W, Doyle K, Rocklin RE (1986) Histamine type I (H$_1$) receptor radioligand binding studies on normal T cell subsets, B cells, and monocytes. J Immunol 136:2116–2120

Camisa C, Eisenstat B, Ragaz A, Weissmann G (1982) The effects of retinoids on neutrophil functions in vitro. J Am Acad Dermatol 6:620–629

Camp RDR, Greaves MW (1987) Inflammatory mediators in the skin. Br Med Bull 43:401–414

Canning PC, Neill JD (1989) Isolation and characterization of interleukin-1 from bovine polymorphonuclear leukocytes. J Leukocyte Biol 45:21–28

Cantin M, Dubois F, Bégin R (1988) Lung exposure to mineral dusts enhances the capacity of lung inflammatory cells to release superoxide. J Leukocyte Biol 43:299–303

Carp H (1982) Mitochondrial N-formylmethionyl proteins as chemoattractants for neutrophils. J Exp Med 155:264–275

Carpentier JL, Dayer JM, Lang U, Silverman R, Orci L, Gorden P (1984) Down-regulation and recycling of insulin receptors. Effect of monensin on IM-9 lymphocytes and U-937 monocyte-like cells. J Biol Chem 259:14190–14196

Casey PJ, Gilman AG (1988) G protein involvement in receptor-effector coupling. J Biol Chem 263:2577–2580

Cassatella MA, Della Bianca V, Berton G, Rossi F (1985) Activation by gamma interferon of human macrophage capability to produce toxic oxygen molecules is accompanied by decreased K_m of the superoxide-generating NADPH oxidase. Biochem Biophys Res Commun 132:908–914

Cassatella MA, Cappelli R, Della Bianca V, Grzeskowiak M, Dusi S, Berton G (1988) Interferon-gamma activates human neutrophil oxygen metabolism and exocytosis. Immunology 63:499–506

Cassatella MA, Berton G, Agostini C, Zambello R, Trentin L, Cipriani A, Semenzato G (1989a) Generation of superoxide anion by alveolar macrophages and sarcoidosis: evidence for the activation of the oxygen metabolism in patients with high-intensity alveolitis. Immunology 66:451–458

Cassatella MA, Hartman L, Trinchieri G (1989b) Tumor necrosis factor and immune interferon synergistically induce cytochrome b_{-245} heavy-chain gene expression and nicotinamide-adenine dinucleotide phosphate dehydrogenase oxidase in human leukemic myeloid cells. J Clin Invest 83:1570–1579

Cassatella MA, Bazzoni F, Flynn RM, Dusi S, Trinchieri G, Rossi F (1990) Molecular basis of interferon-γ and lipopolysaccharide enhancement of phagocyte respiratory burst capability. Studies on the gene expression of several NADPH oxidase components. J Biol Chem 265:20241–20246

Cassidy LF, Lyles DS, Abramson JS (1989) Depression of polymorphonuclear leukocyte functions by purified Influenza virus hemagglutinin and sialic acid-binding lectins. J Immunol 142:4401–4406

Cassidy P, Hoar PE, Kerrick WGL (1979) Irreversible thiophosphorylation and activation of tension in functionally skinned rabbit ileum strips by [^{35}S]ATPγS. J Biol Chem 254:11148–11153

Castagna M, Takai Y, Kaibuchi K, Sano K, Kikkawa U, Nishizuka Y (1982) Direct activation of calcium-activated, phospholipid-dependent protein kinase by tumor-promoting phorbol esters. J Biol Chem 257:7847–7851

Castranova V, Jones GS, Phillips RM, Peden D, Vandyke K (1981) Abnormal responses of granulocytes in chronic granulomatous disease. Biochim Biophys Acta 645:49–53

Celada A, Gray PW, Rinderknecht E, Schreiber RD (1984) Evidence for a gamma-interferon receptor that regulates macrophage tumoricidal activity. J Exp Med 160:55–74

Chabre M (1989) Aluminofluoride action on G-proteins of the adenylate cyclase system is not different from that of transducin. Biochem J 258:931–933

Channon JY, Leslie CC, Johnston RB Jr (1987) Zymosan-stimulated production of phosphatidic acid by macrophages: relationship to release of superoxide anion and inhibition by agents that increase intracellular cyclic AMP. J Leukocyte Biol 41:450–453

Chaplinski TJ, Niedel JE (1982) Cyclic nucleotide-induced maturation of human promyelocytic leukemia cells. J Clin Invest 70:953–964

Chaudhry AN, Santinga JT, Gabig TG (1982) The subcellular particulate NADPH-dependent O_2^--generating oxidase from human blood monocytes: comparison to the neutrophil system. Blood 60:979-983

Chenoweth DE, Hugli TE (1978) Demonstration of specific C5a receptor on intact human polymorphonuclear leukocytes. Proc Natl Acad Sci USA 75:3943–3947

Cherenkevich SN, Vanderkool JM, Restifo R, Daniele RP, Holian A (1982a) The lipid integrity of membranes of guinea pig alveolar macrophages studies by nanosecond fluorescence decay of 1,6-diphenyl-1,3,5-hexatriene: the effect of stimulation by concanavalin A and formyl peptides. Arch Biochem Biophys 214:299-301

Cherenkevich SN, Vanderkool JM, Holian A (1982b) The lipid integrity of membranes of guinea pig alveolar macrophages studied by nanosecond fluorescence decay of 1,6-diphenyl-1,3,5-hexatriene: The influence of temperature and benzyl alcohol. Arch Biochem Biophys 214:305–310

Chiara MD, Bedoya F, Sobrino F (1989) Cyclosporin A inhibits phorbol ester-induced activation of superoxide production in resident mouse peritoneal macrophages. Biochem J 264:21–26

Chiba T, Kaneda M, Fuji M, Clark RA, Nauseef WM, Kakinuma K (1990) Two cytosolic components of the neutrophil NADPH oxidase, p47-*phox* and p67-*phox*, are not flavoproteins. Biochem Biophys Res Commun 173:376–381

Chilton FH, O'Flaherty JTO, Walsh CE, Thomas MJ, Wykle RL, DeChatelet LR, Waite BM (1982) Platelet activating factor. Stimulation of the lipoxygenase pathway in polymorphonuclear leucocytes by 1-*O*-alkyl-2-*O*-acetyl-*sn*-glycero-3-phosphocholine. J Biol Chem 257:5402-5407

Chou S, Kaneko M, Nakaya K, Nakamura Y (1990) Induction of differentiation of human and mouse myeloid leukemia cells by camptothecin. Biochem Biophys Res Commun 166:160–167

Christiansen NO (1988) A time-course study on superoxide generation and protein kinase C activation in human neutrophils. FEBS Lett 239:195–198

Christiansen NO (1990) Pertussis toxin inhibits the FMLP-induced membrane association of protein kinase C in human neutrophils. J Leukocyte Biol 47:60–63

Christiansen NO, Juhl H (1986) Purification and properties of protein kinase C from bovine polymorphonuclear leukocytes. Biochim Biophys Acta 885:170–175

Christiansen NO, Larsen CS, Juhl H, Esmann V (1988a) Membrane-associated protein kinases in superoxide-producing human polymorphonuclear leukocytes. J Leukocyte Biol 44:33–40

Christiansen NO, Larsen CS, Esmann V (1988b) A study on the role of protein kinase C and intracellular calcium in the activation of superoxide generation. Biochim Biophys Acta 961:317–324

Christiansen NP, Skubitz KM (1988) Identification of the major lectin-binding surface proteins of human neutrophils and alveolar macrophages. Blood 71:1624–1632

Chung T, Kim YB (1988) Two distinct cytolytic mechanisms of macrophages and monocytes activated by phorbol myristate acetate. J Leukocyte Biol 44:329–336

Claesson HE, Feinmark SJ (1984) Relationship of cyclic-AMP levels in leukotriene B4-stimulated leukocytes to lysosomal enzyme release and the generation of superoxide anions. Biochim Biophys Acta 804:52–57

Clancy RM, Dahinden CA, Hugli TE (1983) Arachidonate metabolism by human polymorphonuclear leukocytes stimulated by *N*-formyl-Met-Leu-Phe or complement component C5a is independent of phospholipase activation. Proc Natl Acad Sci USA 80:7200–7204

Clark RA (1982) Chemotactic factors trigger their own oxidative inactivation by human neutrophils. J Immunol 129:2725–2728

Clark RA (1990) The human neutrophil respiratory burst oxidase. J Infectious Diseases 161:1140–1147

Clark RA, Leidal KG, Pearson DW, Nauseef WM (1987) NADPH oxidase of human neutrophils. Subcellular localization and characterization of an arachidonate-activatable superoxide-generating system. J Biol Chem 262:4065–4074

Clark RA, Malech HL, Gallin JI, Nunoi H, Volpp BD, Pearson DW, Nauseef WM, Curnutte JT (1989) Genetic variants of chronic granulomatous disease: prevalence of deficiencies of two cytosolic components of the NADPH oxidase system. N Engl J Med 321:647–652

Clark RA, Volpp BD, Leidal KG, Nauseef WM (1990) Two cytosolic components of the human neutrophil respiratory burst oxidase translocate to the plasma membrane during cell activation. J Clin Invest 85:714–721

Clement LT, Lehmeyer JE (1983) Regulation of the growth and differentiation of a human monocytic cell line by lymphokines. I. Induction of superoxide anion production and chemiluminescence. J Immunol 130:2763–2766

Clifford CB, Slauson DO, Neilsen NR, Suyemoto MM, Zwahlen RD, Schlafer DH (1989) Ontogeny of inflammatory cell responsiveness: superoxide anion generation by phorbol ester-stimulated fetal, neonatal, and adult bovine neutrophils. Inflammation 13:221–231

Clifford DP, Rasp FL, Repine JE (1984) Simultaneous measurement of adherence and chemiluminescence by polymorphonuclear leukocytes. Inflammation 8:101–106

Cockcroft S, Stutchfield J (1988) Effect of pertussis toxin and neomycin on G-protein-regulated polyphosphoinositide phosphodiesterase. A comparison between HL60 membranes and permeabilized HL60 cells. Biochem J 256:343-350

Cockcroft S, Stutchfield J (1989a) ATP stimulates secretion in human neutrophils and HL60 cells via a pertussis toxin-sensitive guanine nucleotide-binding protein coupled to phospholipase C. FEBS Lett 245:25–29

Cockcroft S, Stutchfield J (1989b) The receptors for ATP and fMetLeuPhe are independently coupled to phospholipases C and A_2 via G-protein(s). Relationship between phospholipase C and A_2 activation and exocytosis in HL60 cells and human neutrophils. Biochem J 263:715–723

Codina J, Hildebrandt JD, Birnbaumer L, Sekura RP (1984) Effects of guanine nucleotides and Mg on human erythrocyte N_i and N_s, the regulatory components of adenylyl cyclase. J Biol Chem 259:11408–11418

Coffey RG, Davis JS, Djeu JY (1988) Stimulation of guanylate cyclase activity and reduction of adenylate cyclase activity by granulocyte-macrophage colony-stimulating factor in human blood neutrophils. J Immunol 140:2695–2701

Cohen HJ, Chovaniec ME (1978a) Superoxide generation by digitonin-stimulated guinea pig granulocytes. J Clin Invest 61:1081–1087

Cohen HJ, Chovaniec ME (1978b) Superoxide generation by digitonin-stimulated guinea pig granulocytes. The effects of N-ethyl maleimide, divalent cations, and glycolytic and mitochondrial inhibitors on the activation of the superoxide generating system. J Clin Invest 61:1087–1096

Cohen HJ, Chovaniec ME, Davies WA (1980a) Activation of the guinea pig granulocyte NAD(P)H-dependent superoxide generating enzyme: localization in a plasma membrane enriched particle and kinetics of activation. Blood 55:355–363

Cohen HJ, Chovaniec ME, Ellis SE (1980b) Chlorpromazine inhibition of granulocyte superoxide production. Blood 56:23–29

Cohen HJ, Newburger PE, Chovaniec ME, Whitin JC, Simons ER (1981) Opsonized zymosan-stimulated granulocytes—activation and activity of the superoxide-generating system and membrane potential changes. Blood 58:975–982

Cohen HJ, Chovaniec ME, Wilson MK, Newburger PE (1982) Con-A-stimulated superoxide production by granulocytes: reversible activation of NADPH oxidase. Blood 60:1188–1194

Cohen HJ, Whitin JC, Chovaniec ME, Tape EH, Simons ER (1984) Is activation of the granulocyte by concanavalin-A a reversible process? Blood 63:114–120

Cohen HJ, Chovaniec ME, Takahashi K, Whitin JC (1986) Activation of human granulocytes by arachidonic acid: its use and limitations for investigating for granulocyte functions. Blood 67:1103–1109

Cohen MS, Metcalf JA, Root RK (1980) Regulation of oxygen metabolism in human granulocytes: relationship between stimulus binding and oxidative response using plant lectins as probes. Blood 55:1003-1010

Cohen MS, Ryan JL, Root RK (1981) The oxidative metabolism of thioglycollate-elicited mouse peritoneal macrophages: the relationship between oxygen, superoxide and hydrogen peroxide and the effect of monolayer formation. J Immunol 127:1007–1011

Cohen MS, Elliott DM, Chaplinski T, Pike MM, Niedel JE (1982) A defect in the oxidative metabolism of human polymorphonuclear leukocytes that remain in circulation early in hemodialysis. Blood 60:1283–1289

Cohen MS, Mesler DE, Snipes RG, Gray TK (1986) 1,25-Dihydroxyvitamin D_3 activates secretion of hydrogen peroxide by human monocytes. J Immunol 136:1049–1053

Coleman DL, Liu J, Bartiss AH (1989) Recombinant granulocyte-macrophage colony-stimulating factor increases adenylate cyclase activity in murine peritoneal macrophages. J Immunol 143:4134–4140

Collins SJ (1987) The HL-60 promyelocytic leukemia cell line: proliferation, differentiation, and cellular oncogene expression. Blood 70:1233–1244

Collins SJ, Ruscetti FW, Gallagher RE, Gallo RC (1978) Terminal differentiation of human promyelocytic leukemia cells induced by dimethyl sulfoxide and other polar compounds. Proc Natl Acad Sci USA 75:2458–2462

Collins SJ, Ruscetti FW, Gallagher RE, Gallo RC (1979) Normal functional characteristics of cultured human promyelocytic leukemia cells (HL-60) after induction of differentiation by dimethylsulfoxide. J Exp Med 149:969–974

Combadiere C, Hakim J, Giroud JP, Perianin A (1990) Staurosporine, a protein kinase inhibitor, up-regulates the stimulation of human neutrophil respiratory burst by N-formyl peptides and platelet activating factor. Biochem Biophys Res Commun 168:65–70

Conley NS, Yarbro JW, Ferrari HA, Zeidler RB (1986) Bleomycin increases superoxide anion generation by pig peripheral alveolar macrophages. Mol Pharmacol 30:48–52

Conseiller EC, Lederer F (1989) Inhibition of NADPH oxidase by aminoacyl chloromethane protease inhibitors in phorbol-ester-stimulated human neutrophils: a reinvestigation. Are proteases really involved in the activation process? Eur J Biochem 183:107–114

Conseiller EC, Schott D, Lederer F (1990) Inhibition by aminoacyl-chloromethane protease inhibitors of superoxide anion production by phorbol-ester-stimulated human neutrophils. The labeled target is a membrane protein. Eur J Biochem 193:345–350

Cooke E, Hallett MB (1985) The role of C-kinase in the physiological activation of the neutrophil oxidase. Evidence from using pharmacological manipulation of C-kinase activity in intact cells. Biochem J 232:323–327

Cooke E, Al-Mohanna F, Hallett MB (1985) The role of microfilaments in polymorphonuclear leucocyte oxidase activation. Biochem Soc Trans 13:1173–1174

Cooper MR, DeChatelet LR, McCall CE, LaVia MF, Spurr CL, Baehner RL (1972) Complete deficiency of leukocyte glucose-6-phosphate dehydrogenase with defective bactericidal activity. J Clin Invest 51:769–778

Cooper PH, Mayer P, Baggiolini M (1984) Stimulation of phagocytosis in bone marrow-derived mouse macrophages by bacterial lipopolysaccharide: correlation with biochemical and functional parameters. J Immunol 133:913–922

Cope FO (1986) The in vivo inhibition of mouse brain protein kinase-C by retinoic acid. Cancer Lett 30:275–288

Coppi M, Niederman R (1989) Effects of ammonia on human neutrophil N-formyl chemotactic peptide receptor-ligand interaction and cytoskeletal association. Biochem Biophys Res Commun 165:377–383

Corey SJ, Rosoff PM (1989) Granulocyte-macrophage colony-stimulating factor primes neutrophils by activating a pertussis-toxin-sensitive G protein not associated with phosphatidylinositol turnover. J Biol Chem 264:14165–14171

Costa-Casnellie MR, Segel GB, Lichtman MA (1986) Signal transduction in human monocytes: relationship between superoxide production and the level of kinase C in the membrane. J Cell Physiol 129:336–342

Cotgreave IA, Duddy SK, Kass GEN, Thompson D, Moldeus P (1989) Studies on the anti-inflammatory activity of ebselen: ebselen interferes with granulocyte oxidative burst by dual inhibition of NADPH oxidase and protein kinase C? Biochem Pharmacol 38:649–656

Coussens L, Parker PJ, Rhee L, Yang-Feng TL, Chen E, Waterfield MD, Francke U, Ullrich A (1986) Multiple, distinct forms of bovine and human protein kinase C suggest diversity in cellular signaling pathways. Science 233:859–866

Cowen DS, Lazarus HM, Shurin SB, Stoll SE, Dubyak GR (1989) Extracellular adenosine triphosphate activates calcium mobilization in human phagocytic leukocytes and neutrophil/monocyte progenitor cells. J Clin Invest 83:1651–1660

Cox CC, Dougherty RW, Ganong BR, Bell RM, Niedel JE, Snyderman R (1986) Differential stimulation of the respiratory burst and lysosomal enzyme secretion in human polymorphonuclear leukocytes by synthetic diacylglycerols. J Immunol 136:4611–4616

Cox JA, Jeng AY, Sharkey NA, Blumberg PM, Tauber AI (1985) Activation of the human neutrophil nicotinamide adenine dinucleotide phosphate (NADPH)-oxidase by protein kinase C. J Clin Invest 76:1932–1938

Cox JA, Jeng AY, Blumberg PM, Tauber AI (1987) Comparison of subcellular activation of the human neutrophil NADPH-oxidase by arachidonic acid, sodium dodecyl sulfate (SDS), and phorbol myristate acetate (PMA). J Immunol 138:1884–1888

Craighead JE, Mossman BT (1982) The pathogenesis of asbestos-associated diseases. N Engl J Med 306:1446–1455

Crawford DR, Schneider BD (1982) Identification of ubiquinone-50 in human neutrophils and its role in microbicidal events. J Biol Chem 257:6662–6668

Crawford DR, Schneider BD (1983) Ubiquinone content and respiratory burst activity of latex-filled phagolysosomes isolated from human neutrophils and evidence for the probable involvement of a third granule. J Biol Chem 258:5363–5367

Cristol JP, Provencal B, Borgeat P, Sirois P (1988) Characterization of leukotriene B$_4$ binding sites on guinea pig lung macrophages. J Pharmacol Exp Ther 247:1199–1203

Crockett-Torabi E, Fantone JC (1990) Soluble and insoluble immune complexes activate human neutrophil NADPH oxidase by distinct Fcγ receptor-specific mechanisms. J Immunol 145:3026–3032

Cronstein BN, Kramer SB, Weissmann G, Hirschhorn R (1983) Adenosine: a physiological modulator of superoxide anion generation by human neutrophils. J Exp Med 158:1160–1177

Cronstein BN, Rosenstein ED, Kramer SB, Weissmann G, Hirschhorn R (1985) Adenosine; a physiologic modulator of superoxide anion generation by human neutrophils. Adenosine acts via an A$_2$ receptor on human neutrophils. J Immunol 135:1366–1371

Cronstein BN, Kramer SB, Rosenstein ED, Korchak HM, Weissmann G, Hirschhorn R (1988) Occupancy of adenosine receptors raises cyclic AMP alone and in synergy with occupancy of chemoattractant receptors and inhibits membrane depolarization. Biochem J 252:709–715

Cronstein BN, Daguma L, Nichols D, Hutchinson AJ, Williams M (1990) The adenosine/neutrophil paradox resolved: human neutrophils possess both A$_1$ and A$_2$ receptors that promote chemotaxis and inhibit O$_2^-$ generation, respectively. J Clin Invest 85:1150–1157

Crooke ST, Snyder RM, Butt TR, Ecker DJ, Allaudeen HS, Monia B, Mirabelli CK (1986) Cellular and molecular pharmacology of auranofin and related gold complexes. Biochem Pharmacol 35:3423–3431

Cross AR (1987) The inhibitory effects of some iodonium compounds on the superoxide generating system of neutrophils and their failure to inhibit diaphorase activity. Biochem Pharmacol 36:489–493

Cross AR (1990) Inhibitors of the leukocyte superoxide generating oxidase: mechanisms of action and methods for their elucidation. Free Radic Biol Med 8:71–93

Cross AR, Jones OTG (1986) The effect of the inhibitor diphenylene iodonium on the superoxide-generating system of neutrophils. Specific labelling of a component polypeptide of the oxidase. Biochem J 237:111–116

Cross AR, Jones OTG, Garcia R, Segal AW (1983) The subcellular localization of ubiquinone in human neutrophils. Biochem J 216:765–768

Cross AR, Parkinson JF, Jones OTG (1984) The superoxide-generating oxidase of leukocytes: NADPH-dependent reduction of flavin and cytochrome b in solubilized preparations. Biochem J 223:337–344

Cross AR, Parkinson JF, Jones OTG (1985) Mechanism of the superoxide-producing oxidase of neutrophils. O_2 is necessary for the fast reduction of cytochrome b_{-245} by NADPH. Biochem J 226:881–884

Cross AR, Yea CM, Jones OTG (1988) Inhibition of superoxide generation by phagocytic leukocytes. Biochem Soc Trans 16:888–889

Cross AR, Henderson L, Jones OTG, Delpiano MA, Hentschel J, Acker H (1990) Involvement of an NAD(P)H oxidase as a pO_2 sensor protein in the rat carotid body. Biochim J 272:743–747

Csermely P, Szamel M, Resch K, Somogyi J (1988) Zinc can increase the activity of protein kinase C and contributes to its binding to plasma membranes in T lymphocytes. J Biol Chem 263:6487–6490

Cummings NP, Pabst MJ, Johnston RB Jr (1980) Activation of macrophages for enhanced release of superoxide anion and greater killing of *Candida albicans* by injection of muramyl dipeptide. J Exp Med 152:1659–1669

Cunningham CC, DeChatelet LR, Spach PI, Parce JW, Thomas MJ, Lees CJ, Shirley PS (1982) Identification and quantitation of electron-transport components in human polymorphonuclear leukocytes. Biochim Biophys Acta 682:430–435

Curnutte JT (1985) Activation of neutrophil nicotinamide adenine dinucleotide phosphate, reduced (triphosphopyridine nucleotide, reduced) oxidase by arachidonic acid in a cell-free system. J Clin Invest 75:1740–1743

Curnutte JT, Babior BM (1975) Effects of anaerobiosis and inhibitors on O_2^- production by human granulocytes. Blood 45:851–861

Curnutte JT, Tauber AI (1983) Failure to detect superoxide in human neutrophils stimulated with latex particles. Pediatr Res 17:281–284

Curnutte JT, Whitten DM, Babior BM (1974) Defective superoxide production by granulocytes from patients with chronic granulomatous disease. N Engl J Med 290:593–597

Curnutte JT, Kipnes RS, Babior BM (1975) Defect in pyridine nucleotide dependent superoxide production by a particulate fraction from the granulocytes of patients with chronic granulomatous disease. N Engl J Med 293:628–632

Curnutte JT, Babior BM, Karnovsky ML (1979) Fluoride-mediated activation of the respiratory burst in human neutrophils. A reversible process. J Clin Invest 63:637–647

Curnutte JT, Badwey JA, Robinson JM, Karnovsky MJ, Karnovsky ML (1984) Studies on the mechanism of superoxide release from human neutrophils stimulated with arachidonate. J Biol Chem 259:11851–11857

Curnutte JT, Scott PJ, Kuver R, Berkow R (1986) NADPH oxidase activation cofactor (ACF): partial purification and absent activity in a patient with chronic granulomatous disease. Clin Res 34:455A

Curnutte JT, Kuver R, Babior BM (1987a) Activation of the respiratory burst oxidase in a fully soluble system from human neutrophils. J Biol Chem 262:6450–6452

Curnutte JT, Kuver R, Scott PJ (1987b) Activation of neutrophil NADPH oxidase in a cell-free system. Partial purification of compounds and characterization of the activation process. J Biol Chem 262:5563–5569

Curnutte JT, Berkow RL, Roberts RL, Shurin SB, Scott PJ (1988) Chronic granulomatous disease due to a defect in the cytosolic factor required for nicotinamide adenine dinucleotide phosphate oxidase activation. J Clin Invest 81:606–610

Curnutte JT, Scott PJ, Mayo LA (1989a) Cytosolic components of the respiratory burst oxidase: resolution of four components, two of which are missing in complementing types of chronic granulomatous disease. Proc Natl Acad Sci USA 86:825–829

Curnutte JT, Scott P, Babior BM (1989b) Functional defect in neutrophil cytosols from two patients with autosomal recessive cytochrome-positive chronic granulomatous disease. J Clin Invest 83:1236–1240

Currie RW, White FP (1981) Trauma-induced protein in rat tissues: a physiological role for a "heat shock" protein? Science 214:72–73

Dahinden C, Fehr J (1980) Receptor-directed inhibition of chemotactic factor-induced neutrophil hyperactivity by pyrazolon derivatives. Definition of a chemotactic peptide antagonist. J Clin Invest 66:884–891

Dahinden C, Fehr J (1983) Granulocyte activation by endotoxin. II. Role of granulocyte adherence, aggregation, and effect of cytochalasin B, and comparison with formylated chemotactic peptide-induced stimulation. J Immunol 130:863–868

Dahinden C, Galanos C, Fehr J (1983a) Granulocyte activation by endotoxin. I. Correlation between adherence and other granulocyte functions, and role of endotoxin structure and biologic activity. J Immunol 130:857–862

Dahinden C, Fehr J, Hugli TE (1983b) Role of cell surface contact in the kinetics of superoxide production by granulocytes. J Clin Invest 72:113–121

Dahinden CA, Zingg J, Maly FE, de Weck AL (1988) Leukotriene production in human neutrophils primed by recombinant granulocyte/macrophage colony-stimulating factor and stimulated with the complement component C5a and FMLP as second signals. J Exp Med 167:1281–1295

Dahlgren C (1987) Difference in extracellular radical release after chemotactic factor and calcium ionophore activation of the oxygen radical-generating system in human neutrophils. Biochim Biophys Acta 930:33–38

Dahlgren C (1989) The calcium ionophore ionomycin can prime, but not activate, the reactive oxygen generating system in differentiated HL-60 cells. J Leukocyte Biol 46:15–24

Dahlgren C, Follin P (1990) Degranulation in human neutrophils primes the cells for subsequent responsiveness to the chemoattractant N-formylmethionyl-leucylphenylalanine but does not increase the sensitivity of the NADPH oxidase to an intracellular calcium rise. Biochim Biophys Acta 1052:42–46

Dahlgren C, Andersson T, Stendahl O (1987) Chemotactic factor binding and functional capacity: a comparison between human granulocytes and differentiated HL-60 cells. J Leukocyte Biol 42:245–252

Dahlgren C, Johansson A, Orselius K (1989) Difference in hydrogen peroxide release between human neutrophils and neutrophil cytoplasts following calcium ionophore activation. A role of the subcellular granule in activation of the NADPH oxidase in human neutrophils? Biochim Biophys Acta 1010:41–48

Dahm LJ, Roth RA (1990) Differential effects of lithocholate on rat neutrophil activation. J Leukocyte Biol 47:551–560

Dale MM, Penfield A (1984) Synergism between phorbol ester and A23187 in superoxide production by neutrophils. FEBS Lett 175:170–172

Dale MM, Penfield A (1985) Superoxide generation by either 1-oleoyl-2-acetylglycerol or A23187 in human neutrophils is enhanced by indomethacin. FEBS Lett 185:213–217

Dale MM, Penfield A (1987) Comparison of the effects of indomethacin, RHC80267 and R59022 on superoxide production by 1,oleoyl-2,acetyl glycerol and A23187 in human neutrophils. Br J Pharmacol 92:63–68

Dallegri F, Ballestrero A, Ottonello L, Patrone F (1989) Platelets as inhibitory cells in neutrophil-mediated cytolysis. J Lab Clin Med 114:502–509

Damerau B (1987) Biological activites of complement-derived peptides. Rev Physiol Biochem Pharmacol 108:151–206

Damiani G, Kiyotaki C, Soeller W, Sasada M, Peisach J, Bloom BR (1980) Macrophage variants in oxygen metabolism. J Exp Med 152:808–822

Damon M, Vial H, de Paulet AC, Godard P (1988) Phosphoinositide breakdown and superoxide anion release in formyl-peptide-stimulated alveolar macrophages. Comparison between quiescent and activated cells. FEBS Lett 239:169–173

Daniel-Issakani S, Spiegel AM, Strulovici B (1989) Lipopolysaccharide response is linked to the GTP-binding protein, G_{i2}, in the promonocytic cell line U937. J Biol Chem 264:20240–20247

Danley DL, Hilger AE (1981) Stimulation of oxidative metabolism in murine polymorphonuclear leukocytes by unopsonized fungal cells: evidence for a mannose-specific mechanism. J Immunol 127:551–556

Davies BD, Hoss W, Lin JP, Lionetti F (1982) Evidence for noncholinergic receptor on human phagocytic leukocytes. Mol Cell Biochem 44:23–31

Davies P, Allison AC, Ackerman J, Butterfield A, Williams S (1974) Asbestos induces selective release of lysosomal enzymes from mononuclear phagocytes. Nature 251:423–425

Davila DR, Edwards CK III, Arkins S, Simon J, Kelley KW (1990) Interferon-γ-induced priming for secretion of superoxide anion and tumor necrosis factor-α declines in macrophages from aged rats. FASEB J 4:2906–2911

Davis P, Miller CL, Russell AS (1982) Effects of gold compounds on the function of phagocytic cells. I. Suppression of phagocytosis and the generation of chemiluminescence by polymorphonuclear leukocytes. J Rheumatol [Suppl 8] 9:18–24

Davis P, Johnston C, Miller CL, Wong K (1983) Effects of gold compounds on the function of phagocytic cells. II. Inhibition of superoxide radical generation by tripeptide-activated polymorphonuclear leukocytes. Arthritis Rheum 26:82–86

Davis RJ, Czech MP (1985) Amiloride directly inhibits growth factor receptor tyrosine kinase activity. J Biol Chem 260:2543–2551

De Chaffoy de Courcelles D, Roevens P, van Belle H (1985) R 59 022, a diacylglycerol kinase inhibitor. Its effect on diacylglycerol and thrombin-induced C kinase activation in the intact platelet. J Biol Chem 260:15762–15770

DeChatelet LR, Shirley PS, Johnston RB Jr (1976) Effect of phorbol myristate acetate on the oxidative metabolism of human polymorphonuclear leukocytes. Blood 47:545–554

DeChatelet LR, Shirley PS, McPhail LC, Huntley CC, Muss HB, Bass DA (1977) Oxidative metabolism of the human eosinophil. Blood 50:525–533

Deguchi Y, Negoro S, Kishimoto S (1988) Age-related changes of heat shock protein gene transcription in human peripheral blood mononuclear cells. Biochem Biophys Res Commun 157:580–584

Deli E, Kiss Z, Wilson E, Lambeth JD, Kuo JF (1987) Immunocytochemical localization of protein kinase C in resting and activated human neutrophils. FEBS Lett 221:365–369

Dell KR, Severson DL (1989) Effect of cis-unsaturated fatty acids on aortic protein kinase C activity. Biochem J 258:171–175

Della Bianca V, Grzeskowiak M, de Togni P, Cassatella M, Rossi F (1985) Inhibition by verapamil of neutrophil responses to formylmethionylleucylphenylalanine and phorbol myristate acetate. Mechanisms involving Ca^{2+} changes, cyclic AMP and protein kinase C. Biochim Biophys Acta 845:223–236

Della Bianca V, Grzeskowiak M, Cassatella M, Zeni L, Rossi F (1986a) Phorbol 12,myristate 13,acetate potentiates the respiratory burst while inhibits phosphoinositide hydrolysis and calcium mobilization by formyl-methionyl-leucyl-phenylalanine in human neutrophils. Biochem Biophys Res Commun 135:556–565

Della Bianca V, De Togni P, Grzeskowiak M, Vicentini LM, Di Virgilio F (1986b) Cyclic AMP inhibition of phosphoinositide turnover in human neutrophils. Biochim Biophys Acta 836:441–447

Della Bianca V, Grzeskowiak M, Dusi S, Rossi F (1988) Fluoride can activate the respiratory burst independently of Ca^{2+}, stimulation of phosphoinositide turnover and protein kinase C translocation in primed human neutrophils. Biochem Biophys Res Commun 150:955–964

Della Bianca V, Grzeskowiak M, Rossi F (1990) Studies on molecular regulation of phagocytosis and activation of the NADPH oxidase in neutrophils. IgG- and C3b-mediated ingestion and associated respiratory burst independent of phospholipid turnover and Ca^{2+} transients. J Immunol 144:1411–1417

Deme D, Virion A, Hammou NA, Pommier J (1985) NADPH-dependent generation of H_2O_2 in a thyroid particulate fraction requires Ca^{2+}. FEBS Lett 186:107–110

Dent G, Ukena D, Chanez P, Sybrecht G, Barnes P (1989) Characterization of PAF receptors on human neutrophils using the specific antagonist, WEB 2086. FEBS Lett 244:365–368

Deres K, Schild H, Wiesmüller KH, Jung G, Rammensee HG (1989) In vivo priming of virus-specific cytotoxic T lymphocytes with synthetic lipopeptide vaccine. Nature 342:561–564

De Togni P, Cabrini G, Di Virgilio F (1984) Cyclic AMP inhibition of fMet-Leu-Phe-dependent metabolic responses is not due to its effects on cytosolic calcium. Biochem J 224:629–635

De Togni P, Bellavite P, Della Bianca V, Grzeskowiak M, Rossi F (1985a) Intensity and kinetics of the respiratory burst of human neutrophils in relation to receptor occupancy and rate of occupation by formylmethionylleucylphenylalanine. Biochim Biophys Acta 838:12–22

De Togni P, Della Bianca V, Grzeskowiak M, Di Virgilio F, Rossi F (1985b) Mechanism of desensitization of neutrophil response to N-formylmethionylleucylphenylalanine by slow rate of receptor occupancy. Studies on changes in Ca^{2+} concentration and phosphatidylinositol turnover. Biochim Biophys Acta 838:23–31

Dewald B, Baggiolini M (1985) Activation of NADPH oxidase in human neutrophils. Synergism between fMLP and the neutrophil products PAF and LTB_4. Biochem Biophys Res Commun 128:297–304

Dewald B, Baggiolini M (1986) Platelet-activating factor as a stimulus of exocytosis in human neutrophils. Biochim Biophys Acta 888:42–48

Dewald B, Baggiolini M, Curnutte JT, Babior BM (1979) Subcellular localization of the superoxide-forming enzyme in human neutrophils. J Clin Invest 63:21–29

Dewald B, Payne TG, Baggiolini M (1984) Activation of NADPH oxidase of human neutrophils. Potentiation of chemotactic peptice by a diacylglycerol. Biochem Biophys Res Commun 125:367–373

Dewald B, Thelen M, Baggiolini M (1988) Two transduction sequences are necessary for neutrophil activation by receptor agonists. J Biol Chem 263:16179–16184

Dewald B, Thelen M, Wymann MP, Baggiolini M (1989) Staurosporine inhibits the respiratory burst and induces exocytosis in human neutrophils. Biochem J 264:879–884

Diamant M, Henricks PAJ, Nijkamp FP, de Wied D (1989) β-Endorphin and related peptides supress phorbol myristate acetate-induced respiratory burst in human polymorphonuclear leukocytes. Life Sci 45:1537–1545

Diaz P, Jones DG, Kay AB (1979) Histamine-coated particles generate superoxide (O_2^-) and chemiluminescence in alveolar macrophages. Nature 278:454–456

Dickey BF, Pyun HY, Williamson KC, Navarro J (1987) Identification and purification of a novel G protein from neutrophils. FEBS Lett 219:289–292

Didsbury JR, Snyderman R (1987) Molecular cloning of a new human G protein. Evidence for two $G_{i\alpha}$-like protein families. FEBS Lett 219:259–263

Didsbury J, Weber RF, Bokoch GM, Evans T, Snyderman R (1989) rac, a novel ras-related family of proteins that are botulinum toxin substrates. J Biol Chem 264:16378–16382

Dieter P, Schulze-Specking A, Decker K (1986) Differential inhibition of prostaglandin and superoxide production by dexamethasone in primary cultures of rat Kupffer cells. Eur J Biochem 159:451–457

Dieter P, Schulze-Specking A, Karck U, Decker K (1987) Prostaglandin release but not superoxide production in rat Kupffer cells stimulated in vitro depends on Na^+/H^+ exchange. Eur J Biochem 170:201–206

Dieter P, Schulze-Specking A, Decker K (1988) Ca^{2+} requirement of prostanoid but not of superoxide production by rat Kupffer cells. Eur J Biochem 177:61–67

DiGregorio KA, Cilento EV, Lantz RC (1987) Measurement of superoxide release from single pulmonary alveolar macrophages. Am J Physiol 21:C677–C683

DiGregorio KA, Cilento EV, Lantz RC (1989) A kinetic model of superoxide production from single pulmonary alveolar macrophages. Am J Physiol 25:C405–C412

Dillon SB, Murray JJ, Uhing RJ, Snyderman R (1987) Regulation of inositol phospholipid and inositol phosphate metabolism in chemoattractant-activated human polymorphonuclear leukocytes. J Cell Biochem 35:345–359

Dillon SB, Verghese MW, Snyderman R (1988) Signal transduction in cells following binding of chemoattractants to membrane receptors. Virchows Archiv [B] 55:65–80

Dinarello CA, Mier JW (1987) Lymphokines. N Engl J Med 317:940–945

Dinauer MC, Orkin SH, Brown R, Jesaitis AJ, Parkos CA (1987) The glycoprotein encoded by the X-linked chronic granulomatous disease locus is a component of the neutrophil cytochrome b complex. Nature 327:717–720

Dinauer MC, Curnutte JT, Rosen H, Orkin SH (1989) A missense mutation in the neutrophil cytochrome b heavy chain in cytochrome-positive X-linked chronic granulomatous disease. J Clin Invest 84:2012–2016

Dinauer MC, Pierce EA, Bruns GAP, Curnutte JT, Orkin SH (1990) Human neutrophil cytochrome b light chain (p22-phox). Gene structure, chromosomal location, and mutations in cytochrome-negative autosomal recessive chronic granulomatous disease. J Clin Invest 86:1729–1737

Ding AH, Nathan CF (1987) Trace levels of bacterial lipopolysaccharide prevent interferon-γ or tumor necrosis factor-α from enhancing mouse peritoneal macrophage respiratory burst capacity. J Immunol 139:1971–1977

Ding AH, Nathan CF, Stuehr DJ (1988) Release of reactive nitrogen intermediates and reactive oxygen intermediates from mouse peritoneal macrophages. Comparison of activating cytokines and evidence for independent production. J Immunol 141:2407–2412

Di Perri T, Laghi Pasini F, Pasqui AL, Ceccatelli L, Capecchi PL, Orrico A (1984) Polymorphonuclear leukocyte as a model of Ca^{++} and Mg^{++}-dependent cellular activation, effect of flunarizine. Microcir Endothelium Lympathics 1:415–429

DiPersio J, Billing P, Kaufman S, Eghtesady P, Williams RE, Gasson JC (1988) Characterization of the human granulocyte-macrophage colony-stimulating factor receptor. J Biol Chem 263:1834–1841

Di Virgilio F, Lew DP, Pozzan T (1984) Protein kinase C activation of physiological processes in human neutrophils at vanishingly small cytosolic Ca^{2+} levels. Nature 310:691–693

Di Virgilio F, Vicentini LM, Treves S, Riz G, Pozzan T (1985) Inositol phosphate formation in fMet-Leu-Phe-stimulated human neutrophils does not require an increase in the cytosolic free Ca^{2+} concentration. Biochem J 229:361–367

Di Virgilio F, Lew DP, Andersson T, Pozzan T (1987) Plasma membrane potential modulates chemotactic peptide-stimulated cytosolic free Ca^{2+} changes in human neutrophils. J Biol Chem 262:4574–4579

Dolmatch B, Niedel J (1983) Formyl peptide chemotactic receptor. Evidence for an active proteolytic fragment. J Biol Chem 258:7570–7577

Donaldson K, Cullen RT (1984) Chemiluminescence of asbestos-activated macrophages. Br J Exp Pathol 65:81–90

Dorio RJ, Nelson J, Forman HJ (1987) A dual role for calcium in regulation of superoxide generation by stimulated rat alveolar macrophages. Biochim Biophys Acta 928:137–143

Dorio RJ, Hoek JB, Rubin E, Forman HJ (1988) Ethanol modulation of rat alveolar macrophage superoxide production. Biochem Pharmacol 37:3528–3531

Dougherty RW, Niedel JE (1986) Cytosolic calcium regulates phorbol diester binding affinity in intact phagocytes. J Biol Chem 261:4097–4100

Dougherty RW, Godfrey PP, Hoyle PC, Putney JW Jr, Freer RJ (1984) Secretagogue-induced phosphoinositide metabolism in human leukocytes. Biochem J 222:307–314

Doussiere J, Vignais PV (1985) Purification and properties of O_2^--generating oxidase from bovine polymorphonuclear neutrophils. Biochemistry 24:7231–7239

Doussiere J, Vignais PV (1988) Immunological properties of O_2^--generating oxidase from bovine neutrophils. FEBS Lett 234:362–366

Doussiere J, Laporte F, Vignais PV (1986) Photolabeling of an O_2^--generating protein in bovine polymorphonuclear neutrophils by an arylazido $NADP^+$ analog. Biochem Biophys Res Commun 139:85–93

Doussiere J, Pilloud MC, Vignais PV (1988) Activation of bovine neutrophil oxidase in a cell-free system. GTP-dependent formation of a complex between a cytosolic factor and a membrane protein. Biochem Biophys Res Commun 152:993–1001

Doussiere J, Pilloud MC, Vignais PV (1990) Cytosolic factors in bovine neutrophil oxidase activation. Partial purification and demonstration of translocation to a membrane fraction. Biochemistry 29:2225–2232

Downey GP, Chan CK, Grinstein S (1989) Actin assembly in electropermeabilized neutrophils: role of G-proteins. Biochem Biophys Res Commun 164:700–705

Drath DB, Kahan BD (1983) Alterations in rat pulmonary macrophage function by immunosuppressive agents cyclosporine, azathioprine, and prednisolone. Transplantation 35:588–592

Drath DB, Kahan BD (1984) Phagocytic function in response to immunosuppressive therapy. Arch Surg 119:156–160

Drath DB, Karnovsky ML (1975) Superoxide production by phagocytic leukocytes. J Exp Med 141:257–262

Driver AG, Kukoly GA, Bennett TE (1989) Expression of histamine H_1 receptors on cultured histiocytic lymphoma cells. Biochem Pharmacol 38:3083–3091

Dubyak GR, Cowen DS, Meuller LM (1988) Activation of inositol phospholipid breakdown in HL60 cells by P_2-purinergic receptors for extracellular ATP. Evidence for mediation by both pertussis toxin-sensitive and pertussis toxin-insensitive mechanisms. J Biol Chem 263:18108–18117

Dularay B, Elson CJ, Clements-Jewery S, Damais C, Lando D (1990) Recombinant human interleukin-1 beta primes human polymorphonuclear leukocytes for stimulus-induced myeloperoxidase release. J Leukocyte Biol 47:158–163

Dulis BH, Gordon MA, Wilson IB (1979) Identification of muscarinic binding sites in human neutrophils by direct binding. Mol Pharmacol 15:28–34

Duque RE, Phan SH, Sulavik MC, Ward PA (1983) Inhibition by tosyl-L-phenylalanyl chloromethyl ketone of membrane potential changes in rat neutrophils. Correlation with the inhibition of biological activity. J Biol Chem 258:8123–8128

Dusenbery KE, Mendiola JR, Skubitz KM (1988) Evidence for ecto-protein kinase activity on the surface of human neutrophils. Biochem Biophys Res Commun 153:7–13

Eckardt KU, Eckardt H, Harber MJ, Asscher AW (1986) Analysis of polymorphonuclear leukocyte respiratory burst activity in uremic patients using whole-blood chemiluminescence. Nephron 43:274–278

Eckle I, Kolb G, Havemann K (1990) Inhibition of neutrophil oxidative burst by elastase-generated IgG fragments. Biol Chem Hoppe Seyler 371:69–77

Eckstein F (1985) Nucleoside phosphorothioates. Annu Rev Biochem 54:367–402

Eckstein F, Cassel D, Levkovitz H, Lowe M, Selinger Z (1979) Guanosine 5'-O-(2-thiodiphosphate). An inhibitor of adenylate cyclase stimulation by guanine nucleotides and fluoride ions. J Biol Chem 254:9829–9834

Edwards CK III, Ghiasuddin SM, Schepper JM, Yunger LM, Kelley KW (1988) A newly defined property of somatotropin: priming of macrophages for production of super-oxide anion. Science 239:769–771

Edwards SW, Lloyd D (1988) The relationship between superoxide generation, cytochrome b and oxygen in activated neutrophils. FEBS Lett 227:39–42

Edwards SW, Swan TF (1986) Regulation of superoxide generation by myeloperoxidase during the respiratory burst of human neutrophils. Biochem J 237:601–604

Edwards SW, Hallett MB, Lloyd D, Campbell AK (1983) Decrease in apparent K_m for oxygen after stimulation of respiration of rat polymorphonuclear leukocytes. FEBS Lett 161:60–64

Edwards SW, Holden CS, Humprehys JM, Hart CA (1989) Granulocyte-macrophage colony-stimulating factor (GM-CSF) primes the respiratory burst and stimulates protein biosynthesis in human neutrophils. FEBS Lett 256:62–66

Ek B, Heldin CH (1982) Characterization of a tyrosine-specific kinase activity in human fibroblast membranes stimulated by platelet-derived growth factor. J Biol Chem 257:10486–10492

Eklund EA, Gabig TG (1990) Purification and characterization of a lipid thiobis ester from human neutrophil cytosol that reversibly deactivates the O_2^--generating NADPH oxidase. J Biol Chem 265:8426–8430

Elferink JGR (1979) Chlorpromazine inhibits phagocytosis and exocytosis in rabbit polymorphonuclear leukocytes. Biochem Pharmacol 28:965–968

Elferink JGR (1987) The effect of apomorphine on exocytosis and metabolic burst of polymorphonuclear leukocytes. Br J Pharmacol 92:909–913

Elferink JGR (1988) Guanine nucleotides inhibit poly-L-arginine-induced membrane damage in polymorphonuclear leukocytes. Experientia 44:1016–1017

Elferink JGR, Deierkauf M (1984) The effect of verapamil and other calcium antagonists on chemotaxis of polymorphonuclear leukocytes. Biochem Pharmacol 33:35–39

Elferink JGR, Deierkauf M (1985) Involvement of intracellular Ca^{2+} and metabolic burst by neutrophils: the use of antagonists of intracellular Ca^{2+}. Res Commun Chem Pathol Pharmacol 50:67–81

Elferink JGR, Deierkauf M (1988) Felodipine-induced inhibition of polymorphonuclear leukocyte functions. Biochem Pharmacol 37:503–509

Elferink JGR, Deierkauf M (1989a) Activation of the respiratory burst in rabbit neutrophils by a stable guanine nucleotide. In: Biology of cellular transducing signals '89 JY Vanderhoek (Ed) Washington 9th International Washington Spring Symposium. p 279

Elferink JGR, Deierkauf M (1989b) Suppressive action of cobalt on exocytosis and respiratory burst in neutrophils. Am J Physiol 257:C859–864

Elferink JGR, Ebbenhout JL (1988) Asbestos-induced activation of the respiratory burst in rabbit neutrophils. Res Commun Chem Pathol Pharmacol 60:201–211

Elferink JGR, Hoogendijk AM, Riemersma JC (1982) Inhibition of some polymor-phonuclear leukocyte functions by ethacrynic acid. Biochem Pharmacol 31:443–448

Elferink JGR, Boonen GJJC, de Koster BM, Deierkauf M (1990a) Induction of super-oxide production in rabbit neutrophils by a stable analogue of GTP. Agents Actions 29:32–34

Elferink JGR, de Koster BM, Boonen GJJC (1990b) Cytochalasin-induced nitroblue tetrazolium dye (NBT) reduction in polymorphonuclear leukocytes. Res Commun Chem Pathol Pharmacol 68:175–187

Elliott KRF, Leonard EJ (1989) Interactions of formylmethionyl-leucyl-phenyl-alanine, adenosine, and phosphodiesterase inhibitors in human monocytes. Effects on superoxide release, inositol phosphates and cAMP. FEBS Lett 254:94–98

Elliott KRF, Miller PJ, Stevenson HC, Leonard EJ (1986) Synergistic action of adenosine and fMet-Leu-Phe in raising cAMP content of purified human monocytes. Biochem Biophys Res Commun 138:1376–1382

Ellis JA, Mayer SJ, Jones OTG (1988) The effect of the NADPH oxidase inhibitor diphenyleneiodonium on aerobic and anaerobic microbicidal activities of human neutrophils. Biochem J 251:887–891

Ellis JA, Cross AR, Jones OTG (1989) Studies on the electron-transfer mechanism of the human neutrophil NADPH oxidase. Biochem J 262:575–579

Emilsson A, Sundler R (1984) Differential activation of phosphatidylinositol deacylation and a pathway via diphosphoinositide in macrophages responding to zymosan and ionophore A23187. J Biol Chem 259:3111–3116

Engels W, Grauls G, Lemmens PJ, Mullers WJ, Bruggeman CA (1989) Influence of a cytomegalovirus infection on functions and arachidonic acid metabolism of rat peritoneal macrophages. J Leukocyte Biol 45:466–473

English D, Roloff JS, Lukens JN (1981a) Regulation of human polymorphonuclear leukocyte superoxide release by cellular responses to chemotactic peptides. J Immunol 126:165–171

English D, Roloff JS, Lukens JN (1981b) Chemotactic factor enhancement of superoxide release from flouride and phorbol myristate acetate stimulated neutrophils. Blood 58:129–134

English D, Schell M, Siakotos A, Gabig TG (1986) Reversible activation of the neutrophil superoxide generating system by hexachlorocyclohexane: correlation with effects on a subcellular superoxide-generating fraction. J Immunol 137:283–290

English D, Debono DJ, Gabig TG (1987) Relationship of phosphatidylinositol bisphosphate hydrolysis to calcium mobilization and functional activation in fluoride-treated neutrophils. J Clin Invest 80:145–153

English D, Broxmeyer HE, Gabig TG, Akard LP, Williams DE, Hoffman R (1988) Temporal adaptation of neutrophil oxidative responsiveness to n-formylmethionyl-leucyl-phenylalanine. Acceleration by granulocyte-macrophage colony stimulating factor. J Immunol 141:2400–2406

English D, Rizzo MT, Tricot G, Hoffman R (1989) Involvement of guanine nucleotides in superoxide release by flouride-treated neutrophils. Implications for a role of a guanine nucleotide regulatory protein. J Immunol 143:1685–1691

Entman ML, Youker K, Shappell SB, Siegel C, Rothlein R, Dreyer WJ, Schmalstieg FC, Smith CW (1990) Neutrophil adherence to isolated adult canine myocytes. Evidence for a CD18-dependent mechanism. J Clin Invest 85:1497–1506

Eppell BA, Newell AM, Brown EJ (1989) Adenosine receptors are expressed during differentiation of monocytes to macrophages in vitro. Implications for regulation of phagocytosis. J Immunol 143:4141–4145

Ervens J, Seifert R (1991) Differential modulation by N^4, 2'-O-dibutyryl cytidine 3':5'-cyclic monophosphate of neutrophil activation. Biochem Biophys Res Commun 174:258–267

Ervens J, Schultz G, Seifert R (1991) Differential regulation of chemoattractant-induced superoxide formation in human neutrophils by cell-permeant analogues of cyclic AMP and cyclic GMP. Biochem Soc Trans 19:59–64

Evans AT, Sharma P, Ryves WJ, Evans FJ (1990) TPA and resiniferatoxin-mediated activation of NADPH oxidase. A possible role for Rx-kinase augmentation of PKC. FEBS Lett 267:253–256

Exton JH (1988) Mechanisms of action of calcium-mobilizing agonists: some variations on a young theme. FASEB J 2:2670–2676

Ezekowitz RAB, Sim RB, MacPherson GG, Gordon S (1985) Interaction of human monocytes, macrophages, and polymorphonuclear leukocytes with zymosan in vitro. J Clin Invest 76:2368–2376

Ezekowitz RAB, Orkin SH, Newburger PE (1987) Recombinant interferon gamma augments phagocyte superoxide production and X-chronic granulomatous disease gene expression in X-linked variant chronic granulomatous disease. J Clin Invest 80:1009–1016

Ezekowitz RAB, Dinauer MC, Jaffe HS, Orkin SH, Newburger PE (1988) Partial correction of the phagocyte defect in patients with X-linked chronic granulomatous disease by subcutaneous interferon gamma. N Engl J Med 319:146–151

Ezekowitz RA, Sieff CA, Dinauer MC, Nathan DG, Orkin SH, Newburger PE (1990) Restoration of phagocyte function by interferon-γ in X-linked chronic granulomatous disease occurs at the level of a progenitor cell. Blood 76:2443–2448

Falk RJ, Terrell RS, Charles LA, Jennette JC (1990) Anti-neutrophil cytoplasmic autoantibodies induce neutrophils to degranulate and produce oxygen radicals in vitro. Proc Natl Acad Sci USA 87:4115–4119

Fantone JC, Kinnes DA (1983) Prostaglandin E_1 and prostaglandin I_2 modulation of superoxide production by human neutrophils. Biochem Biophys Res Commun 113:506–512

Fantone JC, Marasco WA, Elgas LJ, Ward PA (1983) Anti-inflammatory effects of prostaglandin E_1: in vivo modulation of the formyl peptide chemotactic receptor on the rat neutrophil. J Immunol 130:1495–1497

Fantone JC, Duque RE, Phan SH (1984) Prostaglandin modulation of N-formyl-methionylleucylphenylalanine-induced transmembrane potential changes in rat neutrophils. Biochim Biophys Acta 804:265–274

Farkas G, Enyedi A, Sarkadi B, Gárdos G, Nagy Z, Faragó A (1984) Cyclic AMP-dependent protein kinase stimulates the phosphorylation of phosphatidylinositol to phosphatidylinositol-4-monophosphate in a plasma membrane preparation from pig granulocytes. Biochem Biophys Res Commun 124:871–876

Fearon DT (1980) Identification of the membrane glycoprotein that is the C3b receptor of the human erythrocyte, polymorphonuclear leukocyte, B lymphocyte, and monocyte. J Exp Med 152:20–30

Fearon DT, Collins LA (1983) Increased expression of C3b receptors on polymorphonuclear leukocytes induced by chemotactic factors and by purification procedures. J Immunol 130:370–375

Fehr J, Moser R, Leppert D, Groscurth P (1985) Antiadhesive properties of biological surfaces are protective against stimulated granulocytes. J Clin Invest 76:535–542

Feister AJ, Browder B, Willis HE, Mohanakumar T, Ruddy S (1988) Pertussis toxin inhibits human neutrophil responses mediated by the 42-kilodalton IgG Fc receptor. J Immunol 141:228–233

Fel AOS, Nathan CF, Cohn ZA (1987) Hydrogen peroxide release by alveolar macrophages from sarcoid patients and by alveolar macrophages from normals after exposure to recombinant interferons α, β, and γ and 1,25-dihydroxyvitamin D_3. J Clin Invest 80:381–386

Feltner DE, Smith RH, Marasco WA (1986) Characterization of the plasma membrane bound GTPase from rabbit neutrophils. I. Evidence for an N_i-like protein coupled to the formyl peptide, C5a, and leukotriene B_4 chemotaxis receptors. J Immunol 137:1961–1970

Ferguson KM, Higashijima T, Smigel MD, Gilman AG (1986) The influence of bound GDP on the kinetics of guanine nucleotides binding to G proteins. J Biol Chem 261:7393–7399

Ferrante A, Nandoskar M, Bates EJ, Goh DHB, Beard LJ (1988) Tumor necrosis factor beta (lymphotoxin) inhibits locomotion and stimulates the respiratory burst and degranulation of neutrophils. Immunology 63:507–512

Ferriola PC, Cody V, Middleton E Jr (1989) Protein kinase C inhibition by plant flavonoids. Kinetic mechanisms and structure-activity relationships. Biochem Pharmacol 38:1617–1624

Feuerstein N, Cooper HL (1984) Rapid phosphorylation-dephosphorylation of specific proteins induced by phorbol ester in HL-60 cells. Further characterization of the phosphorylation of 17-kilodalton and 27-kilodalton proteins in myeloid leukemic cells and human monocytes. J Biol Chem 259:2782–2788

Filep J, Földes-Filep E (1989) Effects of C-reactive protein on human neutrophil granulocytes challenged with N-formyl-methionyl-leucyl-phenylalanine and platelet-activating factor. Life Sci 44:517–524

Finch GL, Fisher GL, Hayes TL, Golde DW (1982) Surface morphology and functional studies of human alveolar macrophages from cigarette smokers and nonsmokers J Reticuloendothel Soc 32:1–23

Finkel TH, Pabst MJ, Suzuki H, Guthrie LA, Forehand JR, Phillips WA, Johnston RB Jr (1987) Priming of neutrophils and macrophages for enhanced release of superoxide anion by the calcium ionophore ionomycin. Implications for regulation of the respiratory burst. J Biol Chem 262:12589–12596

Fisher DB, Mueller GC (1971) Gamma-hexachlorocylohexane inhibits the initiation of lymphocyte growth by phytohemagglutinin. Biochem Pharmacol 20:2515–2518

Fisher GL, McNeill KL, Finch GL, Wilson FD, Golde DW (1982) Functional evaluation of lung macrophages from cigarette smokers and nonsmokers. J Reticuloendothel Soc 32:311–321

Flanagan MD, Lin S (1980) Cytochalasins block actin filament elongation by binding to high affinity sites associated with F-actin. J Biol Chem 255:835–838

Fleischmann J, Golde DW, Weisbart RH, Gasson JC (1986) Granulocyte-macrophage colony-stimulating factor enhances phagocytosis of bacteria by human neutrophils. Blood 68:708–711

Fleit HB, Wright SD, Unkeless JC (1982) Human neutrophil Fcγ receptor distribution and structure. Proc Natl Acad Sci USA 79:3275–3279

Fletcher MP (1986) Modulation of the heterogeneous membrane potential response of neutrophils to N-formyl-methionyl-phenylalanine (FMLP) by leukotriene B_4: evidence for cell recruitment. J Immunol 136:4213–4219

Fletcher MP, Gallin JI (1983) Human neutrophils contain an intracellular pool of putative receptors for the chemoattractant N-formyl-methionyl-leucyl-phenylalanine. Blood 62:792–799

Fletcher MP, Seligmann BE (1986) PMN heterogeneity: long-term stability of fluorescent membrane potential responses to the chemoattractant N-formyl-methionyl-leucyl-phenylalanine in healthy adults and correlation with respiratory burst activity. Blood 68:611–618

Fliss H, Weissbach H, Brot N (1983) Oxidation of methionine residues in proteins of activated human neutrophils. Proc Natl Acad Sci USA 80:7160–7164

Foerder CA, Shapiro BM (1977) Release of ovoperoxidase from sea urchin eggs hardens the fertilization membrane with tyrosine crosslinks. Proc Natl Acad Sci USA 74:4214–4218

Foerder CA, Klebanoff SJ, Shapiro BM (1978) Hydrogen peroxide production and chemiluminescence, and the respiratory burst of fertilization: interrelated events in early sea urchin development. Proc Natl Acad Sci USA 75:3183–3187

Follin P, Dahlgren C (1990) Altered O_2^-/H_2O_2 production ratio by in vitro and in vivo primed human neutrophils. Biochem Biophys Res Commun 167:970–976

Ford DA, Miyake R, Glaser PE, Gross RW (1989) Activation of protein kinase C by naturally occurring ether-linked diglycerides. J Biol Chem 264:13818–13824

Ford-Hutchinson AW (1982) Aggregation of rat neutrophils by nucleotide triphosphates. Br J Pharmacol 76:367–371

Ford-Hutchinson AW, Bray MA, Doig MV, Shipley ME, Smith MJH (1980) Leukotriene B, a potent chemokinetic and aggregating substance released from polymorphonuclear leukocytes. Nature 286:264–265

Forehand JR, Pabst MJ, Phillips WA, Johnston RB Jr (1989) Lipopolysaccharide priming of human neutrophils for an enhanced respiratory burst. Role of intracellular free calcium. J Clin Invest 83:74–83

Foreman JC (1987) Peptides and neurogenic inflammation. Br Med Bull 43:386–400

Foreman JC, Jordan CC (1984) Neurogenic inflammation. Trends Pharmacol Sci 5:116–119

Foris G, Medgyesi GA, Hauck M (1986) Bidirectional effect of Met-enkephalin on macrophage effector functions. Mol Cell Biochem 69:127–137

Forman HJ, Thomas MJ (1986) Oxidant production and bactericidal activity of phagocytes. Annu Rev Physiol 48:669–680

Forman HJ, Nelson J, Fisher AB (1980) Rat alveolar macrophages require NADPH for superoxide production in the respiratory burst. Effect of NADPH depletion by paraquat. J Biol Chem 259:9879–9883

Forsberg EJ, Feuerstein G, Shohami E, Pollard HB (1987) Adenosine triphosphate stimulates inositol phospholipid metabolism and prostacyclin formation in adrenal medullary endothelial cells by means of P_2-purinergic receptors. Proc Natl Acad Sci USA 84:5630–5634

Foster JS, Moore RN (1987) Dynorphin and related opioid peptides enhance tumoricidal activity mediated by murine peritoneal macrophages. J Leukocyte Biol 42:171–174

Franson RC, Eisen D, Jesse R, Lanni C (1980) Inhibition of highly purified mammalian phospholipase A_2 by non-steroidal anti-inflammatory agents. Modulation by calcium ions. Biochem J 186:633–636

French JK, Hurst NP, Zalewski PD, Valente L, Forbes IJ (1987) Calcium ionophore A23187 enhances human neutrophil superoxide release, stimulated by phorbol dibutyrate, by converting phorbol ester receptors from a low- to high-affinity state. FEBS Lett 212:242–246

Freyer DR, Boxer LA, Axtell RA, Todd RF III (1988) Stimulation of human neutrophil adhesive properties by adenine nucleotides. J Immunol 141:580–586

Fridkin M, Gottlieb P (1981) Tuftsin, Thr-Lys-Pro-Arg. Anatomy of an immunologically active peptide. Mol Cell Biochem 41:73–97

Fridkin M, Stabinsky Y, Zakuth V, Spirer Z (1977) Tuftsin and some analogs. Synthesis and interaction with human polymorphonuclear leukocytes. Biochim Biophys Acta 496:203–211

Fridovich I (1986) Biological effects of the superoxide radical. Arch Biochem Biophys 247:1–11

Fridovich I (1989) Superoxide dismutases. An adaptation to a paramagnetic gas. J Biol Chem 264:7761–7764

Friedenberg WR, Marx JJ Jr, Hansotia P, Gottschalk PG (1986) Granulocyte dysfunction and myotonic dystrophy. J Neurol Sci 73:1–10

Froscio M, Murray AW, Hurst NP (1989) Inhibition of protein kinase C by the antirheumatic drug auranofin. Biochem Pharmacol 38:2087–2089

Fujimoto S, Smith RM, Curnutte JT, Babior BM (1989) Evidence that activation of the respiratory burst oxidase in a cell-free system from human neutrophils is accomplished in part through an alteration of the oxidase-related 67-kDa cytosolic protein. J Biol Chem 264:21629–21632

Fujita I, Irita K, Takeshige K, Minakami S (1984) Diacylglycerol, 1-oleoyl-2-acetylglycerol, stimulates superoxide-generation from human neutrophils. Biochem Biophys Res Commun 120:318–324

Fujita I, Takeshige K, Minakami S (1986) Inhibition of neutrophil superoxide formation by 1-(5-isoquinolinesulfonyl)-2-methylpiperazine (H-7), an inhibitor of protein kinase-C. Biochem Pharmacol 35:4555–4562

Fujita I, Takeshige K, Minakami S (1987) Characterization of the NADPH-dependent superoxide production activated by sodium dodecyl sulfate in a cell-free system of pig neutrophils. Biochim Biophys Acta 931:41–48

Fukuhara Y, Ise Y, Kakinuma K (1988) Immunological studies on the respiratory burst oxidase of pig blood neutrophils. FEBS Lett 229:150–156

Fülöp T Jr, Jacob MP, Varga Z, Fóris G, Leövey A, Robert L (1986) Effect of elastin peptides on human monocytes: Ca^{2+} mobilization, stimulation of respiratory burst and enzyme secretion. Biochem Biophys Res Commun 141:92–98

Fülöp T Jr, Varga Z, Nagy JT, Fóris G (1988) Studies on opsonized zymosan, FMLP, carbachol, PMA and A_{23187} stimulated respiratory burst of human PMNLs. Biochem Int 17:419–426

Fülöp T Jr, Varga Z, Csongor J, Fóris G, Leövey A (1989) Age related impairment in phosphatidylinositol breakdown of polymorphonuclear granulocytes. FEBS Lett 245:249–252

Furuta R, Junichi Y, Kotani H, Sakamoto F, Fukui T, Matsui Y, Sohmura Y, Yamada M, Yoshimura T, Larsen CG, Oppenheim JJ, Matsushima K (1989) Production and characterization of recombinant human neutrophil chemotactic factor. J Biochem 106:436–441

Gabig TG (1983) The NADPH-dependent O_2^--generating oxidase from human neutrophils. Identification of a flavoprotein component that is deficient in a patient with chronic granulomatous disease. J Biol Chem 258:6352–6356

Gabig TG, Babior BM (1979) The O_2^--forming oxidase responsible for the respiratory burst in human neutrophils. Properties of the solubilized enzyme. J Biol Chem 254:9070–9074

Gabig TG, English D (1986) The arachidonate-dependent activator of the subcellular neutrophil NADPH oxidase acts independently of phosphorylation. Clin Res 34:657A

Gabig TG, Lefker BA (1984a) Deficient flavoprotein component of the NADPH-dependent O_2^--generating oxidase in the neutrophils from three male patients with chronic granulomatous disease. J Clin Invest 73:701–705

Gabig TG, Lefker BA (1984b) Catalytic properties of the resolved flavoprotein and cytochrome B components of the NADPH dependent O_2^- generating oxidase from human neutrophils. Biochem Biophys Res Commun 118:430–436

Gabig TG, Lefker BA (1985) Activation of human neutrophil NADPH oxidase results in coupling of electron carrier function between ubiquinone-10 and cytochrome b_{559}. J Biol Chem 260:3991–3995

Gabig TG, Kipnes RS, Babior BM (1978) Solubilization of the O_2^--forming activity responsible for the respiratory burst in human neutrophils. J Biol Chem 253:6663–6665

Gabig TG, Schervish EW, Santinga JT (1982) Functional relationship of the cytochrome b to the superoxide-generating oxidase of human neutrophils. J Biol Chem 257:4114–4119

Gabig TG, English D, Akard LP, Schell MJ (1987) Regulation of neutrophil NADPH oxidase activation in a cell-free system by guanine nucleotides and fluoride. Evidence for participation of a pertussis and cholera toxin-insensitive G protein. J Biol Chem 262:1685–1690

Gabig TG, Eklund EA, Potter GB, Dykes JR II (1990) A neutrophil GTP-binding protein that regulats cellfree NADPH oxidase activation is located in the cytosolic fraction. J Immunol 145:945–951

Gabler WL (1990) Potential roles of Mg^{2+} and Ca^{2+} in NADPH oxidase dependent superoxide anion synthesis by human neutrophils. Res Commun Chem Pathol Pharmacol 70:213–226

Gabler WL, Creamer HR, Bullock WW (1989) Fluoride activation of neutrophils: similarities to formylmethionyl-leucyl-phenylalanine. Inflammation 13:47–58

Gagnon L, Filion LG, Dubois C, Rola-Pleszczynski M (1989) Leukotrienes and macrophage activation: augmented cytotoxic activity and enhanced interleukin 1, tumor necrosis factor and hydrogen peroxide production. Agents Actions 26:141–147

Gallin JI, Seligmann BE (1984) Mobilization and adaptation of human neutrophil chemoattractant fMet-Leu-Phe receptors. Fed Proc 43:2732–2736

Garbers DL (1989) Molecular basis of fertilization. Annu Rev Biochem 58:719–742

Garcia RC, Segal AW (1988) Phosphorylation of the subunits of cytochrome $b_{.245}$ upon triggering of the respiratory burst of human neutrophils and macrophages. Biochem J 252:901–904

García-Castro I, Mato JM, Vasanthakumar G, Wiesmann WP, Schiffmann E, Chiang PK (1983) Paradoxical effects of adenosine on neutrophil chemotaxis. J Biol Chem 258:4345–4349

García Gil M, Alonso F, Alvarez Chiva V, Sanchez Crespo M, Mato JM (1982) Phospholipid turnover during phagocytosis in human polymorphonuclear leucocytes. Biochem J 206:67–72

Gardner JP, Melnick DA, Malech HL (1986) Characterization of the formyl peptide chemotactic receptor appearing at the phagocytic cell surface after exposure to phorbol myristate acetate. J Immunol 136:1400–1405

Garotta G, Talmadge KW, Pink JRL, Dewald B, Baggiolini M (1986) Functional antagonists between type I and type II interferons on human macrophages. Biochem Biophys Res Commun 140:948–954

Gasson JC, Weisbart RH, Kaufman SE, Clark SC, Hewick RM, Wong GG, Golde DW (1984) Purified human granulocyte-macrophage colony-stimulating factor: direct action on neutrophils. Science 226:1339–1342

Gasson JC, Kaufman SE, Weisbart RH, Tomonaga M, Golde DW (1986) High-affinity binding of granulocyte-macrophage colony-stimulating factor to normal and leukemic human myeloid cells. Proc Natl Acad Sci USA 83:669–673

Gaudry M, Combadiere C, Marquetty C, Sheibani A, El Benna J, Hakim J (1990) Dissimilarities in superoxide anion production by human neutrophils stimulated by phorbol myristate acetate or phorbol dibutyrate. Immunopharmacology 19:23–32

Gaut JR, Carchman RA (1987) A correlation between phorbol diester-induced protein phosphorylation and superoxide anion generation in HL-60 cells during granulocytic maturation. J Biol Chem 262:826–834

Gavioli R, Spisani S, Giuliani A, Traniello S (1987) Protein kinase C mediates human neutrophil cytotoxicity. Biochem Biophys Res Commun 148:1290–1294

Gavison R, Bar-Shavit Z (1989) Impaired macrophage activation in vitamin D_3 deficiency: differential in vitro effects of 1,25-dihydroxyvitamin D_3 on mouse peritoneal macrophage functions. J Immunol 143:3686–3690

Gay JC, Stitt ES (1990) Chemotactic peptide enhancement of phorbol ester-induced protein kinase C activity in human neutrophils. J Leukocyte Biol 47:49–59

Gay JC, Beckman K, Brash AR, Oates JA, Lukens JN (1984) Enhancement of chemotactic factor-stimulated neutrophil oxidative metabolism by leukotriene B_4. Blood 64:780–785

Gay JC, Beckman K, Zaboy KA, Lukens JN (1986) Modulation of neutrophil oxidative responses to soluble stimuli by platelet-activating factor. Blood 67:931–936

Gbarah A, Mhashilkar AM, Boner G, Sharon N (1989) Involvement of protein kinase C in activation of human granulocytes and peritoneal macrophages by type 1 fimbriated (mannose specific) *Escherichia coli*. Biochem Biophys Res Commun 165:1243–1249

Geissler K, Tricot G, Grimm G, Siostrzonek P, Broxmeyer H (1989) Recombinant human colony factor-granulocyte/macrophage and -granulocyte, but not macrophage induce development of a respiratory burst in primary human myeloid leukemic cells in vitro. Blut 59:226–230

Gelas P, Ribbes G, Record M, Terce F, Chap H (189) Differential activation by fMet-Leu-Phe and phobol ester of a plasma membrane phosphatidylcholine-specific phospholipase D in human neutrophil. FEBS Lett 251:213–218

Gennaro R, Pozzan T, Romeo D (1984) Monitoring of cytosolic free Ca^{2+} in C5a-stimulated neutrophils: loss of receptor-modulated Ca^{2+} stores and Ca^{2+} uptake in granule-free cytoplasts. Proc Natl Acad Sci USA 81:1416–1420

Gennaro R, Florio C, Romeo D (1985) Activation of protein kinase C in neutrophil cytoplasts. Localization of protein substrates and possible relationship with stimulus-response coupling. FEBS Lett 180:185–190

Genarro R, Florio C, Romeo D (1986) Co-activation of protein kinase C and NADPH oxidase in the plasma membrane of neutrophil cytoplasts. Biochem Biophys Res Commun 134:305–312

Gentile F, Raptis A, Knipling LG, Wolff J (1988) *Bordetella pertussis* adenylate cyclase. Penetration into host cells. Eur J Biochem 173:447–453

Georgilis K, Schaefer C, Dinarello CA, Klempner MS (1987) Human recombinant interleukin 1β has no effect on intracellular calcium or on functional responses of human neutrophils. J Immunol 138:3403–3407

Gerard C, McPhail LC, Marfat A, Stimler-Gerard NP, Bass DA, McCall CE (1986) Role of protein kinases in stimulation of human polymorphonuclear leukocyte oxidative metabolism by various agonists. Differential effects of a novel protein kinase inhibitor. J Clin Invest 77:61–65

Gerard NP, Hodges MK, Drazen JM, Weller PF, Gerard C (1989) Characterization of a receptor for C5a anaphylatoxin on human eosinophils. J Biol Chem 264:1760–1766

Gespach C, Abita JP (1982) Human polymorphonuclear neutrophils. Pharmacological characterization of histamine receptors mediating the elevation of cyclic AMP. Mol Pharmacol 21:78–85

Gespach C, Saal F, Cost H, Abita JP (1982) Identification and characterization of surface receptors for histamine in the human promyelocytic leukemia cell line HL-60. Comparison with human peripheral neutrophils. Mol Pharmacol 22:547–553

Gespach C, Cost H, Abita JP (1985) Histamine H_2 receptor activity during the differentiation of the human monocytic-like cell line U-937. Comparison with prostaglandins and isoproterenol. FEBS Lett 184:207–213

Gespach C, Courillon-Mallet A, Launay JM, Cost H, Abita JP (1986) Histamine H_2 receptor activity and histamine metabolism in human U-937 monocyte-like cells and human peripheral monocytes. Agents Actions 18:124–128

Ghezzi P, Erroi A, Acero R, Salmona M, Mantovani A (1987) Defensive production of reactive oxygen intermediates by tumor-associated macrophages exposed phorbol ester. J Leukocyte Biol 42:84–90

Gierschik P, Jakobs KH (1987) Receptor-mediated ADP-ribosylation of phospholipase C-stimulating G protein. FEBS Lett 224:219–223

Gierschik P, Falloon J, Milligan G, Pines M, Gallin JI, Spiegel A (1986) Immunochemical evidence for a novel pertussis toxin substrate in human neutrophils. J Biol Chem 261:8058–8062

Gierschik P, Sidiropoulos D, Spiegel A, Jakobs KH (1987) Purification and immunochemical characterization of the major pertussis-toxin-sensitive guanine-nucleotide-binding protein of bovine-neutrophil membranes. Eur J Biochem 165:185–194

Gierschik P, Steisslinger M, Sidiropoulos D, Herrmann E, Jakobs KH (1989a) Dual Mg^{2+} control of formyl-peptide-receptor-G-protein interaction in HL 60 cells. Evidence that the low-agonist-affinity receptor interacts with and activates the G-protein. Eur J Biochem 183:97–105

Gierschik P, Sidiropoulos D, Jakobs KH (1989b) Two distinct G_i-proteins mediate formyl peptide receptor signal transduction in human leukemia (HL-60) cells. J Biol Chem 264:21470–21473

Gilman AG (1987) G proteins: transducers of receptor-generated signals. Annu Rev Biochem 56:615–649

Ginis I, Tauber AI (1990) Activation mechanisms of adherent human neutrophils. Blood 76:1233–1239

Ginsburg I, Ward PA, Varani J (1989) Lysophosphatides enhance superoxide responses of stimulated human neutrophils. Inflammation 13:163–174

Ginsel LA, Onderwater JJM, Fransen JAM, Verhoeven AJ, Roos D (1990) Localization of the Low-M_r subunit of cytochrome b_{558} in human blood phagocytes by immunoelectron microscopy. Blood 76:2105–2116

Glass GA, DeLisle DM, DeTogni P, Gabig TG, Magee BH, Markert M, Babior BM (1986) The respiratory burst oxidase of human neutrophils. Further studies of the purified enzyme. J Biol Chem 261:13247–13251

Gluck WL, Weinberg JB (1987) $1\alpha,25$ Dihydroxyvitamin D_3 and mononuclear phagocytes: enhancement of mouse macrophage and human monocyte hydrogen peroxide production without alteration of tumor cytolysis. J Leukocyte Biol 42:498–503

Godchaux W III, Zimmerman WF (1979) Membrane-dependent guanine nucleotide binding and GTPase activities of soluble protein from bovine rod cell outer segments. J Biol Chem 254:7874–7884

Goetzl EJ, Brash AR, Tauber AI, Oates JA, Hubbard WC (1980) Modulation of human neutrophil function by monohydroxy-eicosatetraenoic acids. Immunology 39:491–501

Goffeau A, Pedersen PL, Lehninger AL (1967) The kinetics and inhibition of the adenosine diphosphate-adenosine triphosphate exchange catalyzed by a purified mitochondrial nucleoside diphosphokinase. J Biol Chem 242:1845–1853

Goldberg ND, Haddox MK (1977) Cyclic GMP metabolism and involvement in biological regulation. Annu Rev Biochem 46:823–896

Goldman DW, Goetzl EJ (1982) Specific binding of leukotriene B_4 to receptors on human polymorphonuclear leukocytes. J Immunol 129:1600–1604

Goldman DW, Gifford LA, Olson DM, Goetzl EJ (1985a) Transduction by leukotriene B_4 receptors of increases in cytosolic calcium in human polymorphonuclear leukocytes. J Immunol 135:525–530

Goldman DW, Chang FH, Gifford LA, Goetzl EJ, Bourne HR (1985b) Pertussis toxin inhibition of chemotactic factor-induced calcium mobilization and function in human polymorphonuclear leukocytes. J Exp Med 162:145–156

Goldman DW, Enkel H, Gifford LA, Chenoweth DE, Rosenbaum JT (1986) Lipopolysaccharide modulates receptors for leukotriene B_4, C5a, and formyl-methionyl-leucyl-phenylalanine on rabbit polymorphonuclear leukocytes. J Immunol 137:1971–1976

Goldman R, Bar-Shavit Z (1983) On the mechanism of the augmentation of the phagocytic capability of phagocytic cells by tuftsin, substance P, neurotensin, and kentsin and the interrelationship between their receptors. Ann NY Acad Sci 419:143–155

Goldsmith P, Gierschik P, Milligan G, Unson CG, Vinitsky R, Malech HL, Spiegel AM (1987) Antibodies directed against synthetic peptides distinguish between GTP-binding proteins in neutrophil and brain. J Biol Chem 262:14683–14688

Goldstein BD, Witz G, Amoruso M, Troll W (1979) Protease inhibitors antagonize the activation of polymorphonuclear leukocyte oxygen consumption. Biochem Biophys Res Commun 88:854–860

Goldstein IM, Roos D, Kaplan HB, Weissmann G (1975) Complement and immunoglobulins stimulate superoxide production by human leukocytes independently of phagocytosis. J Clin Invest 56:1155–1163

Goldstein IM, Kaplan HB, Radin A, Frosch M (1976) Independent effects of IgG and complement upon human polymorphonuclear leukocyte function. J Immunol 117:1282–1287

Goldstein IM, Cerqueira M, Lind S, Kaplan HB (1977) Evidence that the superoxide-generating system of human leukocytes is associated with the cell surface. J Clin Invest 59:249–254

Goldyne ME, Burrish GF, Poubelle P, Borgeat P (1984) Arachidonic acid metabolism among human mononuclear leukocytes. Lipoxygenase-related pathways. J Biol Chem 259:8815–8819

Gomez-Cambronero J, Molski TFP, Becker EL, Sha'afi RI (1987) The diacylglycerol kinase inhibitor R59022 potentiates superoxide production but not secretion induced by fMet-Leu-Phe: effects of leupeptin and the protein kinase C inhibitor H-7. Biochem Biophys Res Commun 148:38–46

Gomez-Cambronero J, Yamazaki M, Metwally F, Molski TFP, Bonak VA, Huang CK, Becker EL, Sha'afi RI (1989a) Granulocyte-macrophage colony-stimulating factor and human neutrophils: role of guanine-nucleotide regulatory proteins. Proc Natl Acad Sci USA 86:3569–3573

Gomez-Cambronero J, Huang CK, Bonak VA, Wang E, Casnellie JE, Shiraishi T, Sha'afi RI (1989b) Trosine phosphorylation in human neutrophil. Biochem Biophys Res Commun 162:1478–1485

Gopalakrishna R, Anderson WB (1989) Ca^{2+}- and phospholipid-independent activation of protein kinase C by selective oxidative modification of the regulatory domain. Proc Natl Acad Sci USA 86:6758–6762

Gord'on DL, Krueger RA, Quie PG, Hostetter MK (1985) Amidation of C3 at the thiolester site: stimulation of chemiluminescence and phagocytosis by a new inflammatory mediator. J Immunol 134:3339–3345

Gordon JL (1986) Extracellular ATP: effects, sources and fate. Biochem J 233:309–319

Gorter A, Hiemstra PS, Leijh PCJ, van der Sluys ME, van der Barselaar LA (1987) IgA- and secretory IgA-opsonized S. aureus induce a respiratory burst and phagocytosis by polymorphonuclear leucocytes. Immunology 61:303–309

Gorter A, Hiemstra PS, Klar-Mohamad N, van Es LA, Daha MR (1988a) Binding, internalization and degradation of soluble aggregates of human secretory IgA by resident rat peritoneal macrophages. Immunology 64:703–708

Gorter A, Hiemstra PS, van der Voort EAM, van Es LA, Daha MR (1988b) Binding of IgA1 to rat peritoneal macrophages. Immunology 64:207–212

Gorter A, Hiemstra PS, Leijh PCJ, van der Sluys ME, van den Barselaar MT (1989) Complement-mediated enhancement of IgA-induced H_2O_2 release by human polymorphonuclear leucocytes. Immunology 67:120–125

Grady PG, Thomas LL (1986) Characterization of cyclic-nucleotide phosphodiesterase activities in resting and N-formylmethionylleucylphenylalanine-stimulated human neutrophils. Biochim Biophys Acta 885:282–293

Grandordy BM, Lacroix H, Mavoungou E, Krilis S, Crea AEG, Spur BW, Lee TH (1990) Lipoxin A_4 inhibits phosphoinositide hydrolysis in human neutrophils. Biochem Biophys Res Commun 167:1022–1029

Gray GR, Klebanoff SJ, Stamatoyannopoulos G, Austin T, Naiman SC, Yoshida A, Kliman MR, Robinson GCF (1973) Neutrophil dysfunction, chronic granulomatous disease, and non-spherolytic haemolytic anaemia caused by complete deficiency of glucose-6-phosphate dehydrogenase. Lancet 2:530–534

Green MJ, Hill HAO, Tew GD, Walton NJ (1984) An opsonised electrode. The direct electrochemical detection of superoxide generated by human neutrophils. FEBS Lett 170:69–72

Green MJ, Hill HAO, Tew GD (1987) The rate of oxygen consumption and superoxide anion formation by stimulated human neutrophils. The effect of particle concentration and size. FEBS Lett 216:31–34

Green TR, Pratt KL (1987) A reassessment of the product specificity of the NADPH:O_2 oxidoreductase of human neutrophils. Biochem Biophys Res Commun 142:213–220

Green TR, Pratt KL (1988) Purification of the solubilized NADPH:O_2 oxidoreductase of human neutrophils. Isolation of its catalytically inactive cytochrome b and flavoprotein redox centers. J Biol Chem 263:5617–5623

Green TR, Schaefer RE (1981) Intrinsic dichlorophenolindophenol reductase activity associated with the superoxide-generating oxidoreductase of human granulocytes. Biochemistry 20:7483–7487

Green TR, Wu DE (1986) The NADPH:O_2 oxidoreductase of human neutrophils. Stoichiometry of univalent and divalent reduction of O_2. J Biol Chem 261:6010–6015

Green TR, Wu DE, Wirtz MK (1983) The O_2^- generating oxidoreductase of human neutrophils: evidence of an obligatory requirement for calcium and magnesium for expression of catalytic activity. Biochem Biophys Res Commun 110:973–978

Greenberg S, Di Virgilio F, Steinberg TH, Silverstein SC (1988) Extracellular nucleotides mediate Ca^{2+} fluxes in J774 macrophages by two distinct mechanisms. J Biol Chem 263:10337–10343

Greening AP, Lowrie DB (1983) Extracellular release of hydrogen peroxide by human alveolar macrophages: the relationship to cigarette smoking and lower respiratory tract infections. Clin Sci 65:661–664

Gresham HD, Zheleznyak A, Mormol JS, Brown EJ (1990) Studies on the molecular mechanisms of human neutrophil Fc receptor-mediated phagocytosis. Evidence that a distinct pathway for activation of the respiratory burst results in reactive oxygen metabolite-dependent amplification of ingestion. J Biol Chem 265:7819–7826

Griese M, Griese S, Reinhardt D (1990) Inhibitory effects of pertussis toxin on the cAMP generating system in human mononuclear leucocytes. Eur J Clin Invest 20:317–322

Grinstein S, Furuya W (1984) Amiloride-sensitive Na^+/H^+ exchange in human neutrophils: mechanisms of activation by chemotactic factors. Biochem Biophys Res Commun 122:755–762

Grinstein S, Furuya W (1986a) Cytoplasmic pH regulation in activated human neutrophils: effects of adenosine and pertussis toxin on Na^+/H^+ exchange and metabolic acidificiation. Biochim Biophys Acta 889:301–309

Grinstein S, Furuya W (1986b) Cytoplasmic pH regulation in phorbol ester-activated human neutrophils. Am J Physiol 251:C55–C65

Grinstein S, Furuya W (1988) Receptor-mediated activation of electropermeabilized neutrophils. Evidence for a Ca^{2+}- and protein kinase C-independent signaling pathway. J Biol Chem 263:1779–1783

Grinstein S, Rothstein A (1986) Mechanisms of regulation of the Na^+/H^+ exchanger. J Membr Biol 90:1–12

Grinstein S, Goetz JD, Furuya W, Rothstein A, Gelfand EW (1984) Amiloride-sensitive Na^+-H^+ exchange in platelets and leukocytes: detection by electronic cell sizing. Am J Physiol 247:C293–C298

Grinstein S, Elder B, Furuya W (1985) Phorbol ester-induced changes of cytoplasmic pH in neutrophils: role of exocytosis in Na^+-H^+ exchange. Am J Physiol 248:C379–C386

Grinstein S, Furuya W, Biggar WD (1986) Cytoplasmic pH regulation in normal and abnormal neutrophils. Role of superoxide generation and Na$^+$/H$^+$ exchange. J Biol Chem 261:512–514

Grinstein S, Furuya W, Nasmith PE, Rotstein OD (1988) Na$^+$/H$^+$ exchange and the regulation of intracellular pH in polymorphonuclear leukocytes. Comp Biochem Physiol 90A:543–549

Grinstein S, Hill M, Furuya W (1989) Activation of electropermeabilized neutrophils by adenosine 5'-[γ-thio]triphosphate (ATP[S]). Role of phosphatases in stimulus-response coupling. Biochem J 261:755–759

Grinstein S, Furuya W, Lu DJ, Mills GB (1990) Vanadate stimulates oxygen consumption and tyrosine phosphorylation in electropermeabilized human neutrophils. J Biol Chem 265:318–327

Groopman JE, Molina JM, Scadden DT (1989) Hematopoietic growth factors. N Engl J Med 321:1449–1459

Grunberger G, Zick Y, Roth J, Gorden P (1983) Protein kinase activity of the insulin receptor in human circulating and cultured mononuclear cells. Biochem Biophys Res Commun 115:560–566

Gryglewski RJ, Palmer RMJ, Moncada S (1986) Superoxide anion is involved in the breakdown of endothelium-derived vascular relaxing factor. Nature 320:454–456

Gryglewski RJ, Szczeklik A, Wandzilak M (1987) The effect of six prostaglandins, prostacyclin and iloprost on generation of superoxide anions by human polymorphonuclear leukocytes stimulated by zymosan or formyl-methionyl-leucyl-phenylalanine. Biochem Pharmacol 36:4209–4213

Grzeskowiak M, Della Bianca V, De Togni P, Papini E, Rossi F (1985) Independence with respect to Ca^{2+} changes of the neutrophil respiratory and secretory response to exogenous phospholipase C and possible involvement of diacylglycerol and protein kinase C. Biochim Biophys Acta 844:81–90

Grzeskowiak M, Della Bianca V, Cassatella MA, Rossi F (1986) Complete dissociation between the activation of phosphoinositide turnover and of NADPH oxidase by formyl-methionyl-leucyl-phenylalanine in human neutrophils depleted of Ca^{2+} and primed by subthreshold doses of phorbol 12,myristate 13, acetate. Biochem Biophys Res Commun 135:785–794

Gschwendt M, Kittstein W, Marks F (1988) The weak immunosupressant cyclosporine D as well as the immunologically inactive cyclosporine H are potent inhibitors in vivo of phorbol ester TPA-induced biological effects in mouse skin and of Ca^{2+}/calmodulin dependent EF-2 phosphorylation in vitro. Biochem Biophys Res Commun 150:545–551

Gudewicz PW, Beezhold DH, Van Alten P, Molnar J (1982) Lack of stimulation of post-phagocytic metabolic activities of polymorphonuclear leukocytes by fibronectin opsonized particles. J Reticuloendothel Soc 32:143–154

Guthrie LA, McPhail LC, Henson PM, Johnston RB Jr (1984) Priming of neutrophils for enhanced release of oxygen metabolites by bacterial lipopolysaccharide. Evidence for increased activity of the superoxide-producing enzyme. J Exp Med 160:1656–1671

Gyllenhammar H, Palmblad J (1989) Linoleic acid-deficient rat neutrophils show decreased bactericidal capacity, superoxide formation and membrane depolarization. Immunology 66:616–620

Hafeman DG, McConnell HM, Gray JW, Dean PN (1982) Neutrophil activation monitored by flow cytometry: stimulation by phorbol diester is an all-or-none event. Science 215:673–675

Hafström I, Seligmann BE, Friedman MM, Gallin JI (1984) Auranofin affects early events in human polymorphonuclear neutrophil activation by receptor-mediated stimuli. J Immunol 132:2007–2014

Hagenlocker BE, Walker BAM, Ward PA (1990) Superoxide responses of immune complex-stimulated rat alveolar macrophages. Intracellular calcium and priming. J Immunol 144:3898–3906

Haines KA, Giedd KN, Rich AM, Korchak HM, Weissmann G (1987) The leukotriene B_4 paradox: neutrophils can, but will not, respond to ligand-receptor interactions by forming leukotriene B_4 or its ω-metabolites. Biochem J 241:55–62

Haines KA, Reibman J, Callegari PE, Abramson SB, Philips MR, Weissmann G (1990) Cocaine and its derivatives blunt neutrophil functions without influence of a 47-kilodalton component of the reduced nicotinamide-adenine dinucleotide phosphate oxidase. J Immunol 144:4757–4764

Hall CL, Munford RS (1983) Enzymatic deacylation of the lipid A moiety of *Salmonella typhimurium* lipopolysaccharides by human neutrophils. Proc Natl Acad Sci USA 80:6671–6675

Hallett MB, Campbell AK (1983) Two distinct mechanisms for stimulation of oxygen-radical production by polymorphonuclear leucocytes. Biochem J 216:459–465

Hallett MB, Campbell AK (1984) Is intracellular Ca^{2+} the trigger for oxygen radical production by polymorphonuclear leucocytes? Cell Calcium 5:1–19

Halliwell B, Gutteridge JMC (1984) Oxygen toxicity, oxygen radicals, transition metals and disease. Biochem J 219:1–14

Halliwell B, Hoult JR, Blake DR (1988) Oxidants, inflammation, and anti-inflammatory drugs. FASEB J 2:2867–2873

Halstensen A, Haneberg B, Glette J, Sandberg S, Solberg CO (1986) Factors important for the measurement of chemiluminescence production by polymorphonuclear leukocytes. J Immunol Methods 88:121–128

Ham EA, Soderman DD, Zanetti ME, Dougherty HW, McCauley E, Kuehl FA Jr (1983) Inhibition by prostaglandins of leukotriene B_4 release from activated neutrophils. Proc Natl Acad Sci USA 80:4349–4553

Hamachi T, Hirata M, Koga T (1984) Effect of cAMP-elevating drugs on Ca^{2+} efflux and actin polymerization in peritoneal macrophages stimulated with *N*-formyl chemotactic peptide. Biochim Biophys Acta 804:230–236

Hamers MN, de Boer M, Meerhof LJ, Weening RS, Roos D (1984) Complementation in monocyte hybrids revealing genetic heterogeneity in chronic granulomatous disease. Nature 307:553–555

Hamilton TA, Adams DO (1987) Molecular mechanisms of signal transduction in macrophages. Immunol Today 8:151–158

Hamilton TA, Becton DL, Somers SD, Gray PW, Adams DO (1985) Interferon-γ modulates protein kinase C activity in murine peritoneal macrophages. J Biol Chem 260:1378–1381

Hammer R, Kobinger W, Pichler L (1980) Binding of an imidazoline (clonidine), an oxazoloazepin (B-HT 933) and a thiazoloazepin (B-HT 920) to rat brain α-adrenoceptors and relation to cardiovascular effects. Eur J Pharmacol 62:277–285

Hammond B, Kontos HA, Hess ML (1985) Oxygen radicals in the adult respiratory distress syndrome, in myocardial ischemia and reperfusion injury, and in cerebral vascular damage. Can J Physiol Pharmacol 63:173–187

Hanahan DJ (1986) Platelet activating factor: a biologically active phosphoglyceride. Annu Rev Biochem 55:483–509

Hancock JT, Jones OTG (1987) The inhibition by diphenyleneiodonium and its analogues of superoxide generation by macrophages. Biochem J 242:103–107

Hancock JT, Maly FE, Jones OTG (1989) Properties of the superoxide-generating oxidase of B-lymphocyte cell lines. Determination of Michaelis parameters. Biochem J 262:373–375

Hancock JT, Henderson LM, Jones OTG (1990) Superoxide generation by EBV-transformed B lymphocytes. Activation by IL-1β, TNF-α and receptor-independent stimuli. Immunology 71:213–217

Hansen BD, Finbloom DS (1990) Characterization of the interaction between recombinant human interferon-γ and its receptor on human polymorphonuclear leukocytes. J Leukocyte Biol 47:64–69

Hanski E (1989) Invasive adenylate cyclase toxin of Bordetella pertussis. Trends Biochem Sci 14:459–463

Hansson A, Serhan CN, Haeggström J, Ingelman-Sundberg M, Samuelsson B (1986) Activation of protein kinase C by lipoxin A and other eicosanoids. Intracellular action of oxygenated products of arachidonic acid. Biochem Biophys Res Commun 134:1215–1222

Hardebo JE, Kahrström J, Owman C, Salford LG (1987) Endothelium-dependent relaxation by uridine tri- and diphosphate in isolated human pial vessels. Blood Vessels 24:150–155

Harper AM, Dunne MJ, Segal AW (1984) Purification of cytochrome b_{-245} from human neutrophils. Biochem J 219:519–527

Harper AM, Chaplin MF, Segal AW (1985) Cytochrome b_{-245} from human neutrophils is a glycoprotein. Biochem J 227:783–788

Harris DN, Asaad MM, Phillips MB, Goldenberg HJ, Antonaccio MJ (1979) Inhibition of adenylate cyclase in human blood platelets by 9-substituted adenine derivatives. J Cyclic Nucleotide Res 5:125–134

Harris PE, Ralph P (1985) Human leukemic models of myelomonocytic development: a review of the HL-60 and U937 cell lines. J Leukocyte Biol 37:407–422

Harris PE, Ralph P, Litcofsky P, Moore MAS (1985) Distinct activities of interferon-γ, lymphokine and cytokine differentiation-inducing factors acting on the human monoblastic leukemia cell line U937. Cancer Res 45:9–13

Harris RE, Boxer LA, Baehner RL (1980) Consequences of vitamin-E deficiency on the phagocytic and oxidative functions of the rat polymorphonuclear leukocyte. Blood 55:338–343

Hartfield PJ, Robinson JM (1990) Fluoride-mediated activation of the respiratory burst in electropermeabilized neutrophils. Biochim Biophys Acta 1054:176–180

Harth M, Keown PA, Orange JF (1983) Monocyte dependent excited oxygen radical generation in rheumatoid arthritis: inhibition by gold sodium thiomalate. J Rheumatol 10:701–707

Hartiala KT, Scott IG, Viljanen MK, Akerman KEO (1987) Lack of correlation between calcium mobilization and respiratory burst activation induced by chemotactic factors in rabbit polymorphonuclear leukocytes. Biochem Biophys Res Commun 144:794–800

Hartshorn KL, Collamer M, White MR, Schwartz JH, Tauber AI (1990a) Characterization of influenza A virus activation of the human neutrophil. Blood 75:218–226

Hartshorn KL, Karnad AB, Tauber AI (1990b) Influenza A virus and the neutrophil: a model of natural immunity. J Leukocyte Biol 47:176–186

Hartung HP (1983) Leukotriene C_4 (LTC₄) elicits a respiratory burst in peritoneal macrophages. Eur J Pharmacol 91:159–160

Hartung HP, Toyka KV (1983) Activation of macrophages by substance P: induction of oxidative burst and thromboxane release. Eur J Pharmacol 89:;301–305

Hartung HP, Wolters K, Toyka KV (1986) Substance P: binding properties and studies on cellular responses in guinea pig macrophages. J Immunol 136:3856–3863

Harvath L, Aksamit RR (1984) Oxidized N-formylmethionyl-leucyl-phenylalanine: effect on the activation of human monocyte and neutrophil chemotaxis and superoxide production. J Immunol 133:1471–1476

Harvath L, McCall CE, Bass DA, McPhail LC (1987) Inhibition of human neutrophil chemotaxis by the protein kinase inhibitor, 1-(5-isoquinolinesulfonyl) piperazine. J Immunol 139:3055–3061

Hashimoto Y, Ziff M, Hurd ER (1982) Increased endothelial cell adherence, aggregation, and superoxide generation by neutrophils incubated in systemic Lupus erythematosus and Felty's syndrome sera. Arthritis Rheum 25:1409–1418

Hassan NF, Campbell DE, Douglas SD (1988) Phorbol myristate acetate induced oxidation of 2',7'-dichlorofluorescin by neutrophils from patients with chronic granulomatous disease. J Leukocyte Biol 43:317–322

Hatch GE, Gardner DE, Menzel DB (1978) Chemiluminescence of phagocytic cells caused by N-formylmethionyl peptides. J Exp Med 147:182–195

Hauschildt S, Hirt W, Bessler W (1988a) Modulation of protein kinase C activity by NaF in bone marrow derived macrophages. FEBS Lett 230:121–124

Hauschildt S, Steffens U, Wagner-Roos L, Bessler WG (1988b) Role of protein-kinase C and phosphatidylinositol metabolism in lipopeptide-induced leukocyte activation as signal transducing mechanism. Mol Immunol 25:1081-1086

Hayakawa H, Umehara K, Myrvik QN (1989) Oxidative responses of rabbit alveolar macrophages: comparative priming activities of MIF/MAF, sera, and serum components. J Leukocyte Biol 45:231–238

Hayakawa T, Suzuki K, Suzuki S, Andrews PC, Babior BM (1986) A possible role for protein phosphorylation in the activation of the respiratory burst in human neutrophils. Evidence from studies with cells from patients with chronic granulomatous disease. J Biol Chem 261:9109–9115

Hecker G, Ney P, Schrör K (1990) Cytotoxic enzyme release and oxygen centered radical formation in human neutrophils are selectively inhibited by E-type prostaglandins but not by PGI_2. Naunyn Schmiedebergs Arch Pharmacol 341:308–315

Heiman DF, Astiz ME, Rackow EC, Rhein D, Berm Kim Y, Weil MH (1990) Monophosphoryl lipid A inhibits neutrophil priming by lipopolysaccharide. J Lab Clin Med 116:237–241

Heinecke JW, Shapiro BM (1989) Respiratory burst oxidase of fertilization. Proc Natl Acad Sci USA 86:1259–1263

Helfand SL, Werkmeister J, Roder JC (1982) Chemiluminescence response of human natural killer cells. I. The relationship between target cell binding, chemiluminescence, and cytolysis. J Exp Med 156:492–505

Helfman DM, Appelbaum BD, Vogler WR, Kuo JF (1983) Phospholipid-sensitive Ca^{2+}-dependent protein kinase and its substrates in human neutrophils. Biochem Biophys Res Commun 111:847–853

Henderson B, Pettipher ER, Higgs GA (1987) Mediators of rheumatoid arthritis. Br Med Bull 43:415–428

Henderson LM, Chappell JB, Jones OTG (1987) The superoxide-generating NADPH oxidase of human neutrophils is electrogenic and associated with an H^+ channel. Biochem J 246:325–329

Henderson LM, Chappell JB, Jones OTG (1988a) Superoxide generation by the electrogenic NADPH oxidase of human neutrophils is limited by the movement of a compensating charge. Biochem J 255:285–290

Henderson LM, Chappell JB, Jones OTG (1988b) Internal pH changes associated with the activity of NADPH oxidase of human neutrophils. Further evidence for the presence of an H^+ conducting channel. Biochem J 251:563–567

Henderson LM, Chappell JB, Jones OTG (1989) Superoxide generation is inhibited by phospholipase A_2 inhibitors. Role for phospholipase A_2 in the activation of the NADPH oxidase. Biochem J 264:249–255

Henderson WR, Kaliner M (1978) Immunologic and nonimmunologic generation of superoxide from mast cells and basophils. J Clin Invest 61:187–196

Hendricks CL, McCall CE, McPhail LC (1987) Cell-free activation of the respiratory burst enzyme: characteristics of the cytosolic co-factor. Fed Proc 46:1034

Henricks PAJ, van Erne-van der Tol ME, Verhoef J (1982) Partial removal of sialic acid enhances phagocytosis and the generation of superoxide and chemiluminescence by polymorphonuclear leukocytes. J Immunol 129:745–750

Herrmann E, Gierschik P, Jakobs KH (1989) Neomycin induces high-affinity agonist binding of G-protein-coupled receptors. Eur J Biochem 185:677–683

Heyneman RA, Vercauteren RE (1984) Activation of a NADPH oxidase from horse polymorphonuclear leucocytes in a cell-free system. J Leukocyte Biol 36:751–759

Heyworth PG, Badwey JA (1990) Continuous phosphorylation of both the 47 and the 49 kDa proteins occurs during superoxide production by neutrophils. Biochim Biophys Acta 1052:299–305

Heyworth PG, Segal AW (1986) Further evidence for the involvement of a phosphoprotein in the respiratory burst oxidase of human neutrophils. Biochem J 239:723–731

Heyworth PG, Shrimpton CF, Segal AW (1989a) Localization of the 47 kDa phosphoprotein involved in the respiratory-burst NADPH oxidase of phagocytic cells. Biochem J 260:243–248

Heyworth PG, Karnovsky ML, Badwey JA (1989b) Protein phosphorylation associated with synergistic stimulation of neutrophils. J Biol Chem 264:14935–14939

Hibbs JB Jr, Taintor RR, Vavrin Z (1987a) Macrophage cytotoxicity: role for L-arginine deiminase and imino nitrogen oxidation to nitrite. Science 235:473–476

Hibbs JB Jr, Vavrin Z, Taintor RR (1987b) L.-Arginine is required for expression of the activated macrophage effector mechanism causing selective metabolic inhibition in target cells. J Immunol 138:550–565

Hibbs JB Jr, Taintor RR, Vavrin Z, Rachlin EM (1988) Nitric oxide: a cytotoxic activated macrophage effector molecule. Biochem Biophys Res Commun 157:87–94

Hidaka H, Inagaki M, Kawamoto S, Sasaki Y (1984) Isoquinolinesulfonamides, novel and potent inhibitors of cyclic nucleotide dependent protein kinase and protein kinase C. Biochemistry 23:5036–5041

Higashijima T, Uzu S, Nakajima T, Ross EM (1988) Mastoparan, a peptide toxin from wasp venom, mimics receptors by activating GTP-binding regulatory proteins (G proteins). J Biol Chem 263:6491–6494

Higson FK, Durbin L, Pavlotsky N, Tauber AI (1985) Studies of cytochrome b_{-245} translocation in the PMA stimulation of the human neutrophil NADPH oxidase. J Immunol 135:519–524

Hingorani VN, Ho YK (1987) A structural model for the α-subunit of transducin. Implications of its role as a molecular switch in the visual signal transduction mechanism. FEBS Lett 220:15–22

Hinsch KD, Rosenthal W, Spicher K, Binder T, Gausepohl H, Frank R, Schultz G, Joost HG (1988) Adipocyte plasma membranes contain two G_i subtypes but are devoid of G_o. FEBS Lett 238:191–196

Hirabayashi Y, Kobayashi T, Nishikawa A, Okazaki H, Aoki T, Takaya J, Kobayashi Y (1988) Oxidative metabolism and phagocytosis of polymorphonuclear leukocytes in patients with chronic renal failure. Nephron 49:305–312

Hladky SB, Haydon DA (1972) Ion transfer across lipid membranes in the presence of gramicidin A. I. Studies of the unit conductance channel. Biochim Biophys Acta 274:294–312

Hoek JB, Thomas AP, Rubin R, Rubin E (1987) Ethanol-induced mobilization of calcium by activation of phosphoinositide-specific phospholipase C in intact hepatocytes. J Biol Chem 262:682–691

Hoffman M, Autor AP (1982) Effect of cyclooxygenase inhibitors and protease inhibitors on phorbol-induced stimulation of oxygen consumption and superoxide production by rat pulmonary macrophages. Biochem Pharmacol 31:775–780

Hoffman M, Weinberg JB (1987) Tumor necrosis factor-α induces increased hydrogen peroxide production and Fc receptor expression, but not increased Ia antigen expression by peritoneal macrophages. J Leukocyte Biol 42:704–707

Hoffman M, Feldman SR, Pizzo SV (1983) α_2-Macroglobulin 'fast' forms inhibit superoxide production by activated macrophages. Biochim Biophys Acta 760:421–423

Hoffstein ST, Gennaro DE, Manzi RM (1985) Surface contact inhibits neutrophil superoxide generation induced by soluble stimuli. Lab Invest 52:515–522

Hohn DC, Lehrer RI (1975) NADPH oxidase deficiency in X-linked chronic granulomatous disease. J Clin Invest 55:707–713

Hokin LE (1985) Receptors and phosphoinositide-generated second messengers. Annu Rev Biochem 54:205–235

Hokin MR, Brown DF (1969) Inhibition by γ-hexachlorocyclohexane of acetylcholine-stimulated phosphatidylinositol synthesis in cerebral cortex slices and of phosphatidic acid-inositol transferase in cerebral cortex particulate fractions. J Neurochem 16:475–483

Holian A (1986) Leukotriene B_4 stimulation of phosphatidylinositol turnover in macrophages and inhibition by pertussis toxin. FEBS Lett 201:15–19

Holian A, Daniele RP (1979) Stimulation of oxygen consumption and superoxide anion production in pulmonary macrophages by N-formyl methionyl peptides. FEBS Lett 108:47–50

Holian A, Daniele RP (1981) Release of oxygen products from lung macrophages by N-formyl peptides. J Appl Physiol 50:736–740

Holian A, Marchiarullo MA, Stickle DF (1984) γ-Hexachlorocyclohexane activation of alveolar macrophage phosphatidylinositol cycle, calcium mobilization and O_2^- production. FEBS Lett 176:151–154

Holian A, Jordan MK, Nguyen HV, Devenyi ZJ (1988) Inhibition of macrophage activation by isoquinolinesulfonamides, phenothiazines, and a naphthalenesulfonamide. J Cell Physiol 137:45–54

Holland JA, Pritchard KA, Pappolla MA, Wolin MS, Rogers NJ, Stemerman MB (1990) Bradykinin induces superoxide anion release from human endothelial cells. J Cell Physiol 143:21–25

Honeycutt PJ, Niedel JE (1986) Cytochalasin B enhancement of the diacylglycerol response in formyl peptide-stimulated neutrophils. J Biol Chem 261:15900–15905

Hoogerwerf M, Weening RS, Hack CE, Roos D (1990) Complement fragments C3b and iC3b coupled to latex induce a respiratory burst in human neutrophils. Mol Immunol 27:159–167

Hopkins NK, Lin AH, Gorman RR (1983) Evidence for mediation of acetyl glyceryl ether phosphorylcholine stimulation of adenosine 3',5'-(cyclic)monophosphate levels in human polymorphonuclear leukocytes by leukotriene B_4. Biochim Biophys Acta 763:276–283

Hopkins NK, Schaub RG, Gorman RR (1984) Acetyl glyceryl ether phosphorylcholine (PAF-acether) and leukotriene B_4-mediated neutrophil chemotaxis through an intact endothelial cell monolayer. Biochim Biophys Acta 805:30–36

Hörl WH, Haag-Waber M, Georgopoulos A, Block LH (1990) Physicochemical characterization of a polypeptide present in uremic serum that inhibits the biological activity of a polymorphonuclear cells. Proc Natl Acad Sci USA 87:6353–6357

Horn W, Karnovsky ML (1986) Features of the translocation of protein kinase C in neutrophils stimulated with the chemotactic peptide fMet-Leu-Phe. Biochem Biophys Res Commun 139:1169–1175

Horton JK, Davies M, Topley N, Thomas D, Williams JD (1990) Activation of the inflammatory response of neutrophils by Tamm-Horsfall glycoprotein. Kidney Int 37:717–726

Houslay MD, Gordon LM (1983) The activity of adenylate cyclase is regulated by the nature of its lipid environment. Current Top Membr Transp 18:179–231

Howard TH, Wang D (1987) Calcium ionophore, phorbol ester, and chemotactic peptide-induced cytoskeleton reorganization in human neutrophils. J Clin Invest 79:1359–1364

Howard TH, Wang D, Berkow RL (1990) Lipopolysaccharide modulates chemotactic peptide-induced actin polymerization in neutrophils. J Leukocyte Biol 47:13–24

Howlett AC, Sternweis PC, Macik BA, van Arsdale PM, Gilman AG (1979) Reconstitution of catecholamine-sensitive adenylate cyclase. Association of a regulatory component of the enzyme with membranes containing the catalytic protein and β-adrenergic receptors. J Biol Chem 254:2287–2295

Hoyle PC, Freer RJ (1984) Isolation and reconstitution of the N-formylpeptide receptor from HL–60 derived neutrophils. FEBS Lett 167:277–280

Hruska KA, Bar-Shavit Z, Malone JD, Teitelbaum S (1988) Ca^{2+} priming during vitamin D-induced monocytic differentiation of a human leukemia cell line. J Biol Chem 263:16039–16044

Huang CK (1987) Partial purification and characterization of formylpeptide receptor from rabbit peritoneal neutrophils. J Leukocyte Biol 41:63–69

Huang CK, Oshana SC (1986) Partial characterization of protein kinase C and inhibitor activity of protein kinase C in rabbit peritoneal neutrophils. J Leukocyte Biol 39:671–678

Huang CK, Hill JM, Bormann BJ, Mackin WM, Becker EL (1983a) Endogenous substrates for cyclic AMP-dependent and calcium-dependent protein phosphorylation in rabbit peritoneal neutrophils. Biochim Biophys Acta 760:126–135

Huang CK, Mackin WM, Bormann BJ, Becker EL (1983b) Cyclic AMP receptor protein and cyclic AMP-dependent protein kinase activity in rabbit peritoneal neutrophils. J Reticuloendothel Soc 34:413–421

Huang CK, Laramee GR, Casnellie JE (1988) Chemotactic factor induced tyrosine phosphorylation of membrane associated proteins in rabbit peritoneal neutrophils. Biochem Biophys Res Commun 151:794–801

Huang CK, Bonak V, Laramee GR, Casnellia JE (1990) Protein tyrosine phosphorylation in rabbit peritoneal neutrophils. Biochem J 269:431–436

Huang SJ, Monk PN, Downes CP, Whetton AD (1988) Platelet-activating factor-induced hydrolysis of phosphatidylinositol 4,5-bisphosphate stimulates the production of reactive oxygen intermediates in macrophages. Biochem J 249:839–845

Huey R, Hugli TE (1985) Characterization of a C5a receptor on human polymorphonuclear leukocytes (PMN). J Immunol 135:2063–2068

Huizinga TWJ, van der Schoot CE, Jost C, Klaassen R, Kleijer M, von dem Borne AEGK, Roos D, Tetteroo PAT (1988) The PI-linked receptor FcRIII is released on stimulation of neutrophils. Nature 333:667–669

Huizinga TWJ, van Kemenade F, Koenderman L, Dolman KM, von dem Borne AEG, Tetteroo PAT, Roos D (1989) The 40-kDa Fcγ receptor (FcRII) on human neutrophils is essential for the IgG-induced respiratory burst and IgG-induced phagocytosis. J Immunol 142:2365–2369

Humphreys JM, Hughes V, Edwards SW (1989) Stimulation of protein synthesis in human neutrophils by γ-interferon. Biochem Pharmacol 38:1241–1246

Hunter T, Cooper JA (1985) Protein-tyrosine kinases. Annu Rev Biochem 54:897–930

Hurst NP, French JK, Bell AL, Nuki G, O'Donnell ML, Betts WH, Cleland LG (1986) Differential effects of mepacrine, chloroquine and hydroxychloroquine on superoxide anion generation, phospholipid methylation and arachidonic acid release by human blood monocytes. Biochem Pharmacol 35:3083–3089

Hyslop PA, Oades ZG, Jesaitis AJ, Painter RG, Cochrane CG, Sklar LA (1984) Evidence for N-formyl chemotactic peptide-stimulated GTPase activity in human neutrophil homogenates. FEBS Lett 166:165–169

Iannone MA, Wolberg G, Zimmerman TP (1989) Chemotactic peptide induces cAMP elevation in human neutrophils by amplification of the adenylate cyclase response to endogenously produced adenosine. J Biol Chem 264:20177–20180

Iglesias G, Pijoan C, Molitor T (1989) Interactions of pseudorabies virus with swine alveolar macrophages: effects of virus infection on cell functions. J Leukocyte Biol 45:410–415

Ignarro LJ, George WJ (1974) Hormonal control of lysosomal enzyme release from human neutrophils: elevation of cyclic nucleotide levels by autonomic neurohormones. Proc Natl Acad Sci USA 71:2027–2031

Iiri T, Tohkin M, Morishima N, Ohoka Y, Ui M, Katada T (1989) Chemotactic peptide receptor-supported ADP-ribosylation of a pertussis toxin substrate GTP-binding protein by cholera toxin in neutrophil-type HL-60 cells. J Biol Chem 264:21394–21400

Iizuka T, Kanegasaki S, Makino R, Tanaka T, Ishimura Y (1985a) Studies on neutrophil b-type cytochrome in situ by low temperature absorption spectroscopy. J Biol Chem 260:12049–12053

Iizuka T, Kanegasaki S, Makino R, Tanaka T, Ishimura Y (1985b) Pyridine and imidazole reversibly inhibit the respiratory burst in porcine and human neutrophils: evidence for the involvement of cytochrome $b558$ in the reaction. Biochem Biophys Res Commun 130:621–626

Imamura K, Sherman ML, Spriggs D, Kufe D (1988) Effect of tumor necrosis factor on GTP binding and GTPase activity in HL-60 and L929 cells. J Biol Chem 263:10247–10253

Ingraham LM, Coates TD, Allen JM, Higgins CP, Baehner RL, Boxer LA (1982) Metabolic, membrane, and functional responses of human polymorphonuclear leukocytes to platelet-activating factor. Blood 59:1259–1266

Irita K, Takeshige K, Minakami S (1984a) Phosphorylation of myosin light chain in intact pig leukocytes stimulated by phorbol 12-myristate 13-acetate. Biochim Biophys Acta 803:21–28

Irita K, Takeshige K, Minakami S (1984b) Protein phosphorylation in intact pig leukocytes. Biochim Biophys Acta 805:44–52

Irita K, Fujita I, Takeshige K, Minakami S, Yoshitake J (1986) Calcium channel antagonist induced inhibition of superoxide production in human neutrophils. Mechanisms independent of antagonizing calcium influx. Biochem Pharmacol 35:3465–3471

Irvine RF, Hemington N, Dawson RMC (1979) The calcium-dependent phosphatidylinositol-phosphodiesterase of rat brain. Mechanisms of suppression and stimulation. Eur J Biochem 99:525–530

Ishida K, Takeshige K, Takasugi S, Minakami S (1989) GTP-dependent and -independent activation of superoxide producing NADPH oxidase in a neutrophil cell-free system. FEBS Lett 243:169–172

Ishida K, Takeshige K, Minakami S (1990) Endothelin-1 enhances superoxide generation of human neutrophils stimulated by the chemotactic peptide N-formyl-methionyl-leucyl-phenylalanine. Biochem Biophys Res Commun 173:496–500

Ishihara Y, Rosolia DL, McKenna PJ, Peters SP, Albertine KH, Gee MH (1990) Calcium is required for PMA induced superoxide release from human neutrophils. J Leukocyte Biol 48:89–96

Ishitoya J, Yamakawa A, Takenawa T (1987) Translocation of diacylglycerol kinase in response to chemotactic peptide and phorbol ester in neutrophils. Biochem Biophys Res Commun 144:1025–1030

Iverson D, CeChatelet LR, Spitznagel JK, Wang P (1977) Comparison of NADH and NADPH oxidase activities in granules isolated from human polymorphonuclear leukocytes with a fluorometric assay. J Clin Invest 59:282–290

Iyengar R, Stuehr DJ, Marletta MA (1987) Macrophage synthesis of nitrite, nitrate, and N-nitrosamines: precursors and role of the respiratory burst. Proc Natl Acad Sci USA 84:6369–6373

Iyer GYN, Islam DMF, Quastel JH (1961) Biochemical aspects of phagocytosis. Nature 192:535–541

Jackowski S, Sha'afi RI (1979) Response of adenosine cyclic 3',5'-monphosphate level in rabbit neutrophils to the chemotactic peptide formyl-methionyl-leucyl-phenylalanine. Mol Pharmacol 16:473–481

Jacob J (1988) Linear gramicidin activates neutrophil functions and the activation is blocked by chemotactic peptide receptor antagonist. FEBS Lett 231:139–142

Jacobs AA, Ward RA, Wellhausen SR, McLeish KR (1989) Polymorphonuclear leukocyte function during hemodialysis: relationship to complement activation. Nephron 52:119–124

Jaconi MEE, Rivest RW, Schlegel W, Wollheim CB, Pittet D, Lew PD (1988) Spontaneous and chemoattractant-induced oscillations of cytosolic free calcium in single adherent human neutrophils. J Biol Chem 263:10557–10560

Jahn H, Nastainczyk W, Röhrkasten A, Schneider T, Hofmann F (1988) Site-specific phosphorylation of the purified receptor for calcium-channel blockers by cAMP- and cGMP-dependent protein kinases, protein kinase C, calmodulin-dependent protein kinase II and casein kinase II. Eur J Biochem 178:535–542

Jakobs KH, Aktories K, Schultz G (1984) Mechanism of pertussis toxin action on the adenylate cyclase system. Inhibition of the turn-on reaction of the inhibitory regulatory site. Eur J Biochem 140:177–181

Jakway JP, DeFranco AL (1986) Pertussis toxin inhibition of B cell and macrophage responses to bacterial lipopolysaccharide. Science 234:743–746

Janco RL, English D (1983) Cyclosporine and human neutrophil function. Transplantation 35:501–503

Jandl RC, André-Schwartz J, Borges-DuBois L, Kipnes RS, McMurrich BJ, Babior BM (1978) Termination of the respiratory burst in human neutrophils. J Clin Invest 61:1176–1185

Jelsema CL, Axelrod J (1987) Stimulation of phospholipase A_2 activity in bovine rod outer segments by the $\beta\gamma$ subunits of transducin and its inhibition by the α subunit. Proc Natl Acad Sci USA 84:3623–3627

Jesaitis AJ, Allen RA (1988) Activation of the neutrophil respiratory burst by chemoattractants: regulation of the N-formyl peptide receptor in the plasma membrane. J Bioenerg Biomembr 20:679–707

Jesaitis AJ, Naemura JR, Sklar LA, Cochrane CG, Painter RG (1984) Rapid modulation of N-formyl chemotactic peptide receptors on the surface of human granulocytes: formation of '. ;h-affinity ligand-receptor complexes in transient association with cytoskeleton. J Cell Biol 98:1378–1387

Jesaitis AJ, Tolley JO, Painter RG, Sklar LA, Cochrane CG (1985) Membrane-cytoskeleton interactions and the regulation of chemotactic peptide-induced activation of human granulocytes: the effects of dihydrocytochalasin B. J Cell Biochem 27:241–253

Jesaitis AJ, Tolley JO, Allen RA (1986) Receptor-cytoskeleton interactions and membrane traffic may regulate chemoattractant-induced superoxide production in human granulocytes. J Biol Chem 261:13662–13669

Johansson A, Dahlgren C (1989) Characterization of the luminol-amplified light-generating reaction induced in human monocytes. J Leukocyte Biol 45:444–451

Johnson RJ, Chenoweth DE (1985) Labeling the granulocyte C5a receptor with a unique photoreactive probe. J Biol Chem 260:7161–7164

Johnston PA, Adams DO, Hamilton TA (1984) Fc-receptor-mediated protein phosphorylation in murine peritoneal macrophages. Biochem Biophys Res Commun 124:197–202

Johnston PA, Adams DO, Hamilton TA (1985) Regulation of the Fc-receptor-mediated respiratory burst: treatment of primed murine peritoneal macrophages with lipopolysaccharide selectively inhibits H_2O_2 secretion stimulated by immune complexes. J Immunol 135:513–518

Johnston RB Jr (1988) Monocytes and macrophages. N Engl J Med 318:747–752

Johnston RB Jr, Kitagawa S (1985) Molecular basis for the enhanced respiratory burst of activated macrophages. Fed Proc 44:2927–2932

Johnston RB Jr, Lehmeyer JE (1976) Elaboration of toxic oxygen by-products by neutrophils in a model of immune complex disease. J Clin Invest 57:836–841

Johnston RB Jr, Lehmeyer JE, Guthrie LA (1976) Generation of superoxide anion and chemiluminescence by human monocytes during phagocytosis and on contact with surface-bound immunoglobulin G. J Exp Med 143:1551–1556

Johnston RB Jr, Godzik CA, Cohn ZA (1978) Increased superoxide anion production by immunologically activated and chemically elicited macrophages. J Exp Med 147:115–127

Johnston RB Jr, Chadwick DA, Cohn ZA (1981) Priming of macrophages for enhanced oxidative metabolism by exposure to proteolytic enzymes. J Exp Med 153:1678–1683

Jones DH, Looney RJ, Anderson CL (1985) Two distinct classes of IgG Fc receptors on a human monocyte line (U937) defined by differences in binding of murine IgG subclasses at low ionic strength. J Immunol 135:3348–3353

Jones GS, Van Dyke K, Castranova V (1981) Transmembrane potential changes associated with superoxide release from human granulocytes. J Cell Physiol 106:75–83

Jones HP, Ghai G, Petrone WF, McCord JM (1982) Calmodulin-dependent stimulation of the NADPH oxidase of human neutrophils. Biochim Biophys Acta 714:152–156

Jones JF (1982) Interactions between human neutrophils and vaccinia virus: induction of oxidative metabolism and virus inactivation. Pediatr Res 16:525–529

Jose PJ (1987) Complement-derived peptide mediators of inflammation. Br Med Bull 43:336–349

Kainoh M, Imai R, Umetsu T, Hattori M, Nishio S (1990) Prostacyclin and beraprost sodium as suppressors of activated rat polymorphonuclear leukocytes. Biochem Pharmacol 39:477–484

Kakinuma K (1974) Effects of fatty acids on the oxidative metabolism of leukocytes. Biochim Biophys Acta 348:76–85

Kakinuma K, Minakami S (1978) Effects of fatty acids on superoxide radical generation in leukocytes. Biochim Biophys Acta 538:50–59

Kakinuma K, Kaneda M, Chiba T, Ohnishi T (1986) Electron spin resonance studies on a flavoprotein in neutrophil plasma membranes. Redox potentials of the flavin and its participation in NADPH oxidase. J Biol Chem 261:9426–9432

Kakinuma K, Fukuhara Y, Kaneda M (1987) The respiratory burst oxidase of neutrophils. Separation of an FAD enzyme and its characterization. J Biol Chem 262:12316–12322

Kamp DW, Dunne M, Weitzman SA, Dunn MM (1989) The interaction of asbestos and neutrophils injures cultured human pulmonary epithelial cells: role of hydrogen peroxide. J Lab Clin Med 114:604–612

Kaneda M, Kakinuma K (1986) Hypertonic glycerol induces a respiratory burst in leukocytes. Biochim Biophys Acta 882:63–70

Kano S, Iizuka T, Ishimura Y, Fujiki H, Sugimura T (1987) Stimulation of superoxide anion formation by the non-TPA-type tumor promoters palytoxin and the thapsigargin in porcine and human neutrophils. Biochem Biophys Res Commun 143:672–677

Kaplan HB, Edelson HS, Friedman R, Weissmann G (1982) The roles of degranulation and superoxide anion generation in neutrophil aggregation. Biochim Biophys Acta 721:55–63

Kaplan HB, Edelson HS, Korchak HM, Given WP, Abramson S, Weissmann G (1984) Effects of non-steroidal anti-inflammatory agents on human neutrophil functions in vitro and in vivo. Biochem Pharmacol 33:371–378

Kaplan L, Weiss J, Elsbach P (1978) Low concentrations of indomethacin inhibit phospholipase A_2 of rabbit polymorphonuclear leukocytes. Proc Natl Acad Sci USA 75:2955–2958

Kaplan SS, Zdziarski UE, Kuhns DB, Basford RE (1988) Inhibition of cell movement and associated changes by hexachlorocyclohexanes due to unregulated intracellular calcium increases. Blood 71:677–683

Kaplan SS, Billiar T, Curran RD, Zdziarski UE, Simmons RL, Basford RE (1989) Inhibition of chemotaxis with N^G-monomethyl-L-arginine: a role for cyclic GMP. Blood 74:1885–1887

Karnad AB, Hartshorn KL, Wright J, Myers JB, Schwartz JH, Tauber AI (1989) Priming of human neutrophils with N-formyl-methionyl-leucyl-phenylalanine by a calcium-independent, pertussis toxin-insensitive pathway. Blood 74:2519–2526

Karnovsky ML (1986) Muramyl peptides in mammalian tissues and their effects at the cellular level. Fed Proc 45:2556–2560

Kase H, Iwahashi K, Nakanishi S, Matsuda Y, Yamada K, Takahashi M, Murakata C, Sato A, Kaneko M (1987) K-252 compounds, novel and potent inhibitors of protein kinase C and cyclic nucleotide-dependent protein kinases. Biochem Biophys Res Commun 142:436–440

Katada T, Gilman AG, Watanabe Y, Bauer S, Jakobs KH (1985) Protein kinase C phosphorylates the inhibitory guanine-nucleotide-binding regulatory component and apparently suppresses its function in hormonal inhibition of adenylate cyclase. Eur J Biochem 151:431–437

Kato H, Ishitoya J, Takenawa T (1986) Inhibition of inositol phospholipids metabolism and calcium mobilization by cyclic AMP-increasing agents and phorbol ester in neutrophils. Biochem Biophys Res Commun 139:1272–1278

Kazura JW, Wenger JD, Salata RA, Budzynski AZ, Goldsmith GH (1989) Modulation of polymorphonuclear leukocyte microbicidal activity and oxidative metabolism by fibrinogen degradation products D and E. J Clin Invest 83:1916–1924

Kemmerich B, Pennington JE (1988) Different calcium and oxidative metabolic responses in human blood monocytes during exposure to various agonists. J Leukocyte Biol 43:125–132

Kharazmi A, Svenson M, Nielsen H, Birgens HS (1985) Effect of cyclosporin A on human neutrophil and monocyte function. Scand J Immunol 21:585–591

Kharazmi A, Nielsen H, Bendtzen K (1988) Modulation of human neutrophil and monocyte chemotaxis and superoxide responses by recombinant TNF-alpha and GM-CSF. Immunobiology 177:363–370

Kharazmi A, Fomsgaard A, Conrad RS, Galanos C, Noiby H (1991) Relationship between chemical composition and biological function of *Pseudomonas aeruginosa* lipopolysaccharide: effect on human neutrophil chemotaxis and oxidative burst. J Leukocyte Biol 49:15–20

Kikuchi A, Kozawa O, Kaibuchi K, Katada T, Ui M, Takai Y (1986) Direct evidence for involvement of a guanine nucleotide-binding protein in chemotactic peptide-stimulated formation of inositol bisphosphate and trisphosphate in differentiated human leukemic (HL-60) cells. Reconstitution with G_i or G_o of the plasma membranes ADP-ribosylated by pertussis toxin. J Biol Chem 261:11558–11562

Kikuchi A, Ikeda K, Kozawa O, Takai Y (1987) Modes of inhibitory action of protein kinase C in the chemotactic peptide-induced formation of inositol phosphates in differentiated human leukemic (HL-60) cells. J Biol Chem 262:6766–6770

Kilpatrick JM, Volanakis JE (1985) Opsonic properties of C-reactive protein. Stimulation by phorbol myristate acetate enables human neutrophils to phagocytize C-reactive protein-coated cells. J Immunol 134:3364–3370

Kilpatrick L, Garty BZ, Lundquist KF, Hunter K, Stanley CA, Baker L, Douglas SD, Korchak HM (1990) Impaired metabolic function and signaling defects in phagocytic cells in glycogen storage disease type 1b. J Clin Invest 86:196–202

Kim UH, Kim JW, Rhee SG (1989) Phosphorylation of phospholipase C-γ by cAMP-dependent protein kinase. J Biol Chem 264:20167–20170

Kimura N, Johnson GS (1983) Increased membrane-associated nucleoside diphosphate kinase activity as a possible basis for enhanced guanine nucleotide-dependent adenylate cyclase activity induced by picolinic acid treatment of simian virus 40-transformed normal rat kidney cells. J Biol Chem 258:12609–12617

Kimura N, Nagata N (1979) Mechanism of glucagon stimulation of adenylate cyclase in the presence of GDP in rat liver plasma membranes. J Biol Chem 254:3451–3457

Kimura N, Shimada N (1983) GDP does not mediate but rather inhibits hormonal signal to adenylate cyclase. J Biol Chem 258:2278–2283

Kimura N, Shimada N (1988) Direct interaction between membrane-associated nucleoside diphosphate kinase and GTP-binding protein(Gs), and its regulation by hormones and guanine nucleotides. Biochem Biophys Res Commun 151:248–256

King CH, Goralnik CH, Kleinhenz PJ, Marino JA, Sedor JR, Mahmoud AAF (1987) Monoclonal antibody characterization of a chymotrypsin-like molecule on neutrophil membrane associated with cellular activation. J Clin Invest 79:1091–1098

Kinsella JE, Bruckner G, Mai J, Shimp J (1981) Metabolism of trans fatty acids with emphasis on the effects of trans, trans-octadecadienoate on lipid composition, essential fatty acid, and prostaglandins: An overview. Am J Clin Nutr 34:2307–2318

Kishimoto A, Mikawa K, Hashimoto K, Yasuda I, Tanaka S, Tominga M, Kuroda T, Nishizuka Y (1989) Limited proteolysis of protein kinase C subspecies by calcium-dependent netural protease (calpain). J Biol Chem 264:4088–4092

Kiss Z, Anderson WB (1989) Phorbol ester stimulates the hydrolysis of phosphatidylethanolamine in leukemic HL-60, NIH 3T3, and by hamster kidney cells. J Biol Chem 264:1483–1487

Kiss Z, Luo Y (1986) Phorbol ester and 1,2-diolein are not fully equivalent activators of protein kinase C in respect to phosphorylation of membrane proteins in vitro. FEBS Lett 198:203–207

Kiss Z, Deli E, Girard PR, Pettit GR, Kuo JF (1987) Comparative effects of polymyxin B, phorbol ester and bryostatin on protein phosphorylation, protein kinase C translocation, phospholipid metabolism and differentiation of HL 60 cells. Biochem Biophys Res Commun 146:208–215

Kitagawa S, Johnston RB Jr (1985) Relationship between membrane potential changes and superoxide-releasing capacity in resident and activated mouse peritoneal macrophages. J Immunol 135:3417–3423

Kitagawa S, Johnston RB Jr (1986) Deactivation of the respiratory burst in activated macrophages: evidence for alteration of signal transduction. J Immunol 136:2605–2612

Kitagawa S, Takaku F (1982) Effect of microtubule-disrupting agents on superoxide production in human polymorphonuclear leukocytes. Biochim Biophys Acta 719:589–598

Kitagawa S, Takaku F, Sakamoto S (1979) Possible involvement of proteases in superoxide production by human polymorphonuclear leukocytes. FEBS Lett 99:275–278

Kitagawa S, Takaku F, Sakamoto S (1980a) Evidence that proteases are involved in superoxide production by human polymorphonuclear leukocytes and monocytes. J Clin Invest 64:74–81

Kitagawa S, Takaku F, Sakamoto S (1980b) A comparison of the superoxide-releasing response in human polymorphonuclear leukocytes and monocytes. J Immunol 125:359–364

Kitagawa S, Takaku F, Sakamoto S (1980c) Serine protease inhibitors inhibit superoxide production by human basophils stimulated by anti-IgE. Biochem Biophys Res Commun 95:801–806

Kitagawa S, Ohta M, Nojiri H, Kakinuma K, Saito M, Takaku F, Miura Y (1984) Functional maturation of membrane potential changes and superoxide-producing capacity during differentiation of human granulocytes. J Clin Invest 73:1062–1071

Kitagawa S, Yuo A, Souza M, Saito M, Miura Y, Takaku F (1987) Recombinant human granulocyte colony-stimulating factor enhances superoxide release in human granulocytes stimulated by the chemotactic peptide. Biochem Biophys Res Commun 144:1143–1146

Kitahara M, Eyre HJ, Lynch RE, Rallison ML, Hill HR (1980) Metabolic activity of diabetic monocytes. Diabetes 29:251–256

Kiyotaki C, Bloom BR (1984) Activation of murine macrophage cell lines. Possible involvement of protein kinases in stimulation of superoxide production. J Immunol 133:923–931

Kiyotaki C, Peisach J, Bloom BR (1984) Oxygen metabolism in cloned macrophage cell lines: glucose dependence of superoxide production, metabolic and spectral analysis. J Immunol 132:857–866

Klauser RJ, Schmer G, Chandler WL, Müller W (1990) Consumption of complement and activation of human neutrophils by an artificial immune complex: polyacrylic acid-IgG-polymer. Biochim Biophys Acta 1052:408–415

Klausner RD, Kleinfeld AM, Hoover RL, Karnovsky MJ (1980) Lipid domains in membranes. Evidence derived from structural perturbations induced by free fatty acids and lifetime heterogeneity analysis. J Biol Chem 255:1286-1295

Klebanoff SJ (1980) Oxygen metabolism and the toxic properties of phagocytes. Ann Intern Med 93:480–489

Klebanoff SJ, Hamon CB (1972) Role of myeloperoxidase-mediated antimicrobial systems in intact leukocytes. J Reticuloendothel Soc 12:170–196

Klebanoff SJ, Vadas MA, Harlan JM, Sparks LH, Gamble JR, Agosti JM, Waltersdorph AM (1986) Stimulation of neutrophils by tumor necrosis factor. J Immunol 136:4220–4225

Klein JB, Schepers TM, Dean WL, Sonnenfeld G, McLeish KR (1990) Role of intracellular calcium concentration and protein kinase C activation in IFN-γ stimulation of U937 cells. J Immunol 144:4305–4311

Kleinberg ME, Rotrosen D, Malech HL (1989) Asparagine-linked glycosylation of cytochrome b_{558} large unit varies in different human phagocytic cells. J Immunol 143:4152–4157

Kleinberg ME, Malech HL, Rotrosen D (1990) The phagocyte 47-kilodalton cytosolic protein is an early reactant in activation of the respiratory burst. J Biol Chem 265:15577–15583

Klempner MS, Rocklin RE (1982) Specific binding of leukocyte inhibitory factor to neutrophil plasma membranes. J Immunol 128:2040–2043

Kloner RA, Przyklenk K, Whittaker P (1989) Deleterious effects of oxygen radicals in ischemia/reperfusion. Resolved and unresolved issues. Circulation 80:1115–1127

Knight DE, Tonge DA, Baker PF (1985) Inhibition of exocytosis in bovine adrenal medullary cells by botulinum toxin type D. Nature 317:719–721

Kobayashi S, Imajoh-Ohmi S, Nakamura M, Kanegasaki S (1990) Occurrence of cytochrome b_{558} in B-cell lineage of human lymphocytes. Blood 75:458–461

Kobzik L, Godleski JJ, Brain JD (1990) Selective down-regulation of alveolar macrophage oxidative response to opsonin-independent phagocytosis. J Immunol 144:4312–4319

Koenderman L, Bruijnzeel PLB (1989) Increased sensitivity of the chemoattractant-induced chemiluminescence in eosinophils isolated from atopic individuals. Immunology 67:534–536

Koenderman L, Yazdanbakhsh M, Roos D, Verhoeven AJ (1989a) Dual mechanisms in priming of the chemoattractant-induced respiratory burst in human granulocytes. A Ca^{2+}-dependent and a Ca^{2+}-independent route. J Immunol 142:623–628

Koenderman L, Tool ATJ, Roos D, Verhoeven AJ (1989b) Accmulation of diacylglycerol modulates, but does not initiate the respiratory burst in human neutrophils and eosinophils. In: Biology of cellular transducing signals '89. Vanderhoek JY (ed) 9th International Washington Spring Symposium, Washington. p 73

Koenderman L, Tool ATJ, Roos D, Verhoeven AJ (1989c) 1,2-Diacylglycerol accumulation in human neutrophils does not correlate with respiratory burst activation. FEBS Lett 243:399–403

Kolbuch-Braddon ME, Peterhans E, Stocker R, Weidemann MJ (1984) Oxygen uptake associated with Sendai-virus-stimulated chemiluminescence in rat thymocytes contains a significant non-mitochondrial component. Biochem J 222:541–551

Koo C, Lefkowitz RJ, Snyderman R (1982) The oligopeptide chemotactic factor receptor on human polymorphonuclear leukocyte membranes exists in two affinity states. Biochem Biophys Res Commun 106:442–449

Koo C, Lefkowitz RJ, Snyderman R (1983) Guanine nucleotides modulate the binding affinity of the oligopeptide chemoattractant receptor on human polymorphonuclear leukocytes. J Clin Invest 72:748–753

Koo C, Sherman JW, Mendelson MA, Harvey J, Goetzl EJ (1989) Diversity of leukotriene receptors on human leukocytes. In: Biology of cellular transducting signals '89. Vanderhoek JY (ed) 9th International Washington Spring Symposium, Washington p 37

Korchak HM, Weissmann G (1978) Changes in membrane potential of human granulocytes antecede the metabolic responses to surface stimulation. Proc Natl Acad Sci USA 75:3818–3822

Korchak HM, Weissmann G (1980) Stimulus-response coupling in the human neutrophil transmembrane potential and the role of extracellular Na^+. Biochim Biophys Acta 601:180–194

Korchak HM, Eisenstat BA, Hoffstein ST, Dunham PB, Weissmann G (1980) Anion channel blockers inhibit lysosomal enzyme secretion from human neutrophils without affecting generation of superoxide anion. Proc Natl Acad Sci USA 77:2721–2725

Korchak HM, Eisenstat BA, Smolen JE, Rutherford LE, Dunham PB, Weissmann G (1982) Stimulus-response coupling in the human neutrophil. The role of anion fluxes in degranulation. J Biol Chem 257:6919–6922

Korchak HM, Roos D, Giedd KN, Wynkoop EM, Vienne K, Rutherford LE, Buyon JP, Rich AM, Weissmann G (1983) Granulocytes without degranulation: neutrophil function in granule-depleted cytoplasts. Proc Natl Acad Sci USA 80:4968–4972

Korchak HM, Rutherford LE, Weissmann G (1984a) Stimulus response coupling in the human neutrophil. I. Kinetic analysis of changes in calcium permeability. J Biol Chem 259:4070–4075

Korchak HM, Vienne K, Rutherford LE, Wilkenfeld C, Finkelstein MC, Weissmann G (1984b) Stimulus response coupling in the human neutrophil. II. Temporal analysis of changes in cytosolic calcium and calcium efflux. J Biol Chem 259:4076–4082

Korchak HM, Wilkenfeld C, Rich AM, Radin AR, Vienne K, Rutherford LE (1984c) Stimulus response coupling in the human neutrophil. Differential requirements for receptor occupancy in neutrophil responses to a chemoattractant. J Biol Chem 259:7439–7445

Korchak HM, Vienne K, Rutherford LE, Weissmann G (1984d) Neutrophil stimulation: receptor, membrane, and metabolic events. Fed Proc 43:2749–2754

Korchak HM, Vosshall LB, Zagon G, Ljubich P, Rich AM, Weissmann G (1988a) Activation of the neutrophil by calcium-mobilizing ligands. I. A chemotactic peptide and the lectin concanavalin A stimulate superoxide anion generation but elicit different calcium movements and phosphoinositide remodeling. J Biol Chem 263:11090–11097

Korchak HM, Vosshall LB, Haines KA, Wilkenfeld C, Lundquist KF, Weissmann G (1988b) Activation of the human neutrophil by calcium-mobilizing ligands. II. Correlation of calcium, diacyl glycerol, and phosphatidic acid generation with superoxide anion generation. J Biol Chem 263:11098–11105

Korn ED (1982) Actin polymerization and its regulation by proteins from nonmuscle cells. Physiol Rev 62:672–737

Kotani S, Tsujimoto M, Koga T, Nagao T, Tanaka A, Kawata S (1986) Chemical structure and biological activity relationship of bacterial cell walls and muramyl peptides. Fed Proc 45:2534–2540

Kownatzki E, Uhrich S (1987) Differential effects of nylon fibre adherence on the production of superoxide anion by human polymorphonuclear neutrophilic granulocytes stimulated with chemoattractants, ionophore A23187 and phorbol myristate acetate. Clin Exp Immunol 69:213–220

Kownatzki E, Uhrich S, Grüninger G (1988a) Functional properties of a novel neutrophil chemotactic factor derived from human monocytes. Immunobiology 177:352–362

Kownatzki E, Kapp A, Uhrich S (1988b) Modulation of human neutrophilic granulocyte functions by recombinant human tumor necrosis factor and recombinant human lymphotoxin. Promotion of adherence, inhibition of chemotactic migration and superoxide anion release from adherent cells. Clin Exp Immunol 74:143–148

Kownatzki E, Neumann M, Uhrich S (1989) Stimulation of human neutrophilic granulocytes by two monocyte-derived cytokines. Agents Actions 26:180–182

Kraft AS, Anderson WB (1983) Phorbol esters increase the amount of Ca^{2+}, phospholipid-dependent protein kinase associated with plasma membrane. Nature 301:621–623

Kraft AS, Berkow RL (1987) Tyrosine kinase and phosphotyrosine phosphatase activity in human promyelocytic leukemia cells and human polymorphonuclear leukocytes. Blood 70:356–362

Kraft AS, Smith JB, Berkow RL (1986) Bryostatin, an activator of the calcium phospholipid-dependent protein kinase, blocks phorbol ester-induced differentiation of human promyelocytic leukemia cells HL-60. Proc Natl Acad Sci USA 83:1334–1338

Kramer IM, van der Bend RL, Verhoeven AJ, Roos D (1988a) The 47-kDa protein involved in the $NADPH:O_2$ oxidoreductase activity of human neutrophils is phosphorylated bycyclic AMP-dependent protein kinase without induction of a respiratory burst. Biochim Biophys Acta 971:189–196

Kramer IM, Verhoeven AJ, van der Bend RL, Weening RS, Roos D (1988b) Purified protein kinase C phosphorylates a 47 kDa protein in control neutrophil cytoplasts but not in neutrophil cytoplasts from patients with the autosomal form of chronic granulomatous disease. J Biol Chem 263:2352–2357

Kramer IM, van der Bend RL, Tool ATJ, van Blitterswijk WJ, Roos D, Verhoeven AJ (1989) 1-O-Hexadecyl-2-O-methylglycerol, a novel inhibitor of protein kinase C, inhibits the respiratory burst in human neutrophils. J Biol Chem 264:5876–5884

Kraus JL, DiPaola A, Belleau B (1984) Cyclic tetrameric clusters of chemotactic peptides as superactive activators of lysozyme release from human neutrophils. Biochem Biophys Res Commun 124:939–944

Krause KH, Lew PD (1987) Subcellular distribution of Ca^{2+} pumping sites in human neutrophils. J Clin Invest 80:107–116

Krause KH, Schlegel W, Wollheim CB, Andersson T, Waldvogel FA, Lew PD (1985) Chemotactic peptide activation of human neutrophils and HL-60 cells. Pertussis toxin reveals correlation between inositol triphosphate generation, calcium ion transients, and cellular activation. J Clin Invest 76:1348–1354

Kreisle RA, Parker CW (1983) Specific binding of leukotriene B_4 to a receptor on human polymorphonuclear leukocytes. J Exp Med 157:628:641

Krishnamurthi S, Patel Y, Kakkar VV (1989) Weak inhibition of protein kinase C coupled with various non-specific effects make sphingosine an unsuitable tool in patelet signal transduction studies. Biochim Biophys Acta 1010:258–264

Kroegel C, Yukawa T, Dent G, Venge P, Chung KF, Barnes PJ (1989) Stimulation of degranulation from human eosinophils by platelet-activating factor. J Immunol 142:3518–3526

Kuhns DB, Kaplan SS, Basford RE (1986) Hexachlorocyclohexanes, potent stimuli of O_2^- production and calcium release in human polymorphonuclear leukocytes. Blood 68:535–540

Kuhns DB, Wright DG, Nath J, Kaplan SS, Basford RE (1988) ATP induces transient elevations of $[Ca^{2+}]_i$ in human neutrophils and primes these cells for enhanced O_2^- generation. Lab Invest 58:448–453

Kukreja RC, Weaver AB, Hess ML (1989) Stimulated human neutrophils damage cardiac sarcoplasmic reticulum function by generation of oxidants. Biochim Biophys Acta 990:198–205

Kuroki M, Minakami S (1989) Extracellular ATP triggers superoxide production in human neutrophils. Biochem Biophys Res Commun 162:377–380

Kuroki M, Kamo N, Kobatake Y, Okimasu E, Utsumi K (1982) Measurement of membrane potential in polymorphonuclear leukocytes and its changes during surface stimulation. Biochim Biophys Acta 693:326–334

Kusner DJ, King CH (1989) Protease-modulation of neutrophil superoxide response. J Immunol 143:1696–1702

Lacal PM, Balsinde J, Cabanas C, Bernabeau C, Sanchez-Madrid F, Mollinedo F (1990) The CD11c antigen couples concanavalin A binding to generation of superoxide anion in human phagocytes. Biochem J 268:707–712

Lackie JM, Lawrence AJ (1987) Signal response transduction in rabbit neutrophil leucocytes. The effects of exogenous phospholipase A_2 suggest two pathways exist. Biochem Pharmacol 36:1941–1945

Lad PM, Glovsky MM, Smiley PA, Klempner M, Reisinger DM, Richards JH (1984) The β-adrenergic receptor in the human neutrophil plasma membrane: Receptor-cyclase uncoupling is associated with amplified GTP activation. J Immunol 132:1466–1471

Lad PM, Goldberg BJ, Smiley PA, Olson CV (1985a) Receptor-specific threshold effects of cyclic AMP are involved in the regulation of enzyme release and superoxide production from human neutrophils. Biochim Biophys Acta 846:286–295

Lad PM, Olson CV, Smiley PA (1985b) Association of the N-formy-Met-Leu-Phe receptor in human neutrophils with a GTP-binding protein sensitive to pertussis toxin. Proc Natl Acad Sci USA 82:869–873

Lad PM, Olson CV, Grewal IS (1985c) Platelet-activating factor mediated effects on human neutrophil function are inhibited by pertussis toxin. Biochem Biophys Res Commun 129:632–638

Lad PM, Olson CV, Grewal IS, Scott SJ (1985d) A pertussis toxin-sensitive GTP-binding protein in the human neutrophil regulates multiple receptors, calcium mobilization, and lectin-induced capping. Proc Natl Acad Sci USA 82:8643–8647

Lad PM, Glovsky MM, Richards JH, Smiley PA, Backstrom B (1985e) Regulation of human neutrophil guanylate cyclase by metal ions, free radicals and the muscarinic cholinergic receptor. Mol Immunol 22:731–739

Lad PM, Olson CV, Grewal IS (1986) Role of a pertussis toxin substrate in the control of lectin-induced cap formation in human neutrophils. Biochem J 238:29–36

Lagunoff D, Martin TW, Read G (1983) Agents that release histamine from mast cells. Annu Rev Pharmacol Toxicol 23:331–351

Laine E, Couderc R, Roch-Arveiller M, Vasson MP, Giroud JP, Raichvarg D (1990) Modulation of human polymorphonuclear neutrophil functions by α_1-acid glycoprotein. Inflammation 14:1–9

Lambeth JD (1988) Activation of the respiratory burst oxidase in neutrophils: on the role of membrane-derived second messengers, Ca^{++}, and protein kinase C. J Bioenerg Biomembr 20:709–733

Lambeth JD, Burnham DN, Tyagi SR (1988) Sphinganine effects on chemoattractant-induced diacylglycerol generation, calcium fluxes, superoxide production, and on cell viability in the human neutrophil. Delivery of sphinganine with bovine serum albumin minimizes cytotoxicity without affecting inhibition of the respiratory burst. J Biol Chem 263:3818–3822

Laporte F, Doussiere J, Vignais PV (1990) Respiratory burst of rabbit paritoneal neutrophils. Transition from an NADPH diaphorase activity to an O_2^- generating oxidase activity. Eur J Biochem 194:301–308

Largent BL, Walton KM, Hoppe CA, Lee YC, Schnaar RL (1984) Carbohydrate-specific adhesion of alveolar macrophages to mannose-dervatized surfaces. J Biol Chem 259:1764–1769

Larrick JW, Graham D, Toy K, Lin LS, Senyk G, Fendly BM (1987) Recombinant tumor necrosis factor causes activation of human granulocytes. Blood 69:640–644

Larsen NE, Enelow RI, Simons ER, Sullivan R (1985) Effect of bacterial endotoxin on the transmembrane electrical potential and plasma membrane fluidity of human monocytes. Biochim Biophys Acta 815:1–8

Laskin DL, Kimura T, Sakakibara S, Riley DJ, Berg RA (1986) Chemotactic activity of collagen-like polypeptides for human peripheral blood neutrophils. J Leukocyte Biol 39:255–266

Laskin DL, Sirak AA, Pilaro AM, Laskin JD (1988) Functional and biochemical properties of rat Kupffer cells and peritoneal macrophages. J Leukocyte Biol 44:71–78

Laudanna C, Miron S, Berton G, Rossi F (1990) Tumor necrosis factor-α/cachectin activates the O_2^--generating system of human neutrophils independently of the hydrolysis of phosphoinositides and the release of arachidonic acid. Biochem Biophys Res Commun 166:308–315

Lazzari KG, Proto PJ, Simons ER (1986) Simultaneous measurement of stimulus-induced changes in cytoplasmic Ca^{2+} and in membrane potential of human neutrophils. J Biol Chem 261:9710–9713

Leb L, Beatson P, Fortier N, Newburger PE, Snyder LM (1985) Modulation of mononuclear phagocyte cytotoxicity by alpha-tocopherol (vitamin E). J Leukocyte Biol 37:449–459

Leca G, Joly F, Vazquez A, Galanaud P, Ninio E (1990) Paf-acether-induced superoxide anion generation in human B cell line. FEBS Lett 269:171–173

Lee GHD, Kaptein JS, Scott SJ, Niedzin H, Kalunta CI, Lad PM (1989) Desensitization of calcium mobilization and cell function in human neutrophils. Biochem J 262:165–172

Lefkowitz RJ, Caron MG (1986) Regulation of adrenergic receptor function by phosphorylation. Current Top Cell Regul 28:209–231

Lehmeyer JE, Johnston RB Jr (1978) Effect of anti-inflammatory drugs and agents that elevate intracellular cyclic AMP on the release of toxic oxygen metabolites by phagocytes: Studies in a model of tissue-bound IgG. Clin Immunol Immunopathol 9:482–490

Lehmeyer JE, Snyderman R, Johnston RB Jr (1979) Stimulation of neutrophil oxidative metabolism by chemotactic peptides. Influence of calcium ion concentration and cytochalasin B and comparison with stimulation by phorbol myristate acetate. Blood 54:35–45

Lehn M, Weiser WY, Engelhorn S, Gillis S, Remold HG (1989) IL-4 inhibits H_2O_2 production and antileishmanial capacity of human cultured monocytes mediated by IFN-γ. J Immunol 143:3020–3024

Lehrer RI, Cohen L (1981) Receptor-mediated regulation of superoxide production in human neutrophils stimulated by phorbol myristate acetate. J Clin Invest 68:1314–1320

Lengyel P (1982) Biochemistry of interferons and their actions. Annu Rev Biochem 51:251–282

Lepay DA, Steinman RM, Nathan CF, Murray HW, Cohn ZA (1985) Liver macrophages in murine listeriosis. Cell-mediated immunity is correlated with an influx of macrophages capable of generating reacting oxygen intermediates. J Exp Med 161:1503–1512

Lepoivre M, Tenu JP, Petit JF (1982) Transmembrane potential variations accompanying the PMA-triggered O_2^- and H_2O_2 release by mouse peritoneal macrophages. FEBS Lett 149:233–239

Leslie CC, Detty DM (1986) Arachidonic acid turnover in response to lipopolysaccharide and opsonized zymosan in human monocyte-derived macrophages. Biochem J 236:251–259

Leto TL, Lomax KJ, Sechler JMG, Nunoi H, Gallin JI, Malech HL (1989) Molecular cloning of the 65 kD cytosolic factor absent in a rare form of autosomal recessive chronic granulomatous disease. Clin Res 37:547A

Leto TL, Lomax KJ, Volpp BD, Nunoi H, Sechler JMG, Nauseef WM, Clark RA, Gallin JI, Malech HL (1990) Cloning of a 67-kD neutrophil oxidase factor with similarity to a noncatalytic region of p60$^{c\text{-}src}$. Science 248:727–730

Levy R, Malech HL, Retrosen D (1990a) Production of myeloid cell cytosols functionally and immunochemically deficient in the 47-kDa or 67-kDa NADPH oxidase cytosolic factors. Biochem Biophys Res Commun 170:1114–1120

Levy R, Rotrosen D, Nagauker D, Leto TL, Malech HL (1990b) Induction of the respiratory burst in HL-60 cells. Correlation of function and protein expression. J Immunol 145:2596–2601

Lew PD, Stossel TP (1981) Effect of calcium on superoxide production by phagocytic vesicles from rabbit alveolar macrophages. J Clin Invest 67:1–9

Lew PD, Southwick FS, Stossel TP, Whitin JC, Simons E, Cohen HJ (1981) A variant of chronic granulomatous disease: deficient oxidative metabolism due to a low-affinity NADPH oxidase. N Engl J Med 305:1329–1333

Lew PD, Dayer JM, Wollheim CB, Pozzan T (1984a) Effect of leukotriene B_4, prostaglandin E_2 and arachidonic acid on cytosolic-free calcium in human neutrophils. FEBS Lett 166:44–48

Lew PD, Wollheim CB, Waldvogel FA, Pozzan T (1984b) Modulation of cytosolic-free calcium transients by changes in intracellular calcium-buffering capacity: correlation with exocytosis and O_2^- production in human neutrophils. J Cell Biol 99:1212–1220

Lew PD, Favre L, Waldvogel FA, Vallotton MB (1985) Cytosolic free calcium and intracellular calcium stores in neutrophils from hypertensive subjects. Clin Sci 69:227–230

Lew PD, Monod A, Waldvogel FA, Pozzan T (1987) Role of cytosolic free calcium and phospholipase C in leukotriene-B_4-stimulated secretion in human neutrophils. Comparison with the chemotactic peptide formyl-methionyl-leucyl-phenylalanine. Eur J Biochem 162:;161–168

Lewis SL, Van Epps DE (1987) Neutrophil and monocyte alterations in chronic dialysis patients. Am J Kidney Dis 9:381–395

Lichtenstein A (1987) Stimulation of the respiratory burst of murine peritoneal inflammatory neutrophils by conjugation with tumor cells. Cancer Res 47:2211–2217

Lieners C, Redl H, Schlag G, Hammerschmidt DE (1989) Inhibition by halothane, but not by isoflurane, of oxidative response to opsonized zymosan in whole blood. Inflammation 13:621–630

Ligeti E, Doussiere J, Vignais PV (1988) Activation of the O_2^--generating oxidase in plasma membranes from bovine polymorphonuclear neutrophils by arachidonic acid, a cytosolic factor of protein nature, and nonhydrolyzable analogues of GTP. Biochemistry 27:193–200

Ligeti E, Tardif M, Vignais PV (1989) Activation of O_2^--generating oxidase of bovine neutrophils in a cell-free system. Interaction of a cytosolic factor with the plasma membrane and control by G nucleotides. Biochemistry 28:7116–7123

Light DR, Walsh C, O'Callaghan AM, Goetzl EJ, Tauber AI (1981) Characteristics of cofactor requirements for the superoxide-generating NADPH oxidase of human polymorphonuclear leukocytes. Biochemistry 20:1468–1476

Lim LK, Hunt NH, Weidemann MJ (1983) Reactive oxygen production, arachidonate metabolism and cyclic AMP in macrophages. Biochem Biophys Res Commun 114:549–555

Lin AH, Ruppel PL, Gorman RR (1984) Leukotriene B_4 binding to human neutrophils. Prostaglandins 28:837–849

Lin AH, Ruppel PL, Gorman RR (1985) Detergent solubilization of human neutrophil leukotriene B_4 receptors. Biochem Biophys Res Commun 128:878–883

Linden DJ, Murakami K, Routtenberg A (1986) A newly discovered protein kinase C activator (oleic acid) enhances long-term potentiation in the intact hippocampus. Brain Res 379:358–363

Lindley I, Aschauer H, Seifert JM, Lam C, Brunowsky W, Kownatzki E, Thelen M, Peveri P, Dewald B, von Tscharner V, Walz A, Baggiolini M (1988) Synthesis and expression in *Escherichia coli* of the gene encoding monocyte-derived neutrophil-activating factor: biological equivalence between natural and recombinant neutrophil-activating factor. Proc Natl Acad Sci USA 85:9199–9203

Lindquist S (1986) The heat-shock response. Annu Rev Biochem 55:1151–1191

Lis H, Sharon N (1986) Lectins as molecules and as tools. Annu Rev Biochem 55:35–67

Lochner JE, Badwey JA, Horn W, Karnovsky ML (1986) all-*trans*-Retinal stimulates superoxide release and phospholipase C activity in neutrophils without significantly blocking protein kinase C. Proc Natl Acad Sci USA 83:7673–7677

Lock R, Johansson A, Orseilus K, Dahlgren C (1988) Analysis of horseradish peroxidase-amplified chemiluminescence produced by human neutrophils reveals a role for the superoxide anion in the light emitting reaction. Anal Biochem 173:450–455

Lohr KM, Snyderman R (1982) Amphotericin B alters the affinity and functional activity of the oligopeptide chemotactic factor receptor on human polymorphonuclear leukocytes. J Immunol 129:1594–1599

Lomax KJ, Leto TL, Nunoi H, Gallin JI, Malech HL (1989) Recombinant 47-kilodalton cytosol factor restores NADPH oxidase in chronic granulomatous disease. Science 245:409–412

Looney RJ, Ryan DH, Takahashi K, Fleit HB, Cohen HJ, Abraham GN, Anderson CL (1986) Identification of a second class of IgG Fc receptors on human neutrophils. A 40 kilodalton molecule also found on eosinophils. J Exp Med 163:826–836

Lopez AF, Nicola NA, Burgess AW, Metcalf D, Battye FL, Sewell WA, Vadas M (1983) Activation of granulocyte cytotoxic function by purified mouse colony-stimulating factors. J Immunol 131:2983–2988

Lopez AF, Williamson DJ, Gamble JR, Begley CG, Harlan JM, Klebanoff SJ, Waltersdorph A, Wong G, Clark SC, Vadas MA (1986) Recombinant human granulocyte-macrophage colony-stimulating factor stimulates in vitro mature human neutrophil and eosinophil function, surface receptor expression and survival. J Clin Invest 78:1220–1228

Lopker A, Abood LG, Hoss W, Lionetti FJ (1980) Stereoselective muscarinic acetylcholine and opiate receptors in human phagocytic leukocytes. Biochem Pharmacol 29:1361–1365

Love JT Jr, Padula SJ, Lingenheld EG, Amin JK, Sgroi DC, Wong RL, Sha'afi RI, Clark RB (1989) Effects of H-7 are not exclusively mediated through protein kinase C or the cyclic nucleotide-dependent kinases. Biochem Biophys Res Commun 162:138–143

Lu DJ, Grinstein S (1989) Concanavalin A stimulation of O_2 consumption in electropermeabilized neutrophils via a pertussis toxin-insensitive G protein. FEBS Lett 253:151–156

Lu DJ, Grinstein S (1990) ATP and guanine nucleotide dependence of neutrophil activation. Evidence for the involvement of two distinct GTP-binding proteins. J Biol Chem 265:13721–13729

Lucchi L, Cappelli G, Acerbi MA, Spattini A, Lusvarghi E (1989) Oxidative metabolism of polymorphonuclear leukocytes and serum opsonic activity in chronic renal failure. Nephron 51:44–50

Lüderitz O, Galanos C, Lehmann V, Mayer H, Rietschel ET, Weckesser J (1978) Chemical structure and biological activities of lipid A's from various bacterial families. Naturwissenschaften 65:578–585

Lukey PT, Anderson R, Dippenaar UH (1988) Benoxaprofen activates membrane-associated oxidative metabolism in human polymorphonuclear leukocytes by apparent modulation of protein kinase C. Br J Pharmacol 93:289–294

Lund-Johansen F, Olweus J, Aarli A, Bjerkens R (1990) Signal transduction in human monocytes and granulocytes through the PI-linked antigen CD14. FEBS Lett 273:55–58

Lunec J, Griffiths HR, Blake DR (1987) Oxygen radical in inflammation. In: Grimloade AM (editor and publisher) Philadelphia Pharmacology. pp 45–48 (ISI atlas of science)

Lutter R, van Zwieten R, Weening RS, Hamers MN, Roos D (1984) Cytochrome b, flavins, and ubiquinone-50 in enucleated human neutrophils (polymorphonuclear leukocyte cytoplasts). J Biol Chem 259:9603–9606

Lutter R, van Schaik MLJ, van Zwieten R, Wever R, Roos D, Hamers MN (1985) Purification and partial characterization of the b-type cytochrome from human polymorphonuclear leukocytes. J Biol Chem 260:2237–2244

Macdonald G, Assef R, Guiffre A, Lo E (1984) Vasoconstrictor effects of uridine and its nucleotides and their inhibition by adenosine. Clin Exp Pharmacol Physiol 11:381–384

Machoczek K, Fischer M, Söling HD (1989) Lipocortin I and lipocortin II inhibit phosphoinositide- and polyphosphoinositide-specific phospholipase C. The effect results from interaction with the substrate. FEBS Lett 251:207–212

Mack JA, Nielson CP, Stevens DL, Vestal RE (1986) β-Adrenoceptor-mediated modulation of calcium ionophore activated polymorphonuclear leukocytes. Br J Pharmacol 88:417–423

Mahadevappa VG (1988) [³H]Phosphatidic acid formed in response to FMLP is not inhibited by R59 022, a diacylglycerol kinase inhibitor. Biochem Biophys Res Commun 153:1097–1104

Mahoney CW, Azzi A (1988) Vitamin E inhibits protein kinase C activity. Biochem Biophys Res Commun 154:694–697

Mahoney CW, Hensey CE, Azzi A (1989) Auranofin, gold thiomalate, and gold thioglucose inhibit protein kinase C. Biochem Pharmacol 38:3383–3386

Mahoney CW, Azzi A, Huang KP (1990) Effects of suramin, and anti-human immunodeficiency virus reverse transcriptase agent, on protein kinase C. Differential activation and inhibition of protein kinase C isozymes. J Biol Chem 265:5424–5428

Majander A, Wikström M (1989) The plasma membrane potential of human neutrophils. Role of ion channels and the sodium/potassium pump. Biochim Biophs Acta 980:139–145

Majerus PW, Connolly TM, Deckmyn H, Ross TS, Bross TE, Ishii H, Bansal VS, Wilson DB (1986) The metabolism of phosphoinositide-derived messenger molecules. Science 234:1519–1526

Makowske M, Ballester R, Cayre Y, Rosen OM (1988) Immunochemical evidence that three protein kinase C isozymes increase in abundance during HL-60 differentiation induced by dimethyl sulfoxide and retinoic acid. J Biol Chem 263:3402–3410

Malawista SE, van Blaricom G (1987) Cytoplasts made from human blood polymorphonuclear leukocytes with our without heat: preservation of both motile function and respiratory burst oxidase activity. Proc Natl Acad Sci USA 84:454–458

Malech HL, Gallin MD (1987) Neutrophils in human diseases. N Engl J Med 317:687–694

Malech HL, Gardner JP, Heiman DF, Rosenzweig SA (1985) Asparagine-linked oligosaccharides on formyl peptide chemotactic receptors of human phagocytic cells. J Biol Chem 260:2509–2514

Mallery SR, Zeligs BJ, Ramwell PW, Bellanti JA (1986) Gender-related variations and interaction of human neutrophil cyclooxygenase and oxidative burst metabolites. J Leukocyte Biol 40:133–146

Maly FE, Cross AR, Jones OTG, Wolf-Vorbeck G, Walker C, Dahinden CA, De Weck AL (1988) The superoxide generating system of B cell lines. Structural homology with the phagocytic oxidase and triggering via surface Ig. J Immunol 140:2334–2339

Maly FE, Nakamura M, Gauchat JF, Urwyler A, Walker C, Dahinden CA, Cross AR, Jones OTG, De Weck AL (1989) Superoxide-dependent nitroblue tetrazolium reduction and expression of cytochrome b_{-245} components by human tonsillar B lymphocytes and B cell lines. J Immunol 142:1260–1267

Manara FS, Schneider DL (1985) The activation of the human neutrophil respiratory burst occurs only at temperature above 17°C : evidence that activation requires membrane fusion. Biochem Biophys Res Commun 132:696–701

Marasco WA, Showell HJ, Becker EL (1981) Substance P binds to the formylpeptide chemotaxis receptor on the rabbit neutrophil. Biochem Biophys Res Commun 99:1065–1072

Marasco WA, Phan SH, Krutzsch H, Showell HJ, Feltner DE, Nairn R, Becker EL, Ward PA (1984) Purification and identification of formyl-methionyl-leucyl-phenylalanine as the major peptide neutrophil chemotactic factor produced by Escherichia coli. J Biol Chem 259:5430–5439

Marasco WA, Becker KM, Feltner DE, Brown CS, Ward PA, Nairn R (1985) Covalent affinity labeling, detergent solubilization, and fluid-phase characterization of the rabbit neutrophil formyl peptide chemotaxis receptor. Biochemistry 24:2227–2236

Maridonneau-Parini I, Tauber AI (1986) Activation of NADPH oxidase by arachidonic acid involves phospholipase A_2 in intact human neutrophils but not in the cell-free system. Biochem Biophys Res Commun 138:1099–1105

Maridonneau-Parini I, Tringale SM, Tauber AI (1986) Identification of distinct activation pathways of the human neutrophil NADPH oxidase. J Immunol 137:2925–2929

Maridonneau-Parini I, Clerc J, Polla BS (1988) Heat shock inhibits NADPH oxidase in human neutrophils. Biochem Biophys Res Commun 154:179–186

Maridonneau-Parini I, Errasfa M, Russo-Marie F (1989) Inhibition of O_2^- generation by dexamethasone is mimicked by lipocortin I in alveolar macrophages. J Clin Invest 83:1936–1940

Markert M, Andrews PC, Babior BM (1984) Measurement of O_2^- production by human neutrophils. The preparation and assay of NADPH oxidase-containing particles from human neutrophils. Methods Enzymol 105:358–365

Markert M, Glass GA, Babior BM (1985) Respiratory burst oxidase from human neutrophils: purification and some properties. Proc Natl Acad Sci USA 82:3144–3148

Marletta MA, Yoon PS, Iyengar R, Leaf CD, Wishnok JS (1988) Macrophage oxidation of L-arginine to nitrite and nitrate: nitric oxide is an intermediate. Biochemistry 27:8706–8711

Marodi L, Kalmar A, Karmazsin L (1990) Stimulation of the respiratory burst and promotion of bacterial killing in human granulocytes by intravenous immunoglobulin preparations. Clin Exp Immunol 79:164–169

Martin M, Resch K (1988) Interleukin 1: more than a mediator between leukocytes. Trends Pharmacol Sci 9:171–177

Martin MA, Nauseef WM, Clark RA (1988) Depolarization blunts the oxidative burst of human neutrophils. Parallel effects of monoclonal antibodies, depolarizing buffers, and glycolytic inhibitors. J Immunol 140:3928–3935

Masucci G, Szigeti R, Stevens D, Masucci MG, Klein E, Petersen J, Bendtzen K, Klein G (1984) Production of leukocyte migration inhibitory factory (LIF) in human lymphocyte subsets exposed to polyclonal activators. Cell Immunol 85:511–518

Matsubara T, Ziff M (1986a) Superoxide anion release by human endothelial cells: Synergism between a phorbol ester and a calcium ionophore. J Cell Physiol 127:207–210

Matsubara T, Ziff M (1986b) Increased superoxide anion release from human endothelial cells in response to cytokines. J Immunol 137:3295–3298

Matsuda I, Oka Y, Taniguchi N, Furuyama M, Kodama S, Arashima S, Mitsuyama T (1976) Leukocyte glutathione peroxidase deficiency in a male patient with chronic granulomatous disease. J Pediatr 88:581–583

Matsui H, Wada O, Ushijima Y, Mizuta T (1983a) Inhibition of oxidative metabolism in rabbit polymorphonuclear leukocytes by triphenyltin chloride. Arch Toxicol 54:227–233

Matsui H, Wada O, Ushijima Y, Akuzawa T (1983b) Triphenyltin chloride inhibits superoxide production by human neutrophils stimulated with a surface active agents. FEBS Lett 164:251–254

Matsumoto T, Takeshige K, Minakami S (1979) Inhibition of phagocytotic metabolic changes of leukocytes by an intracellular calcium-antagonist 8-(N,N-diethylamino)-octyl-3,4,5-trimethoxybenzoate. Biochem Biophys Res Commun 88:974–979

Matsumoto T, Molski TFP, Volpi M, Pelz C, Kanaho Y, Becker EL, Feinstein MB, Naccache PH, Sha'afi RI (1986) Treatment of rabbit neutrophils with phorbol esters results in increased ADP-ribosylation catalyzed by pertussis toxin and inhibition of the GTPase stimulated by fMet-Leu-Phe. FEBS Lett 198:295–300

Matsumoto T, Molski TFP, Kanaho Y, Becker EL, Sha'afi RI (1987) G-Protein dissociation, GTP-GDP exchange and GTPase activity in control and PMA treated neutrophils stimulated by fMet-Leu-Phe. Biochem Biophys Res Commun 143:489–498

Matsumoto T, Tao W, Sha'afi RI (1988) Demonstration of calcium-dependent phospholipase A_2 activity in membrane preparation of rabbit neutrophils. Absence of activation by fMet-Leu-Phe, phorbol 12-myristate 13-acetate and A-kinase. Biochem J 250:343–348

Matsuno T, Orita K, Edashige K, Kobuchi H, Sato EF, Inouye B, Inoue M, Utsumi K (1990) Inhibition of active oxygen generation in guinea pig neutrophils by bis-coclaurine alkaloids. Biochem Pharmacol 39:1255–1259

Matsuo S, Nakagawara A, Ikeda K, Mitsuyama M (1985) Enhanced release of reactive oxygen intermediates by immunologically activated rat Kupffer cells. Clin Exp Immunol 59:203–209

Matsuoka I, Sakuma H, Syuto B, Moriishi K, Kubo S, Kurihara K (1989) ADP-ribosylation of 24-26-kDa GTP-binding proteins localized in neuronal and non-neuronal cells by botulinum neurotoxin D. J Biol Chem 264:706–712

May JM, de Haen C (1979) Insulin-stimulated intracellular hydrogen peroxide production in rat epididymal fat cells. J Biol Chem 254:2214–2220

Mazzei GJ, Katoh N, Kuo JF (1982) Polymyxin B is a more selective inhibitor for phospholipid-sensitive Ca^{2+}-dependent protein kinase than for calmodulin-sensitive Ca^{2+}-dependent protein kinase. Biochem Biophys Res Commun 109:1129–1133

McCarthy SA, Hallam TJ, Merritt JE (1989) Activation of protein kinase C in human neutrophils attenuates agonist-stimulated rises in cytosolic free Ca^{2+} concentrations by inhibiting bivalent-cation influx and intracellular Ca^{2+} release in addition to stimulating Ca^{2+} efflux. Biochem J 264:357–364

McColl SR, Kreis C, Di Persio JF, Borgeat P, Naccache PH (1989) Involvement of guanine nucleotide binding proteins in neutrophil activation and priming by GM-CSF. Blood 73:588–591

McColl SR, Beauseigle O, Gilbert C, Naccache PH (1990) Priming of the human neutrophil respiratory burst by granulocyte-macrophage colony-stimulating factor and tumor necrosis factor-a involves regulation of a post-cell surface receptor level. Enhancement of the effect of agents which directly activate G proteins. J Immunol 145:3047–3053

McCord JM (1985) Oxygen-derived free radicals in postischemic tissue injury. N Engl J Med 312:159–163

McGarrity ST, Hyers TM, Webster RO (1988a) Inhibition of neutrophil functions by platelets and platelet-derived products: description of multiple inhibitory properties. J Leukocyte Biol 44:93–100

McGarrity ST, Stephenson AH, Hyers TM, Webster RO (1988b) Inhibition of neutrophil superoxide anion generation by platelet products: role of adenine nucleotides. J Leukocyte Biol 44:411–421

McGarrity ST, Stephenson AH, Webster RO (1989) Regulation of human neutrophil functions by adenine nucleotides. J Immunol 142:1986–1994

McIntyre TM, Reinhold SL, Prescott SM, Zimmerman GA (1987) Protein kinase C activity appears to be required for the synthesis of platelet-activating factor and leukotriene B_4 by human neutrophils. J Biol Chem 262:15370–15376

McLeish KR, Gierschik P, Schepers T, Sidiropoulos D, Jakobs KH (1989a) Evidence that activation of a common G-protein by receptors for leukotriene B_4 and N-formylmethionyl-leucyl-phenylalanine in HL-60 cells occurs by different mechanisms. Biochem J 260:427–434

McLeish KR, Gierschik P, Jakobs KH (1989b) Desensitization uncouples the formyl peptide receptor-guanine nucleotide-binding protein interaction in HL60 cells. Mol Pharmacol 36:384–390

McNeely TB, Turco SJ (1987) Inhibition of protein kinase C activity by the Leishmania donovani lipophosphoglycan. Biochem Biophys Res Commun 148:653–657

McPhail LC, Snyderman R (1983) Activation of the respiratory burst enzyme in human polymorphonuclear leukocytes by chemoattractants and other soluble stimuli. Evidence that the same oxidase is activated by different transductional mechanisms. J Clin Invest 72:192–200

McPhail LC, Snyderman R (1984) Mechanisms of regulating the respiratory burst in leukocytes. In: Snyderman R (ed) Regulation of leukocyte function. Plenum, New York, pp 247–281

McPhail LC, DeChatelet LR, Shirley PS (1976) Further characterization of NADPH oxidase activity of human polymorphonuclear leukocytes. J Clin Invest 58:774–780

McPhail LC, Henson PM, Johnston RB Jr (1981) Respiratory burst enzyme in human neutrophils. Evidence for multiple mechanisms of activation. J Clin Invest 67:710–716

McPhail LC, Clayton CC, Snyderman R (1984a) The NADPH oxidase of human polymorphonuclear leukocytes. Evidence for regulation by multiple signals. J Biol Chem 259:5768–5775

McPhail LC, Clayton CC, Snyderman R (1984b) A potential second messenger role for unsaturated fatty acids: activation of Ca^{2+}-dependent protein kinase. Science 224:622–625

McPhail LC, Shirley PS, Clayton CC, Snyderman R (1985) Activation of the respiratory burst enzyme from human neutrophils in a cell-free system. Evidence for a soluble cofactor. J Clin Invest 75:1735–1739

McPhail LC, Hendricks CL, Seede MC, Bass DA (1986) A possible role for calmodulin in activation of the respiratory burst enzyme of human neutrophils. Clin Res 34:465A

Meade CJ, Harvey J, Boot JR, Turner GA, Bateman PE, Osborne DJ (1984) γ-Hexachlorocyclohexane stimulation of macrophage phospholipid hydrolysis and leukotriene production. Biochem Pharmacol 33:289–293

Meade CJ, Turner GA, Bateman PE (1986) The role of polyphosphoinositides and their breakdown produts in A23187-induced release of arachidonic acid from rabbit polymorphonuclear leucocytes. Biochem J 238:425–436

Mege JL, Capo C, Benoliel AM, Bongrand P (1986) Self-limitation of the oxidative burst of rat polymorphonuclear leukocytes. J Leukocyte Biol 39:599–616

Mege JL, Volpi M, Becker EL, Sha'afi RI (1988a) Effect of botulinum D toxin on neutrophils. Biochem Biophys Res Commun 152:926–932

Mege JL, Pouget J, Capo C, Andre P, Benoliel AM, Serratrice G, Bongrand P (1988b) Myotonic dystrophy: defective respiratory burst of polymorphonuclear leukocytes. J Leukocyte Biol 44:180–186

Mege JL, Tao W, Molski TFP, Gomez-Cambronero J, Huang CK, Becker EL, Sha'afi RI (1988c) Diacylglycerol kinase inhibitor R59022 and stimulated neutrophil responses. Am J Physiol 24:C589–C594

Mege JL, Gomez-Cambronero J, Molski TFP, Becker EL, Sha'afi RI (1989) Effect of granulocyte-macrophage colony-stimulating factor on superoxide production in cytoplasts and intact human neutrophils: role of protein kinase and G-proteins. J Leukocyte Biol 46:161–168

Mehta JL, Lawson D, Mehta P (1988) Modulation of human neutrophil superoxide production by selective thromboxane synthetase inhibitor U63,557A. Life Sci 43:923–928

Meier B, Radeke HH, Selle S, Younes M, Sies H, Resch K, Habermehl GG (1989) Human fibroblasts release reactive oxygen species in response to interleukin-1 or tumour necrosis factor-α. Biochem J 263:539–545

Melchers F, Braun V, Galanos C (1975) The lipoprotein of the outer membrane of Escherichia coli: a B-lymphocyte mitogen. J Exp Med 142:473–482

Melinn M, McLaughlin H (1987) Nitroblue tetrazolium reduction in lymphocytes. J Leukocyte Biol 41:325–329

Melloni E, Pontremoli S, Michetti M, Sacco O, Sparatore B, Salamino F, Horecker BL (1985) Binding of protein kinase C to neutrophil membranes in the presence of Ca^{2+} and its activation by a Ca^{2+}-requiring proteinase. Proc Natl Acad Sci USA 82:6435–6439

Melloni E, Pontremoli S, Michetti M, Sacco O, Sparatore B, Horecker BL (1986a) The involvement of calpain in the activation of protein kinase C in neutrophils stimulated by phorbol myristic acid. J Biol Chem 261:4101–4105

Melloni E, Pontremoli S, Salamino F, Sparatore B, Michetti M, Sacco O, Horecker BL
(1986b) ATP induces the release of a neutral serine proteinase and enhances the
production of superoxide anion in membranes from phorbol ester-activated
neutrophils. J Biol Chem 261:11437–11439

Merrill AH Jr, Stevens VL (1989) Modulation of protein kinase C and diverse cell
functions by sphingosine—a pharmacologically interesting compound linking sphin-
golipids and signal transduction. Biochim Biophys Acta 1010:131–139

Meshulam DM, Blair HE, Wong BL, Charm S, Minowada J, Rocklin RE (1982)
Purification of a lymphoid cell line product with leukocyte inhibitory factor activity.
Proc Natl Acad Sci USA 79:602–605

Meshulam T, Diamond RD, Lyman CA, Wysong DR, Melnick DA (1988) Temporal
association of calcium mobilization, inositol triphosphate generation, and superoxide anion release by human neutrophils activated by serum opsonized and
nonopsonized particulate stimuli. Biochem Biophys Res Commun 150:532–539

Messner RP, Jelinek J (1970) Receptors for human γG globulin on human neutrophils.
J Clin Invest 49:2165–2171

Metcalf D (1985) The granulocyte-macrophage colony-stimulating factors. Science
229:16–22

Metcalf D (1986) The molecular biology and functions of the granulocyte-macrophage
colony-stimulating factors. Blood 67:257–267

Metzger Z, Hoffeld JT, Oppenheim JJ (1981) Regulation by PGE_2 of the production of
oxygen intermediates by LPS-activated macrophages. J Immunol 127:1109–1113

Meurer R, MacIntyre DE (1989) Lack of effect of pertussis toxin on TNF-α-induced
formation of reactive oxygen intermediates by human neutrophils. Biochem
Biophys Res Commun 159:763–769

Meurs H, Kauffman HF, Timmermans A, Van Amsterdam FTM, Koeter GH, De Vries
K (1986) Phorbol 12-myristate 13-acetate induces beta-adrenergic receptor uncoupling and non-specific desensitization of adenylate cyclase in human mononuclear
leukocytes. Biochem Pharmacol 35:4217–4222

Michell RH (1975) Inositol phospholipids and cell surface receptor function. Biochim
Biophys Acta 415:81–147

Middleton E Jr (1984) The flavonoids. Trends Pharmacol Sci 5:335–338

Millard JA, Gerard KW, Schneider DL (1979) The isolation from rat peritoneal
leukocytes of plasma membrane enriched in alkaline phosphatase and a b-type
cytochrome. Biochem Biophys Res Commun 90:312–319

Milligan G (1988) Techniques used in the identification and analysis of function of
pertussis toxin-sensitive guanine nucleotide binding proteins. Biochem J 255:1–13

Minta JO, Williams D (1986) Interactions of antirheumatic drugs with the superoxide
generation system of activated human polymorphonuclear leukocytes. J Rheumatol
13:498–504

Misaki N, Imaizumi T, Watanabe Y (1989) Cyclic AMP-dependent protein kinase
interferes with GTPγS stimulated IP_3 formation in differentiated HL-60 cell
membranes. Life Sci 45:1671–1678

Mitsuhashi M, Mitsuhashi T, Payan DG (1989) Multiple signaling pathways of
histamine H_2 receptors. Identification of an H_2 receptor-dependent Ca^{2+} mobilization pathway in human HL-60 promyelocytic leukemia cells. J Biol Chem 264:18356–
18362

Miura Y, Matsui H (1989) Inhibition by triphenyltin chloride of cytosolic free calcium
mobilization in human neutrophils stimulated by a chemotactic factor. Res Commun
Chem Pathol Pharmacol 64:331–341

Miyahara M, Watanabe S, Okimasu E, Utsumi K (1987) Charge-dependent regulation
of NADPH oxidase activity in guinea-pig polymorphonuclear leukocytes. Biochim
Biophys Acta 929:253–262

Miyahara M, Okimasu E, Uchida H, Sato EF, Yamamoto M, Utsumi K (1988) Charge-dependent regulation of NADPH oxidase activities in intact and subcellular systems of polymorphonuclear leukocytes. Biochim Biophys Acta 971:46–54

Miyake R, Tanaka Y, Tsuda T, Kaibuchi K, Kikkawa U, Nishizuka Y (1984) Activation of protein kinase C by non-phorbol tumor promoter, mezerein. Biochem Biophys Res Commun 121:649–656

Miyanoshita A, Takahashi T, Endou H (1989) Inhibitory effect of cyclic AMP on phorbol ester-stimulated production of reactive oxygen intermediates in rat glomeruli. Biochem Biophys Res Commun 165:519–525

Mizel SB (1989) The interleukins. FASEB J 3:2379–2388

Molski TFP, Naccache PH, Marsh ML, Kermode J, Becker EL, Sha'afi RL (1984) Pertussis toxin inhibits the rise in the intracellular concentration of free calcium that is induced by chemotactic factors in rabbit neutrophils: possible role of the "G proteins" in calcium mobilization. Biochem Biophys Res Commun 124:644–650

Molski TFP, Ford C, Weisman SJ, Sha'afi RI (1986) Cell alkalinization is not necessary and increased sodium influx is not sufficient for stimulated superoxide production. FEBS Lett 203:267–272

Monboisse JC, Bellon G, Dufer J, Randoux A, Borel JP (1987) Collagen activates superoxide anion production by human polymorphonuclear neutrophils. Biochem J 246:599–603

Mong S, Chi-Rosso G, Miller J, Hoffman K, Razgaitis KA, Bender P, Crooke ST (1986) Leukotriene B_4 induces formation of inositol phosphates in rat peritoneal polymorphonuclear leukocytes. Mol Pharmacol 30:235–242

Moon BC, Girotti MJ, Wren SFG, Dawson R, Brar D (1986) Effect of antibiotics and sedatives on normal neutrophil nicotinamide-adenine dinucleotide phosphate-reduced oxidase activity. Arch Surg 121:73–76

Morel F, Vignais PV (1984) Examination of the oxidase function of the b-type cytochrome in human polymorphonuclear leucocytes. Biochim Biophys Acta 764:213–225

Morel F, Dianoux AC, Vignais PV (1988) Superoxide anion measurement by sulfonated phenyl isothiocyanate cytochrome C. Biochem Biophys Res Commun 156:1175–1181

Morell GP, Niaudet P, Jean G, Descamps-Latscha B (1985) Altered oxidative metabolism, motility, and adherence in phagocytic cells from cystinotic children. Pediatr Res 19:1318–1321

Morikawa K, Sato T, Nakazawa H, Fujita M (1986) Calcium-independent induction of cytocidal activity of polymorphonuclear leukocytes by phorbol myristate acetate-like tumor promoters. Life Sci 38:1073–1079

Morikawa M, Abe M, Yamauchi Y, Inoue M, Tsuboi M (1990) Priming effect of 2,3-dibenzylbutane-1,4-diol (mammalian lignan) on superoxide production in human neutrophils. Biochem Biophys Res Commun 168:194–199

Morimoto YM, Sato E, Nobori K, Takahashi R, Utsumi K (1986) Effect of calcium ion on fatty acid-induced generation of superoxide in guinea pig neutrophils. Cell Struct Funct 11:143–155

Morin MJ, Kreutter D, Rasmussen H, Sartorelli AC (1987) Disparate effects of activators of protein kinase C on HL-60 promyelocytic leukemia cell differentiation. J Biol Chem 262:11758–11763

Morstyn G, Burgess AW (1988) Hemopoietic growth factors: a review. Cancer Res 48:5624–5637

Morstyn G, Lieschke GJ, Sheridan W, Layton J, Cebon J (1989) Pharmacology of the colony-stimulating factors. Trends Pharmacol Sci 10:154–159

Mossman B, Light W, Wei E (1983) Asbestos: mechanisms of toxicity and carcinogenicity in the respiratory tract. Annu Rev Pharmacol Toxicol 23:595–615

Motulsky HJ, Smith D, Terman BI, Feldman BI (1987) Regulation of hormone-stimulated cyclic AMP accumulation in intact human mononuclear leukocytes by blood plasma. J Cyclic Nucleotide Protein Phosphor Res 11:329–343

Mousli M, Bronner C, Landry Y, Bockaert J, Rouot B (1990) Direct activation of GTP-binding regulatory proteins (G-proteins) by substance P and compound 48/80. FEBS Lett 259:260–262

Moy JN, Gleich GJ, Thomas LT (1990) Noncytotoxic activation of neutrophils by eosinophil granule major basic protein. Effect on superoxide anion generation and lysosomal enzyme release. J Immunol 145:2626–2632

Mrowietz U, Konter U, Traut R, Schröder JM, Christophers E (1988) Atopic dermatitis: influence of bacterial infections on human monocyte and neutrophil granulocyte functional activities. J Allergy Clin Immunol 82:1027–1037

Mueller H, Fehr J (1986) Binding of C-reactive protein to human polymorphonuclear leukocytes: evidence for association of binding sites with Fc receptors. J Immunol 136:2202–2207

Mueller H, Sklar LA (1989) Coupling of antagonistic signalling pathways in modulation of neutrophil function. J Cell Biochem 40:287–294

Mueller H, Motulsky HJ, Sklar LA (1988) The potency and kinetics of the β-adrenergic receptors on human neutrophils. Mol Pharmacol 34:347–353

Muid RE, Penfield A, Dale MM (1987) The diacylglycerol inhibitor, R59022, enhances the superoxide generation from human neutrophils induced by stimulation of fMet-Leu-Phe, IgG and C3b receptors. Biochem Biophys Res Commun 143:630–637

Muid RE, Twomey B, Dale MM (1988) The effect of inhibition of both diacylglycerol metabolism and phospholipase A_2 activity on superoxide generation by human neutrophils. FEBS Lett 234:235–240

Müller-Peddinghaus R, Wurl M (1987) The amplified chemiluminescence test to characterize antirheumatic drugs as oxygen radical scavengers. Biochem Pharmacol 36:1125–1132

Mukherjee SP, Mukherjee C (1982) Similar activities of nerve growth factor and its homologue proinsulin in intracellular hydrogen peroxide production and metabolism in adipocytes. Transmembrane signalling relative to insulin-mimicking cellular effects. Biochem Pharmacol 31:3163–3172

Munford RS, Hall CL (1986) Detoxification of bacterial lipopolysaccharides (endotoxins) by a human neutrophil enzyme. Science 234:203–205

Murakami K, Routtenberg A (1985) Direct activation of purified protein kinase C by unsaturated fatty acids (oleate and arachidonate) in the absence of phospholipids and Ca^{2+}. FEBS Lett 192:189–193

Murakami K, Chan SY, Routtenberg A (1986) Protein kinase C activation by cis-fatty acid in the absence of Ca^{2+} and phospholipids. J Biol Chem 261:15424–15429

Murakami K, Whiteley MK, Routtenberg A (1987) Regulation of protein kinase C activity by cooperative interaction of Zn^{2+} and Ca^{2+}. J Biol Chem 262:13902–13906

Murakami M, Nakamura M, Minakami S (1986) NADPH oxidase of guinea-pig macrophages catalyses the reduction of ubiquinone-1 under anaerobic conditions. Biochem J 237:541-545

Murayama T, Ui M (1987) Phosphatidic acid may stimulate membrane receptors mediating adenylate cyclase inhibition and phospholipid breakdown in 3T3 fibroblasts. J Biol Chem 262:5522–5528

Murphy PM, Eide B, Goldsmith P, Brann M, Gierschik P, Spiegel A, Malech HL (1987) Detection of multiple forms of $G_{i\alpha}$ in HL60 cells. FEBS Lett 221:81–86

Murray HW (1981) Susceptibility of Leishmania to oxygen intermediates and killing by normal macrophages. J Exp Med 153:1302–1315

Murray HW, Cartelli DM (1983) Killing of intracellular *Leishmania donovani* by human mononuclear phagocytes. Evidence for oxygen-dependent and -independent leishmanicidal activity. J Clin Invest 72:32–44

Murray HW, Cohn ZA (1980) Macrophage oxygen-dependent antimicrobial activity. III. Enhanced oxidative metabolism as an expression of macrophage activation. J Exp Med 152:1596–1609

Murray HW, Spitalny GL, Nathan CF (1985a) Activation of mouse peritoneal macrophages in vitro and in vivo by interferon-γ. J Immunol 134:1619–1622

Murray HW, Rubin BY, Carriero SM, Harris AM, Jaffee EA (1985b) Human mononuclear phagocyte antiprotozoal mechanisms: oxygen-dependent vs oxygen-independent activity against intracellular *Toxoplasma gondii*. J Immunol 134:1982–1988

Murray HW, Gellene RA, Libby DM, Rothermel CD, Rubin BY (1985c) Activation of tissue macrophages from AIDS patients: in vitro response of AIDS alveolar macrophages to lymphokines and interferon-γ. J Immunol 135:2374–2377

Murrell GAC, Francis MJO, Bomley L (1990) Modulation of fibroblast proliferation by oxygen free radicals. Biochem J 265:659–665

Musson RA, McPhail L, Shafran H, Johnston RB Jr (1982) Differences in the ability of human peripheral blood monocytes and in vitro monocyte-derived macrophages to produce superoxide anion: studies with cells from normals and patients with chronic granulomatous disease. J Reticuloendothel Soc 31:261–266

Myers MA, McPhail LC, Snyderman R (1985) Redistribution of protein kinase C activity in human monocytes: correlation with activation of the respiratory burst. J Immunol 135:3411–3416

Naccache PH, Volpi M, Showell HJ, Becker EL, Sha'afi RI (1979) Chemotactic factor-induced release of membrane calcium in rabbit neutrophils. Science 203:461–463

Naccache PH, Molski TFP, Borgeat P, Sha'afi RI (1984) Association of leukotriene B$_4$ with the cytoskeleton of rabbit neutrophils. Effect of chemotactic factor and phorbol esters. Biochem Biophys Res Commun 124:963–969

Naccache PH, Molski TFP, Borgeat P, White JR, Sha'afi RI (1985a) Phorbol esters inhibit the fMet-Leu-Phe- and leukotriene B$_4$-stimulated calcium mobilization and enzyme secretion in rabbit neutrophils. J Biol Chem 260:2125–2131

Naccache PH, Molski MM, Volpi M, Becker EL, Sha'afi RI (1985b) Unique inhibitory profile of platelet activating factor induced calcium mobilization, polyphosphoinositide turnover and granule enzyme secretion in rabbit neutrophils towards pertussis toxin and phorbol ester. Biochem Biophys Res Commun 130:677–684

Naccache PH, Molski MM, Volpi M, Shefcyk J, Molski TFP, Loew L, Becker EL, Sha'afi RI (1986) Biochemical events associated with the stimulation of rabbit neutrophils by platelet-activating factor. J Leukocyte Biol 40:533–548

Naccache PH, Faucher N, Borgeat P, Gasson JC, DiPersio JF (1988a) Granulocyte-macrophage colony-stimulating factor modulates the excitation-response coupling sequence in human neutrophils. J Immunol 140:3541–3546

Naccache PH, Faucher N, Caon AC, McColl SR (1988b) Propionic acid-induced calcium mobilization in human neutrophils. J Cell Physiol 136:118–124

Naccache PH, Therrien S, Caon AC, Liao N, Gilbert C, McColl SR (1989) Chemoattractant-induced cytoplasmic pH changes and cytoskeletal reorganization in human neutrophils. Relationship to the stimulated calcium transients and oxidative burst. J Immunol 142:2438–2444

Naccache PH, Gilbert C, Caon AC, Huang CK, Bonak VA, Umezawa K, McColl SR (1990) Selective inhibition of human neutrophil functional responsiveness by erbstatin, an inhibitor of tyrosine protein kinase. Blood 76:2098–2104

Nagashima K, Koike M, Saito K, Yasuhara T, Tsukamoto Y, Mori M, Fujino M, Nakajima T (1990) Role of lysine residue at 7th position of wasp chemotactic peptides. Biochem Biophys Res Commun 168:844–849

Nagy JT, Foris G, Fulop T Jr, Paragh G, Plotnikoff NP (1988) Activation of the lipoxygenase pathway in the methionine enkephalin induced respiratory burst in human polymorphonuclear leukocytes. Life Sci 42:2299–2306

Najjar VA (1983) Tuftsin, a natural activator of phagocyte cells: an overview. Ann NY Acad Sci 419:1–11

Najjar VA, Nishioka K (1970) "Tuftsin:" a natural phagocytosis stimulating peptide. Nature 228:672–673

Naka M, Nishikawa M, Adelstein RS, Hidaka H (1983) Phorbol ester-induced activation of human platelets is associated with protein kinase C phosphorylation of myosin light chains. Nature 306:490–492

Nakagawara A, Minakami S (1975) Generation of superoxide anions by leucocytes treated with cytochalasin E. Biochem Biophys Res Commun 64:760–767

Nakagawara A, Nathan CF, Cohn ZA (1981) Hydrogen peroxide metabolism in human monocytes during differentiation in vitro. J Clin Invest 68:1243–1252

Nakagawara M, Takeshige K, Sumimoto H, Yoshitake J, Minakami S (1984) Superoxide release and intracellular free calcium of calcium-depleted human neutrophils stimulated by N-formy-methionyl-leucyl-phenylalanine. Biochim Biophys Acta 805:97–103

Nakagawara M, Takeshige K, Takamatsu J, Takahashi S, Yoshitake J, Minakami S (1986) Inhibition of superoxide production and Ca^{2+} mobilization in human neutrophils by halothane, enflurane, and isoflurane. Anesthesiology 64:4–12

Nakamura Y, Ogihara S, Ohtaki S (1987) Activation by ATP of calcium-dependent NADPH oxidase generating hydrogen peroxide in thyroid plasma membranes. J Biochem 102:1121–1132

Nakamura Y, Ohtaki S, Makino R, Tanaka T, Ishimura Y (1989) Superoxide anion is the initial product in the hydrogen peroxide formation catalyzed by NADPH oxidase in porcine thyroid plasma membrane. J Biol Chem 264:4759–4761

Nakashima S, Suganuma A, Sato M, Tohmatsu T, Nozawa Y (1989) Mechanism of arachidonic acid liberation in platelet-activating factor-stimulated human polymorphonuclear neutrophils. J Immunol 143:1295–1302

Naqui A, Chance B, Cadenas E (1986) Reactive oxygen intermediates in biochemistry. Annu Rev Biochem 55:137–166

Nasmith PE, Grinstein S (1986) Impairment of Na^+/H^+ exchange underlies inhibitory effects of Na^+-free media on leukocyte function. FEBS Lett 202:79–85

Nasmith PE, Grinstein S (1987a) Phorbol ester-induced changes in cytoplasmic Ca^{2+} in human neutrophils. Involvement of a pertussis toxin-sensitive G protein. J Biol Chem 262:13558–13566

Nasmith PE, Grinstein S (1987b) Are Ca^{2+} channels in neutrophils activated by a rise in cytosolic free Ca^{2+}? FEBS Lett 221:95–100

Nasmith PE, Mills GB, Grinstein S (1989) Guanine nucleotides induce tyrosine phosphorylation and activation of the respiratory burst in neutrophils. Biochem J 257:893–897

Nasrallah VN Jr, Shirley PS, Myrvik Q, Waite M (1983) The use of acetylated cytochrome c in detecting superoxide anion production in rabbit alveolar macrophages. J Immunol 131:2104–2106

Nastainczyk W, Röhrkasten A, Sieber M, Rudolph C, Schächtele C, Marmé E, Hofmann F (1987) Phosphorylation of the purified receptor for calcium channel blockers by cAMP kinase and protein kinase C. Eur J Biochem 169:137–142

Nath J, Powledge A (1988) Temperature-dependent inhibition of fMet-Leu-Phe-stimulated superoxide generation by C-I and H-7 in human neutrophils. Biochem Biophys Res Commun 156:1376–1382

Nath J, Powledge A, Wright DG (1989) Studies of signal transduction in the respiratory burst-associated stimulation of fMet-Leu-Phe-induced tubulin tyrosinolation and phorbol 12-myristate 13-acetate-induced posttranslational incorporation of tyrosine into multiple proteins in activated neutrophils and HL-60 cells. J Biol Chem 264:848–855

Nathan CF (1982) Secretion of oxygen intermediates: role in effector functions of activated macrophages. Fed Proc 41:2206–2211

Nathan CF (1987) Neutrophil activation on biological surfaces. Massive secretion of hydrogen peroxide in response to products of macrophages and lymphocytes. J Clin Invest 80:1550–1560

Nathan CF (1989) Respiratory burst in adherent human neutrophils: triggering by colony-stimulating factors CSF-GM and CSF-G. Blood 73:301–306

Nathan C, Sanchez E (1990) Tumor necrosis factor and CD11/CD18 (β2) integrins act synergistically to lower cAMP in human neutrophils. J Cell Biol 111:2171–2181

Nathan CF, Murray HW, Cohn ZA (1980) The macrophage as an effector cell. N Engl J Med 303:622–626

Nathan CF, Murray HW, Wiebe ME, Rubin BY (1983) Identification of interferon-γ as the lymphokine that activates human macrophage oxidative metabolism and antimicrobial activity. J Exp Med 158:670–689

Nathan CF, Prendergast TJ, Wiebe ME, Stanley ER, Platzer E, Remold HG, Welte K, Rubin BY, Murray HW (1984) Activation of human macrophages. Comparison of other cytokines with interferon-γ. J Exp Med 160:600–605

Nathan CF, Horowitz CR, De la Harpe J, Vadhan-Raj S, Sherwin SA, Oettgen HF, Krown SE (1985) Administration of recombinant interferon γ to cancer patients enhances monocyte secretion of hydrogen peroxide. Proc Natl Acad Sci USA 82:8686–8690

Nathan DG, Baehner RL, Weaver DK (1969) Failure of nitro blue tetrazolium reduction in the phagocytic vacuoles of leukocytes in chronic granulomatous disease. J Clin Invest 48:1895–1904

Naum CC, Kaplan SS, Basford RE (1991) Platelets and ATP prime neutrophils for enhanced O_2^- generation at low concentrations but inhibit O_2^- generation at high concentration. J Leukocyte Biol 49:83–89

Nauseef WM, Root RK, Newman SL, Malech HL (1983a) Inhibition of zymosan activation of human neutrophil oxidative metabolism by a mouse monoclonal antibody. Blood 62:635–644

Nauseef WM, Metcalf JA, Root RK (1983b) Role of myeloperoxidase in the respiratory burst of human neutrophils. Blood 61:483–492

Nayar R, Mayer LD, Hope MJ, Cullis PR (1984) Phosphatidic acid as a calcium ionophore in large unilamellar vesicle systems. Biochim Biophys Acta 777:343–346

Neal TM, Vissers MCM, Winterbourn CC (1987a) Inhibition by nonsteroidal antiinflammatory drugs of superoxide production and granule enzyme release by polymorphonuclear leukocytes stimulated with immune complexes or formyl-methionyl-leucyl-phenylalanine. Biochem Pharmacol 36:2511–2517

Neal TM, Winterbourn CC, Vissers MCM (1987b) Inhibition of neutrophil degranulation and superoxide production by sulfasalazine. Comparison with 5-aminosalicylic acid, sulfapyridine and olsalazine. Biochem Pharmacol 36:2765–2768

Needleman P, Turk J, Jakschik BA, Morrison AR, Lefkowith JB (1986) Arachidonic acid metabolism. Annu Rev Biochem 55:69–102

Neer EJ, Clapham DE (1988) Roles of G protein subunits in transmembrane signalling. Nature 333:129–134

Neumann M, Kownatzki E (1989) The effect of adherence on the generation of reactive oxygen species by human neutrophilic granulocytes. Agents Actions 26:183–185

Newburger PE, Chovaniec ME, Greenberger JS, Cohen HJ (1979) Functional changes in human leukemic cell line HL-60. A model for myeloid differentiation. J Cell Biol 82:315–322

Newburger PE, Chovaniec ME, Cohen HJ (1980a) Activity and activation of the granulocyte superoxide-generating system. Blood 55:85–92

Newburger PE, Pagano JS, Greenberger JS, Karpas A, Cohen JH (1980b) Dissociation of opsonized particle phagocytosis and respiratory burst activity in an Epstein-Barr virus-infected myeloid cell line. J Cell Biol 85:549–557

Newburger PE, Robinson JM, Pryzwansky KB, Rosoff PM, Greenberger JS, Tauber AI (1983) Human neutrophil dysfunction with giant granules and defective activation of the respiratory burst. Blood 61:1247–1257

Newburger PE, Speier C, Borregaard N, Walsh CE, Whitin JC, Simons ER (1984) Development of the superoxide-generating system during differentiation of the HL-60 human promyelocytic leukemia cell line. J Biol Chem 259:3771–3776

Newburger PE, Luscinskas FW, Ryan T, Beard CJ, Wright J, Platt OS, Simons ER, Tauber AI (1986) Variant chronic granulomatous disease: modulation of the neutrophil defect by severe infection. Blood 68:914–919

Newburger PE, Ezekowitz RAB, Whitney C, Wright J, Orkin SH (1988) Induction of phagocyte cytochrome b heavy chain gene expression by interferon γ. Proc Natl Acad Sci USA 85:5215–5219

Ng DS, Wong K (1986) GTP regulation of platelet-activating factor binding to human neutrophil membranes. Biochem Biophys Res Commun 141:353–359

Ng DS, Wong K (1988) Specific binding of platelet-activating factor (PAF) by human peripheral blood mononuclear leukocytes. Biochem Biophys Res Commun 155:311–316

Ng DS, Wong K (1989) Effect of platelet-activating factor (PAF) on cytosolic free clacium human peripheral blood mononuclear leukocytes. Res Commun Chem Pathol Pharmacol 64:351–354

Nguyen AT, Lethias C, Zingraff J, Herbelin A, Naret C, Descamps-Latscha B (1985) Hemodialysis membrane-induced activation of phagocyte oxidative metabolism detected in vivo and in vitro within microamounts of whole blood. Kidney Int 28:158–167

Nicola NA, Peterson L (1986) Identification of distinct receptors for two hemopoietic growth factors (granulocyte colony-stimulating factor and multipotential colony-stimulating factor) by chemical cross-linking. J Biol Chem 261:12384–12389

Niedel J, Wilkinson S, Cuatrecasas P (1979) Receptor-mediated uptake and degradation of ^{125}I-chemotactic peptide by human neutrophils. J Biol Chem 254:10700–10706

Niedel J, Davis J, Cuatrecasas P (1980) Covalent affinity labeling of the formyl peptide chemotactic receptor. J Biol Chem 255:7063–7066

Nielson CP, Vestal RE (1989) Effects of adenosine on polymorphonuclear leucocyte function, cyclic 3′:5′-adenosine monophosphate, and intracellular calcium. Br J Pharmacol 97:882–888

Nilsson E, Palmblad J (1988) Effects of ethanol on mechanisms for secretory and aggregatory responses of human granulocytes. Biochem Pharmacol 37:3237–3243

Nishihira J, McPhail LC, O'Flaherty JT (1986) Stimulus-dependent mobilization of protein kinase C. Biochem Biophys Res Commun 134:587–594

Nishioka K, Constantopoulos A, Satoh PS, Najjar VA (1972) The characteristics, isolation and synthesis of the phagocytosis stimulating peptide tuftsin. Biochem Biophys Res Commun 47:172–179

Nishioka K, Constantopoulos A, Satosh PS, Mitchell WM, Najjar VA (1973a) Characteristics and isolation of the phagocytosis-stimulating peptide, tuftsin. Biochim Biophys Acta 310:217–229

Nishioka K, Satoh PS, Constantopoulos A, Najjar VA (1973b) The chemical synthesis of the phagocytosis-stimulating tetrapeptide tuftsin (Thr-Lys-Pro-Arg) and its biological properties. Biochim Biophys Acta 310:230–237

Nishizuka Y (1984) The role of protein kinase C in cell surface signal transduction and tumour promotion. Nature 308:693–698

Nishizuka Y (1986) Studies and perspectives of protein kinase C. Science 233:305–312

Nishizuka Y (1988) The molecular heterogeneity of protein kinase C and its implications for cellular regulation. Nature 334:661–665

Nishizuka Y (1989) The family of protein kinase C for signal transduction. JAMA 262:1826–1833

Niwa Y, Mizushima Y (1990) Neutrophil-potentiating factors released from stimulated lymphocytes; special reference to the increase in neutrophil-potentiating factors from streptococcus-stimulated lymphocytes of patients with Behçet's disease. Clin Exp Immunol 79:353–360

Niwa Y, Miyake S, Sakane T, Shingu M, Yokoyama M (1982) Auto-oxidative damage in Behçet's disease—endothelial cell damage following the elevated oxygen radicals generated by stimulated neutrophils. Clin Exp Immunol 49:247–255

Niwa Y, Kano T, Taniguchi S, Miyachi Y, Sakane T (1986) Effect of cyclosporin A on the membrane-associated events in human leukocytes with special reference to the similarity with dexamethasone. Biochem Pharmacol 35:947–951

Niwa Y, Kasama T, Miyachi Y, Kanoh T (1989) Neutrophil chemotaxis, phagocytosis and parameters of reactive oxygen species in human aging: cross-sectional and longitudinal studies. Life Sci 44:1655–1664

Norgauer J, Kownatzki E, Seifert R, Aktories K (1988) Botulinum C2 toxin ADP-ribosy-lates actin and enhances O_2^- production and secretion but inhibits migration of activated human neutrophils. J Clin Invest 82:1376–1382

Norgauer J, Just I, Aktories K, Sklar LA (1989) Influence of botulinum C2 toxin on F-actin and N-formyl peptide receptor dynamics in human neutrophils. J Cell Biol 109:1133–1140

Northover AM (1985) The effects of some non-steroidal anti-inflammatory agents on membrane-associated calcium in rabbit peritoneal neutrophils. Biochem Pharmacol 34:3123–3129

Nozaki M, Takeshige K, Sumimoto H, Minakami S (1990) Reconstitution of the partially purified membrane component of the superoxide-generating NADPH oxidase of pig neutrophils with phospholipid. Eur J Biochem 187:335–340

Nozawa R, Kato H, Yokota T (1988) Induction of cytosolic activation factor for NADPH oxidase in differentiated HL-60 leukemia cells. J Biochem (Tokyo) 103:43–47

Nugent JHA, Gratzer W, Segal AW (1989) Identification of the haem-binding subunit of cytochrome b_{-245}. Biochem J 264:921–924

Nunoi H, Rotrosen D, Gallin JI, Malech HL (1988) Two forms of autosomal chronic granulomatous disease lack distinct neutrophil cytosol factors. Science 242:1298–1301

Odell EW, Segal AW (1988) The bactericidal effects of the respiratory burst and the myeloperoxidase system isolated in neutrophil cytoplasts. Biochim Biophys Acta 971:266–274

Offermans S, Schäfer R, Hoffmann B, Bombien E, Spicher K, Hinsch KD, Schultz G, Rosenthal W (1990) Agonist-sensitive binding of a photoreactive GTP analog to a G-protein α-subunit in membranes of HL-60 cells. FEBS Lett 260:14–18

O'Flaherty JT, Nishihira J (1987) 5-Hydroxyicosatetraenoate promotes Ca^{2+} and protein kinase C mobilization in neutrophils. Biochem Biophys Res Commun 148:575–581

O'Flaherty JT, Dechatelet LR, McCall CE, Bass DA (1980) Neutrophil aggregation: evidence for a different mechanism of action by phorbol myristate acetate. Proc Soc Exp Biol Med 165:225–232

O'Flaherty JT, Schmitt JD, Wykle RL (1985a) Interactions of arachidonate metabolism and protein kinase C in mediating neutrophil function. Biochem Biophys Res Commun 127:916–923

O'Flaherty JT, Schmitt JD, Wykle RL, Redman JF Jr, McCall CE (1985b) Diacylglycerols and mezerein activate neutrophils by a phorbol myristate acetate-like mechanism. J Cell Physiol 125:192–199

O'Flaherty JT, Surles JR, Redman J, Jacobson D, Piantados C, Wykle RL (1986) Binding and metabolism of platelet-activating factor by human neutrophils. J Clin Invest 78:381–388

O'Flaherty JT, Jacobson DP, Redman JF (1989) Bidirectional effects of protein kinase C activators. Studies with human neutrophils and platelet-activating factor. J Biol Chem 264:6836–6843

Oh SK, Pavlotsky N, Tauber AI (1990) Specific binding of haptoglobin to human neutrophils and its functional consequences. J Leukocyte Biol 47:142–148

Ohashi Y, Narumiya S (1987) ADP-ribosylation of a M_r 21,000 membrane protein by type D botulinum toxin. J Biol Chem 262:1430–1433

Ohkubo S, Yamada E, Endo T, Itoh H, Hidaka H (1984) Vitamin A acid-induced activation of Ca^{2+}-activated, phospholipid-dependent protein kinase from rabbit retina. Biochem Biophys Res Commun 118:460–466

Ohlstein EH, Nichols AJ (1989) Rabbit polymorphonuclear neutrophils elicit endothelium-dependent contraction in vascular smooth muscle. Circ Res 65:917–924

Ohno S, Kawasaki H, Imajoh S, Suzuki K, Inagaki M, Yokokura H, Sakoh T, Hidaka H (1987) Tissue-specific expression of three distinct types of rabbit protein kinase C. Nature 325:161–166

Ohno Y, Seligmann BE, Gallin JI (1985) Cytochrome b translocation to human neutrophil plasma membranes and superoxide release. Differential effects of N-formylmethionylleucylphenylalanine, phorbol myristate acetate, and A23187. J Biol Chem 260:2409–2414

Ohno Y, Buescher ES, Roberts R, Metcalf JA, Gallin JI (1986) Reevaluation of cytochrome b and flavin adenine dinucleotide in neutrophils from patients with chronic granulomatous disease and description of a family with probable autosomal recessive inheritance of cytochrome b deficiency. Blood 67:1132–1138

Ohsaka A, Saito M, Suzuki I, Miura Y, Takaku F, Kitagawa S (1988) Phorbol myristate acetate potentiates superoxide release and membrane depolarization without affecting an increase in cytoplasmic free calcium in human granulocytes stimulated by the chemotactic peptide, lectins and the calcium ionophore. Biochim Biophys Acta 941:19–30

Ohsaka A, Kitagawa S, Sakamoto S, Miura Y, Takanashi N, Takaku F, Saito M (1989) In vivo activation of human neutrophil functions by administration of recombinant human granulocyte colony-stimulating factor in patients with malignant lymphoma. Blood 74:2743–2748

Ohta H, Okajima F, Ui M (1985) Inhibition by islet-activating protein of a chemotactic peptide-induced early breakdown of inositol phospholipids and Ca^{2+} mobilization in guinea pig neutrophils. J Biol Chem 260:15771–15780

Ohtsuka T, Okamura N, Ishibashi S (1986) Involvement of protein kinase C in the phosphorylation of 46 kDa proteins which are phosphorylated in parallel with activation of NADPH oxidase in intact guinea-pig polymorphonuclear leukocytes. Biochim Biophys Acta 888:332–337

Ohtsuka T, Ozawa M, Okamoto T, Uchida M, Okamura N, Ishibashi S (1987) Significance of phosphorylation/dephosphorylation of 46K protein(s) in regulation of superoxide anion production in intact guinea pig polymorphonuclear leukocytes. J Biochem (Tokyo) 101:897–903

Ohtsuka T, Ozawa M, Okamura N, Ishibashi S (1989) Stimulatory effects of a short chain phosphatidate on superoxide anion production in guinea pig polymorphonuclear leukocytes. J Biochem (Tokyo) 106:259–263

Ohtsuka T, Hiura M, Ozawa M, Okamura N, Nakamura M, Ishibashi S (1990a) Involvement of membrane charges in constituting the active form of NADPH oxidase in guinea pig polymorphonuclear leukocytes. Arch Biochem Biophys 280:74–79

Ohtsuka T, Hiura M, Yoshida K, Okamura H, Ishibashi S (1990b) A diacylglycerol kinase inhibitor, R 59 022, potentiates superoxide anion production and 46-kDa protein phosphorylation in guinea pig polymorphonuclear leukocytes. J Biol Chem 265:15148–15423

Ohtsuki K, Yokoyama M (1987) Direct activation of guanine nucleotide binding proteins through a high-energy phosphate-transfer by nucleoside-diphosphate kinase. Biochem Biophys Res Commun 148:300–307

Ohtsuki K, Ikeuchi T, Yokoyama M (1986) Characterization of nucleoside-diphosphate kinase-associated guanine nucleotide-binding proteins from HeLa S3 cells. Biochim Biophys Acta 882:322–330

Ohtsuki K, Yokoyama M, Uesaka H (1987) Physiological correlation between nucleoside-diphosphate kinases and the 21-kDa guanine-nucleotide-binding proteins copurified with the enzymes from the cell membrane fractions of Ehrlich ascites tumor cells. Biochim Biophys Acta 929:231-238

Oinuma M, Katada T, Ui M (1987) A new GTP-binding protein in differentiated human leukemic (HL-60) cells serving as the specific substrate of islet-activating protein, pertussis toxin. J Biol Chem 262:8347–8353

Okajima F, Ui M (1984) ADP-ribosylation of the specific membrane protein by islet-activating protein, pertussis toxin, associated with inhibition of a chemotactic peptide-induced arachidonate release in neutrophils. A possible role of the toxin substrate in Ca^{2+}-mobilizing biosignaling. J Biol Chem 259:13863–13871

Okajima F, Katada T, Ui M (1985) Coupling of the guanine nucleotide regulatory protein to chemotactic peptide receptors in neutrophil membranes and its uncoupling by islet-activating protein, pertussis toxin. A possible role of the toxin substrate in Ca^{2+}-mobilizing receptor-mediated signal transduction. J Biol Chem 260:6761–6768

Okamura N, Hanakura K, Kodakari M, Ishibashi S (1980) Cooperation of cytochalasin C and anti-microtubular agents in stimulating superoxide anion production in polymorphonuclear leukocytes. J Biochem (Tokyo) 88:139–144

Okamura N, Uchida M, Ohtsuka T, Kawanishi M, Ishibashi S (1985) Diverse involvements of Ni protein in superoxide anion production in polymorphonuclear leukocytes depending on the type of membrane stimulants. Biochem Biophys Res Commun 130:939–944

Okamura N, Curnutte JT, Roberts RL, Babior BM (1988a) Relationship of protein phosphorylation to the activation of the respiratory burst in human neutrophils. Defects in the phosphorylation of a group of closely related 48-kDa proteins in two forms of chronic granulomatous disease. J Biol Chem 263:6777–6782

Okamura N, Malawista SE, Roberts RL, Rosen H, Ochs HD, Babior BM, Curnutte JT (1988b) Phosphorylation of the oxidase-related 48K phosphoprotein family in the unusual autosomal cytochrome-negative and X-linked cytochrome-positive types of chronic granulomatous disease. Blood 72:811–816

Olson SC, Raj Tyagi S, Lambeth JD (1990) Fluoride activates diradylglycerol and superoxide generation in human neutrophils via PLD/PA phosphohydrolase-dependent and -independent pathways. FEBS Lett 272:19–24

Omann GM, Lakowicz JR (1982) Interactions of chlorinated hydrocarbon insecticides with membranes. Biochim Biophys Acta 684:83–95

Omann GM, Allen RA, Bokoch GM, Painter RG, Traynor AE, Sklar LA (1987a) Signal transduction and cytoskeletal activation in the neutrophil. Physiol Rev 67:285–322

Omann GM, Traynor AE, Harris AL, Sklar LA (1987b) LTB$_4$-induced activation signals and responses in neutrophils are short-lived compared to formylpeptide. J Immunol 138:2626–2632

Omann GM, Swann WN, Oades ZG, Parkos CA, Jesaitis AJ, Sklar LA (1987c) N-For-myl-peptide-receptor dynamics, cytoskeletal activation, and intracellular calcium response in human neutrophil cytoplasts. J Immunol 139:3447–3455

Omann GM, Porasik MM, Sklar LA (1989) Oscillating actin polymeriza-tion/depolymerization responses in human polymorphonuclear leukocytes. J Biol Chem 264:16355–16358

Otero AS, Breitwieser GE, Szabo G (1988) Activation of muscarinic potassium currents by ATPγS in atrial cells. Science 242:443–445

Owen DAA (1987) Inflammation — histamine and 5-hydroxytryptamine. Br Med Bull 43:256–269

Oyanagui Y (1978) Inhibition of superoxide anion production in non-stimulated guinea pig peritoneal exudate cells by anti-inflammatory drugs. Biochem Pharmacol 27:777–782

Ozaki Y, Iwata J, Ohashi T (1984a) Mechanism of neutrophil chemiluminescence induced by wheat germ agglutinin: partial characterization of the antigens recog-nized by wheat germ agglutinin. Blood 64:1094–1102

Ozaki Y, Kume S, Ohashi T (1984b) Effects of histamine agonists and antagonists on luminol-dependent chemiluminescence of granulocytes. Agents Actions 15:182–188

Ozaki Y, Kume S, Ohashi T, Niwa Y (1986a) Partial dependency of 1-oleoyl-2-acetyl-glycerol-induced superoxide production by human neutrophils on calcium ions and cytochalasin B. Biochem Biophys Res Commun 134:690–697

Ozaki Y, Ohashi T, Niwa Y (1986b) A comparative study on the effects of inhibitors of the lipoxygenase pathway on neutrophil function. Inhibitory effects on neutrophil functions may not be attributed to inhibition of the lipoxygenase pathway. Biochem Pharmacol 35:3481–3488

Ozaki Y, Ohashi T, Kume S (1987) Potentiation of neutrophil function by recombinant DNA-produced interleukin 1a. J Leukocyte Biol 42:621–627

Ozawa M, Ohtsuka T, Okamura N, Ishibashi S (1989) Synergism between protein kinase C activator and fatty acids in stimulating superoxide anion production in guinea pig polymorphonuclear leukocytes. Arch Biochem Biophys 273:491–496

Pabst MJ, Johnston RB Jr (1980) Increased production of superoxide anion by macro-phages exposed in vitro to muramyl dipeptide or lipopolysaccharide. J Exp Med 151:101–114

Pabst MJ, Hedegaard HB, Johnston RB Jr (1982) Cultured human monocytes require exposure to bacterial products to maintain an optimal oxygen radical response. J Immunol 128:123–128

Paganelli Parker K, Benjamin WR, Kaffka KL, Kilian PL (1989) Presence of IL-1 receptors on human and murine neutrophils. Relevance to IL-1-mediated effects in inflammation. J Immunol 142:537–542

Pagonis C, Tauber AI, Pavlotsky N, Simons ER (1986) Flavonoid impairment of neutrophil response. Biochem Pharmacol 35:237–245

Pai JK, Siegel MI, Egan RW, Billah MM (1988a) Activation of phospholipase D by chemotactic peptide in HL-60 granulocytes. Biochem Biophys Res Commun 150:355–364

Pai JK, Siegel MI, Egan RW, Billah MM (1988b) Phospholipase D catalyzes phos-pholipid metabolism in chemotactic peptide-stimulated HL-60 granulocytes. J Biol Chem 263:12472–12477

Pai JK, August J, Dobek E (1989) The source of the chemotactic peptide-induced diacylglycerol in human neutrophils is via phospholipase D activation. In: Biology of cellular transducing signals '89 Vanderhoek JY (Ed) Washington 9th Internation-al Washington Spring Symposium. p. 106

Painter RG, Sklar LA, Jesaitis AJ, Schmitt M, Cochrane CG (1984) Activation of neutrophils by N-formyl chemotactic peptides. Fed Proc 43:2737–2742

Painter RG, Dukes R, Sullivan J, Carter R, Erdös EG, Johnson AR (1988) Function of neutrophil endopeptidase on the cell membrane of human neutrophils. J Biol Chem 262:9456–9461

Palmblad J, Gyllenhammar H, Lindgren JA, Malmsten CL (1984) Effects of leukotrienes and fMet-Leu-Phe on oxidative metabolism of neutrophils and eosinophils. J Immunol 132:3041–3045

Palmblad J, Gyllenhammar H, Ringertz B, Serhan CN, Samuelsson B, Nicolaou KC (1987) The effects of lipoxin A and lipoxin B on functional responses of human granulocytes. Biochem Biophys Res Commun 145:168–175

Palmblad J, Gyllenhammar H, Ringertz B, Nilsson E, Cottell B (1988a) Leukotriene B_4 triggers highly characteristic and specific functional responses in neutrophils: studies of stimulus specific mechanisms. Biochim Biophys Acta 970:92–102

Palmblad J, Wannemacher RW, Salem N Jr, Kuhns DB, Wright DG (1988b) Essential fatty acid deficiency and neutrophil function: studies of lipid-free total parenteral nutrition in monkeys. J Lab Clin Med 111:634–644

Panosian JO, Marinetti GV (1983) α_2-Adrenergic receptors in human polymorphonuclear leukocyte membranes. Biochem Pharmacol 32:2243–2247

Panus PC, Jones HP (1987) Inhibition of neutrophil response by mepacrine. Biochem Pharmacol 36:1281–1284

Papini E, Grzeskowiak M, Bellavite P, Rossi F (1985) Protein kinase C phosphorylates a component of NADPH oxidase of neutrophils. FEBS Lett 190:204–208

Parente JE, Tyers MD, Harley CB, Davis P, Wong K (1986a) Gold compounds may alter transduction of inflammatory signals at the level of protein kinase C. In: Signal transduction in biological systems. Adelstein R et al. (Eds) Bethesda 6th International Conference on Cyclic Nucleotides, Calcium and Protein Phosphorylation. p 196

Parente JE, Wong K, Davis P, Burka JF, Percy JS (1986b) Effects of gold compounds on leukotriene B_4, leukotriene C_4 and prostaglandin E_2 production by polymorphonuclear leukocytes. J Rheumatol 13:47–51

Parente JE, Walsh MP, Girard PR, Kuo JF, Ng DS, Wong K (1989) Effects of gold coordination complexes on neutrophil function are mediated via inhibition of protein kinase C. Mol Pharmacol 35:26–33

Parker KP, Benjamin WR, Kaffka KL, Kilian PL (1989) Presence of IL-1 receptors on human and murine neutrophils. Relevance to IL-1-mediated effects in inflammation. J Immunol 142:537–542

Parkinson JF, Gabig TG (1988) Phagocyte NADPH-oxidase: Studies with flavin analogues as active site probes in Triton X-100-solubilized preparations. J Biol Chem 263:8859–8863

Parkinson JF, Akard LP, Schell MJ, Gabig TG (1987) Cell-free activation of phagocyte NADPH-oxidase: tissue and differentiation-specific expression of cytosolic cofactor activity. Biochem Biophys Res Commun 145:1198–1204

Parkos CA, Cochrane CG, Schmitt M, Jesaitis AJ (1985) Regulation of the oxidative response of human granulocytes to chemoattractants. No evidence for stimulated traffic of redox enzymes between endo and plasma membranes. J Biol Chem 260:6541–6547

Parkos CA, Allen RA, Cochrane CG, Jesaitis AJ (1987) Purified cytochrome b purified from human granulocyte plasma membrane is comprised of two polypeptides with relative molecular weights of 91,000 and 22,000. J Clin Invest 80:732–742

Parkos CA, Allen RA, Cochrane CG, Jesaitis AJ (1988a) The quaternary structure of the plasma membrane b-type cytochrome of human granulocytes. Biochim Biophys Acta 932:71–83

Parkos CA, Dinauer MC, Walker LE, Allen RA, Jesaitis AJ, Orkin SH (1988b) Primary structure and unique expression of the 22-kilodalton light chain of human neutrophil cytochrome b. Proc Natl Acad Sci USA 85:3319–3323

Parkos CA, Dinauer MC, Jesaitis AJ, Orkin SH, Curnutte JT (1989) Absence of both the 91kD and 22kD subunits of human neutrophil cytochrome *b* in two genetic forms of chronic granulomatous disease. Blood 73:1416–1420

Parks RE, Agarwal RP (1973) Nucleoside diphosphokinases. In: Boyer PD (ed) The enzymes, vol 8A, 3rd edn. Academic, New York, pp 307–333

Parnham MJ, Bittner C, Lambrecht G (1989) Antagonism of platelet activating factor-induced chemiluminescence in guinea-pig peritoneal macrophages in differing states of activation. Br J Pharmacol 98:574–580

Parries GS, Hokin-Neaverson M (1985) Inhibition of phosphatidylinositol synthase and other membrane-associated enzymes by stereoisomers of hexachlorocyclohexane. J Biol Chem 260:2687–2693

Patriarca P, Cramer R, Marussi M, Rossi F, Romeo D (1971) Mode of activation of granule-bound NADPH oxidase in leucocytes during phagocytosis. Biochim Biophys Acta 237:335–338

Pearson RD, Steigbigel RT (1981) Phagocytosis and killing of the protozoan *Leishmania donovani* by human polymorphonuclear leukocytes. J Immunol 127:1438–1443

Pearson RD, Symes P, Conboy M, Weiss AA, Hewlett EL (1987) Inhibition of monocyte oxidative responses by *Bordetella pertussis* adenylate cyclase toxin. J Immunol 139:2749–2754

Pelham HRB (1986) Speculations on the functions of the major heat shock and glucose-regulated proteins. Cell 46:959–961

Pember SO, Heyl BL, Kinkade JM Jr, Lambeth JD (1984) Cytochrome b_{558} from (bovine) granulocytes. Partial purification from Triton X-114 extracts and properties of the isolated cytochrome. J Biol Chem 259:10590–10595

Penfield A, Dale MM (1984) Synergism between A23187 and 1-oleoyl-2-acetyl-glycerol in superoxide production by human neutrophils. Biochem Biophys Res Commun 125:332–336

Penfield A, Dale MM (1985) Prostaglandin E_1 and E_2 enhance the stimulation of superoxide release by 1-oleoyl-2-acetylglycerol from human neutrophils. FEBS Lett 181:335–338

Pennington JE, Kemmerich B, Kazanjian PH, Marsh JD, Boerth LW (1986) Verapamil impairs human neutrophil chemotaxis by a non-calcium-mediated mechanism. J Lab Clin Med 108:44–52

Perez HD, Ong RR, Elfman F (1985) Removal or oxidation of surface membrane sialic acid inhibits formyl-peptide-induced polymorphonuclear leukocyte chemotaxis. J Immunol 134:1902–1908

Perez HD, Elfman F, Chenoweth D, Hooper C (1986) Preparation and characterization of a derivative of wheat germ agglutinin which specifically inhibits polymorphonuclear leukocyte chemotaxis to the synthetic chemotactic peptide N-formyl-methionyl-leucyl-phenylalanine. J Immunol 136:1813–1819

Perianin A, Snyderman R (1989) Analysis of calcium homeostasis in activated human polymorphonuclear leukocytes: evidence for two distinct mechanisms for lowering cytosolic calcium. J Biol Chem 264:1005–1009

Perianin A, Torres M, Labro MT, Hakim J (1983) The inhibitory effects of phenylbutazone on soluble and particle stimulation of human neutrophil oxidative burst. Biochem Pharmacol 32:2819–2822

Perianin A, Gougerot-Pocidalo MA, Giroud JP, Hakim J (1985) Diclofenac sodium, a negative chemokinetic factor for neutrophil locomotion. Biochem Pharmacol 34:3433–3438

Perianin A, Gougerot-Pocidalo MA, Giroud JP, Hakim J (1987) Diclofenac binding to human polymorphonuclear neutrophils: effect on respiratory burst and N-formylated peptide binding. Biochem Pharmacol 36:2609–2615

Perianin A, Snyderman R, Malfroy B (1989) Substance P primes human neutrophil activation: a mechanism for neurological regulation of inflammation. Biochem Biophys Res Commun 161:520–524

Pernow B (1983) Substance P. Pharmacol Rev 35:85–141

Peroutka SJ (1988a) Serotonin receptor subtypes. In: Pharmacology. pp 1–4 (ISI atlas of science) Grimwade AM (Editor and Publisher) Philadelphia.

Peroutka SJ (1988b) 5-Hydroxytryptamine receptor subtypes. Annu Rev Neurosci 11:45–60

Perussia B, Kobayashi M, Rossi ME, Anegon I, Trinchieri G (1987) Immune interferon enhances functional properties of human granulocytes: role of Fc receptors and effect of lymphotoxin, tumor necrosis factor, and granulocyte-macrophage colony-stimulating factor. J Immunol 138:765–774

Peters SP, Cerasoli F, Albertine KH, Gee MH, Berd D, Ishihara Y (1990) "Autoregulation" of human neutrophil activation in vitro: regulation of phorbol myristate acetate-induced neutrophil activation by cell density. J Leukocyte Biol 47:457–474

Peters-Golden M, McNish RW, Brieland JK, Fantone JC (1990) Diminished protein kinase C-activated arachidonate metabolism accompanies rat macrophage differentiation in the lung. J Immunol 144:4320–4326

Petreccia DC, Nauseef WM, Clark RA (1987) Respiratory burst of normal human eosinophils. J Leukocyte Biol 41:283–288

Peveri P, Walz A, Dewald B, Baggiolini M (1988) A novel neutrophil-activating factor produced by human mononuclear phagocytes. J Exp Med 167:1547–1559

Pfefferkorn LC, Fanger MW (1989a) Cross-linking of the high affinity Fc receptor for human immunoglobulin G1 triggers transient activation of NADPH oxidase activity. Continuous oxidase activation requires continuous de novo receptor cross-linking. J Biol Chem 264:14112–14120

Pfefferkorn LC, Fanger MW (1989b) Transient activation of NADPH oxidase through FcγRI. Oxidase deactivation precedes internalization of cross-linked receptors. J Immunol 143:2640–2649

Philip R, Epstein LB (1986) Tumour necrosis factor as immunomodulator and mediator of monocyte cytotoxicity induced by itself, γ-interferon and interleukin-1. Nature 323:86–89

Phillips WA, Hamilton JA (1989) Phorbol ester-stimulated superoxide production in murine bone marrow-derived macrophages requires preexposure to cytokines. J Immunol 142:2445–2449

Philips WA, Hamilton JA (1990) Colony stimulating factor-1 is a negative regulator of the macrophage respiratory burst. J Cell Physiol 144:190–196

Philips WA, Croatto M, Hamilton JA (1990) Priming the macrophage respiratory burst with IL-4: Enhancement with TNF-α but inhibition by IFN-γ. Immunology 70:498–503

Phillips WA, Fujiki T, Rossi MW, Korchak HM, Johnston RB Jr (1989) Influence of calcium on the subcellular distribution of protein kinase C in human neutrophils. Extraction conditions determine partitioning of histone-phosphorylating activity and immunoreactivity between cytosol and particulate fractions. J Biol Chem 264:8361–8365

Pick E (1986) Microassays for superoxide and hydrogen peroxide production and nitroblue tetrazolium reduction using an enzyme immunoassay microplate reader. Methods Enzymol 132:407–421

Pick E, Gadba R (1988) Certain lymphoid cells contain the membrane-associated component of the phagocyte-specific NADPH oxidase. J Immunol 140:1611–1617

Pick E, Keisari Y (1981) Superoxide anion and hydrogen peroxide production by chemically elicited peritoneal macrophages — induction by multiple nonphagocytic stimuli. Cell Immunol 59:301–318

Pick E, Bromberg Y, Shpungin S, Gadba R (1987) Activation of the superoxide forming NADPH oxidase in a cell-free system by sodium dodecyl sulfate. Characterization of the membrane-associated component. J Biol Chem 262:16476–16483

Pick E, Kroizman T, Abo A (1989) Activation of the superoxide-forming NADPH oxidase of macrophages requires two cytosolic components-one of them is also present in certain nonphagocytic cells. J Immunol 143:4180–4187

Picq M, Dubois M, Munari-Silem Y, Prigent AF, Pacheco H (1989) Flavonoid modulation of protein kinase C activation. Life Sci 44:1563–1571

Pike MC, Snyderman R (1982) Transmethylation reactions regulate affinity and functional activity of chemotactic factor receptors on macrophages. Cell 28:107–114

Pike MC, Kredich NM, Snyderman R (1979) Phospholipid methylation in macrophages is inhibited by chemotactic factors. Proc Natl Acad Sci USA 76:2922–2926

Pike MC, Jakoi L, McPhail LC, Snyderman R (1986) Chemoattractant-mediated stimulation of the respiratory burst in human polymorphonuclear leukocytes may require appearance of protein kinase activity in the cells' particulate fraction. Blood 67:909–913

Pike MC, Wicha MS, Yoon P, Mayo L, Boxer LA (1989) Laminin promotes the oxidative burst in human neutrophils via increased chemoattractant receptor expression. J Immunol 142:2004–2011

Pilloud MC, Doussiere J, Vignais PV (1989a) Parameters of activation of the membrane-bound O_2^- generating oxidase from bovine neutrophils in a cell-free system. Biochem Biophys Res Commun 159:783–790

Pilloud MC, Doussiere J, Vignais PV (1989b) The O_2^- generating oxidase activation of bovine neutrophils. Evidence for synergism of multiple cytosolic factors in a cell-free system. FEBS Lett 257:167–170

Pincus SH, Schooley WR, DiNapoli AM, Broder S (1981) Metabolic heterogeneity of eosinophils from normal and hypereosinophilic patients. Blood 58:1175–1181

Piper PJ, Samhoun MN (1987) Leukotrienes. Br Med Bull 43:297–311

Pittet D, Krause KH, Wollheim CB, Bruzzone B, Lew DP (1987) Nonselective inhibition of neutrophil functions by sphinganine. J Biol Chem 262:10072–10076

Pittet D, Lew DP, Mayr GW, Monod A, Schlegel W (1989) Chemoattractant receptor promotion of Ca^{2+} influx across the plasma membrane of HL-60 cells. A role for cytosolic free calcium elevations and inositol 1,3,4,5-tetrakiphosphate production. J Biol Chem 264:7251–7261

Poitevin B, Roubin R, Benveniste J (1984) Paf-acether generates chemiluminescence in human neutrophils in the absence of cytochalasin B. Immunopharmacology 7:135–144

Polakis PG, Uhing RJ, Snyderman R (1988) The formylpeptide chemoattractant receptor copurifies with a GTP-binding protein containing a distinct 40-kDa pertussis toxin substrate. J Biol Chem 263:4969–4976

Polakis PG, Evans T, Snyderman R (1989) Multiple chromatographic forms of the N-formylpeptide chemoattractant receptor and their relationship to GTP-binding proteins. Biochem Biophys Res Commun 161:276–283

Polla BS, Healy AM, Wojno WC, Krane SM (1987) Hormone 1α,25-dihydroxyvitamin D3 modulates heat shock response in monocytes. Am J Physiol 252:C640–C649

Polla BS, Werlen G, Clerget M, Pittet D, Rossier MF, Capponi AM (1989) 1,25-Dihydroxyvitamin D3 induces responsiveness to the chemotactic peptide fMet-Leu-Phe in the human monocytic line U937: dissociation between calcium and oxidative metabolic responses. J Leukocyte Biol 45:381–388

Pontremoli S, Melloni E, Michetti M, Salamino F, Sparatore B, Sacco O, Horecker BL (1986a) Differential mechanisms of translocation of protein kinase C to plasma membranes in activated human neutrophils. Biochem Biophys Res Commun 136:228–234

Pontremoli S, Melloni E, Michetti M, Sacco O, Salamino F, Sparatore B, Horecker BL (1986b) Biochemical responses in activated human neutrophils mediated by protein kinase C and a Ca^{2+}-requiring proteinase. J Biol Chem 261:8309–8313

Pontremoli S, Melloni E, Salamino F, Sparatore B, Michetti M, Sacco O, Horecker BL (1986c) Phosphorylation of proteins in human neutrophils activated with phorbol myristate acetate or with chemotactic factor. Arch Biochem Biophys 250:23–29

Pontremoli S, Melloni E, Salamino F, Sparatore B, Michetti M, Sacco O, Horecker BL (1986d) Activation of NADPH oxidase and phosphorylation of membrane proteins in human neutrophils: coordinate inhibition by a surface antigen-directed monoclonal antibody. Biochem Biophys Res Commun 140:1121–1126

Pontremoli S, Melloni E, Damiani G, Salamino F, Sparatore B, Michetti M, Horecker BL (1988) Effects of a monoclonal anti-calpain antibody on responses of stimulated human neutrophils. Evidence for a role for proteolytically modified protein kinase C. J Biol Chem 263:1915–1919

Pontremoli S, Salamino F, De Tullio R, Patrone M, Tizianello A, Melloni E (1989) Enhanced activation of the respiratory burst oxidase in neutrophils from hypertensive patients. Biochem Biophys Res Commun 158:966–972

Pontremoli S, Melloni E, Sparatore B, Michetti M, Salamino F, Horecker BL (1990) Isozymes of protein kinase C in human neutrophils and their modification by two endogenous proteinases. J Biol Chem 265:706–712

Poubelle PE, De Medicis R, Naccache PH (1987) Monosodium urate and calcium pyrophosphate crystals differentially activate the excitation-response coupling sequence of human neutrophils. Biochem Biophys Res Commun 149:649–657

Pozzan T, Lew DP, Wollheim CB, Tsien RY (1983) Is cytosolic ionized calcium regulating neutrophil activation? Science 221:1413–1415

Prescott SM, Zimmerman GA, Seeger AR (1984) Leukotriene B_4 is an incomplete agonist for the activation of human neutrophils. Biochem Biophys Res Commun 122:535–541

Prigent AF, Fonlupt P, Dubois M, Nemoz G, Timouyasse L, Pacheco H, Pacheco Y, Biot N, Perrin-Fayolle M (1990) Cyclic nucleotide phosphodiesterases and methyltransferases in purified lymphocytes, monocytes and polymorphonuclear leucocytes from healthy donors and asthmatic patients. Eur J Clin Invest 20:323–329

Prin I, Charon J, Capron M, Gosset P, Taelman H, Tonnel AB, Capron A (1984) Heterogeneity of human eosinophils. II. Variability of respiratory burst activity related to cell density. Clin Exp Immunol 57:735–742

Prpic V, Weiel JE, Somers SD, DiGuiseppi J, Gonias SL, Pizzo SV, Hamilton TA, Herman B, Adams DO (1987) Effects of bacterial lipopolysaccharide on the hydrolysis of phosphatidylinositol-4,5-biphosphate in murine peritoneal macrophages. J Immunol 139:526–533

Pryzwansky KB, Steiner AL, Spitznagel JK, Kapoor CL (1981) Compartmentalization of cyclic AMP during phagocytosis by human neutrophilic granulocytes. Science 211:407–410

Pryzwansky KB, Wyatt TA, Nicols H, Lincoln TM (1990) Compartmentalization of cyclic GMP-dependent protein kinase in formyl-peptide stimulated neutrophils. Blood 76:612–618

Puustinen T, Uotila P (1984) Thromboxane formation in human polymorphonuclear leukocytes is inhibited by prednisolone and stimulated by leukotrienes B_4, C_4, D_4 and histamine. Prostaglandins Leukotriene Med 14:161–167

Pyne NJ, Murphy GJ, Milligan G, Houslay MD (1989) Treatment of intact hepatocytes with either the phobol ester TPA or glucagon elicits the phosphorylation and functional inactivation of the inhibitory guanine nucleotide regulatory protein G_i. FEBS Lett 243:77–82

Quilliam LA, Lacal JC, Bokoch GM (1989) Identification of rho as a substrate for botulinum toxin C_3-catalyzed ADP-ribosylation. FEBS Lett 247:221–226

Quinn MT, Parkos CA, Walker L, Orkin SH, Dinauer MC, Jesaitis AJ (1989) Association of a *ras*-related protein with cytochrome *b* of human neutrophils. Nature 342:198–200

Radeke HH, Meier B, Topley N, Flöge J, Habermehl GG, Resch K (1990) Interleukin 1-α and tumor necrosis factor-α induce oxygen radical production in mesangial cells. Kidney Int 37:767–775

Rajkovic IA, Williams R (1985a) Rapid microassays of phagocytosis, bacterial killing, superoxide and hydrogen peroxide production by human neutrophils in vitro. J Immunol Meth 78:35–47

Rajkovic IA, Williams R (1985b) Inhibition of neutrophil function by hydrogen peroxide. Effect of SH-group-containing compounds. Biochem Pharmacol 34:2083–2090

Rall TW, Sutherland EW (1958) Formation of a cyclic adenine ribonucleotide by tissue particles. J Biol Chem 232:1065–1076

Randriamampita C, Trautmann A (1989) Biphasic effect increase in intracellular calcium induced by platelet-activating factor in macrophages. FEBS Lett 249:199–206

Rao KMK, Castranova V (1988) Phenylmethylsulfonyl fluoride inhibits chemotactic peptide-induced actin polymerization and oxidative burst activity in human neutrophils by an effect unrelated to its anti-proteinase activity. Biochim Biophys Acta 969:131–138

Rao KMK, Currie MS, Cohen HJ, Weinberg JB (1989) Chemotactic peptide receptor-cytoskeletal interactions and functional correlations in differentiated HL-60 cells and human polymorphonuclear leukocytes. J Cell Physiol 141:119–125

Rasmussen H, Waisman DM (1983) Modulation of cell function in the calcium messenger system. Rev Physiol Biochem Pharmacol 95:111–148

Rebut-Bonneton C, Bailly S, Pasquier C (1988) Superoxide anion production in glass-adherent polymorphonuclear leukocytes and its relationship to calcium movement. J Leukocyte Biol 44:402–410

Reed SG, Nathan CF, Pihl DL, Rodricks P, Shanebeck K, Conlon PJ, Grabstein KH (1987) Recombinant granulocyte/macrophage colony-stimulating factor activates macrophages to inhibit *Trypanosoma cruzi* and release hydrogen peroxide. Comparison with interferon γ. J Exp Med 166:1734–1746

Regoli D, Drapeau G, Dion S, d'Orléans-Juste P (1987) Pharmacological receptors for substance P and neurokinins. Life Sci 40:109–117

Reibman J, Korchak HM, Vosshall LB, Haines KA, Rich AM, Weissmann G (1988) Changes in diacylglycerol labeling, cell shape, and protein phosphorylation distinguish "triggering" from "activation" of human neutrophils. J Biol Chem 263:6322–6328

Reinhold SL, Prescott SM, Zimmerman GA, McIntyre TM (1990) Activation of human neutrophil phospholipase D by three separable mechanisms. FASEB J 4:208–214

Reitermann A, Metzger J, Wiesmüller KH, Jung G, Bessler WG (1989) Lipopeptide derivatives of bacterial lipoprotein constitute potent immune adjuvants combined with or covalently coupled to antigen or hapten. Biol Chem Hoppe Seyler 370:343–352

Rellstab P, Schaffner A (1989) Endotoxin suppresses the generation of O_2^- and H_2O_2 by "resting" and lymphokine-activated human blood-derived macrophages. J Immunol 142:2813–2820

Remaley AT, Kuhns DB, Basford RE, Glew RH, Kaplan SS (1984) Leishmanial phosphatase blocks neutrophil O_2^- production. J Biol Chem 259:11173–11175

Renkema TEJ, Postma DS, Noordhoek JA, Faber H, Sluiter HJ, Kauffman HF (1989) Association between nonspecific bronchial hyperreactivity and superoxide anion production by polymorphonuclear leukocytes in chronic airflow obstruction. Agents Actions 26:52–54

Rest RF, Farrell CF, Naids FL (1988) Mannose inhibits the human neutrophil oxidative burst. J Leukocyte Biol 43:158–164

Rhee MS, McGoldrick MD, Meuwissen HJ (1986) Serum factor from patients with chronic renal failure enhances polymorphonuclear leukocyte oxidative metabolism. Nephron 42:6–13

Richards FM, Watson A, Hickman JA (1988) Investigation of the effects of heat shock and agents which induce a heat shock response on the induction of differentiation of HL-60 cells. Cancer Res 48:6715–6720

Riches DWH, Channon JY, Leslie CC, Henson PM (1988) Receptor-mediated signal transduction in mononuclear phagocytes. Prog Allergy 42:65–122

Richter J, Andersson T, Olsson I (1989) Effect of tumor necrosis factor and granulocyte/macrophage colony-stimulating factor on human neutrophil degranulation. J Immunol 142:3199–3205

Rider LG, Niedel JE (1987) Diacylglycerol accumulation and superoxide anion production in stimulated human neutrophils. J Biol Chem 262:5603–5608

Rider LG, Dougherty RW, Niedel JE (1988) Phorbol diesters and dioctanoylglycerol stimulate accumulation of both diacylglycerols and alkylacylglycerols in human neutrophils. J Immunol 140:200–207

Rieder H, Ramadori G, Meyer zum Büschenfelde KH (1988a) Guinea pig Kupffer cells can be activated in vitro to an enhanced superoxide response. I. Comparison with peritoneal macrophages. J Hepatol 7:338–344

Rieder H, Ramadori G, Meyer zum Büschenfelde KH (1988b) Guinea pig Kupffer cells can be activated in vitro to an enhanced superoxide response. II. Involvement of eicosanoids. J Hepatol 7:345–351

Rimele TJ, Sturm RJ, Adams LM, Henry DE, Heaslip RJ, Weichman BM, Grimes D (1988) Interactions of neutrophils with vascular smooth muscle: identification of a neutrophil-derived relaxing factor. J Pharmacol Exp Ther 245:102–111

Rivkin I, Rosenblatt J, Becker EL (1975) The role of cyclic AMP in the chemotactic responsiveness and spontaneous motility of rabbit peritoneal neutrophils. The inhibition of neutrophil movement and the elevation of cyclic AMP levels by catecholamines, prostaglandins, theophylline and cholera toxin. J Immunol 115:1126–1134

Roberts PA, Newby AC, Hallett MB, Campbell AK (1985) Inhibition by adenosine of reactive oxygen metabolite production by human polymorphonuclear leukocytes. Biochem J 227:669–674

Roberts PA, Knight J, Campbell AK (1987) Pholasin — a bioluminescent indicator for detecting activation of single neutrophils. Anal Biochem 160:139–148

Roberts PJ, Cross AR, Jones OTG, Segal AW (1982) Development of cytochrome b and an active oxidase system in association with maturation of a human promyelocytic (HL-60) cell line. J Cell Biol 95:720–726

Roberts PJ, Cummins D, Bainton AL, Walshe KJ, Fisher-Hoch SP, McCormick JB, Gribben JG, Machin SJ, Linch DC (1989) Plasma from patients with severe Lassa fever profoundly modulates fMet-Leu-Phe induced superoxide generation in neutrophils. Br J Haematol 73:152–157

Robertson FM, Beavis AJ, Oberyszyn TM, O'Connell SM, Dokidos A, Leskin DL, Laskin JD, Reiners JJ Jr (1990) Production of hydrogen peroxide by murine epidermal kerationocytes following treatment with the tumor promoter 12-O-tetradecanoyl-13-acetate. Cancer Res 50:6062–6067

Robinson JM, Badwey JA, Karnovksy ML, Karnovsky MJ (1984) Superoxide release by neutrophils: synergistic effects of a phorbol ester and a calcium ionophore. Biochem Biophys Res Commun 122:734–739

Robinson JM, Badwey JA, Karnovsky ML, Karnovsky MJ (1985) Release of superoxide and change in morphology by neutrophils in response to phorbol esters: antagonism by inhibitors of calcium-binding proteins. J Cell Biol 101:1052–1058

Robinson JM, Heyworth PG, Badwey JA (1990) Utility of staurosporine in uncovering differences in the signal transduction pathways for superoxide production in neutrophils. Biochim Biophys Acta 1055:55–62

Roccatello D, Mazzucco G, Coppo R, Piccoli G, Rollino C, Scalzo B, Guerra MG, Cavalli G, Giachino O, Amore A, Malavasi F, Sena LM (1989) Functional changes of monocytes due to dialysis membranes. Kidney Int 35:622–631

Rocklin RE, Blidy A, Butler J, Minowada J (1981) Partial characterization of a lymphoid cell line (Reh) product with leukocyte inhibitory factor (LIF) activity. J Immunol 127:534–539

Roghani M, da Silva C, Castagna M (1987) Tumor promoter chloroform is a potent protein kinase C activator. Biochem Biophys Res Commun 142:738–744

Rokutan K, Hosokawa T, Nakamura K, Koyama K, Aoike A, Kawai K (1988) Increased superoxide anion production and glutathione peroxidase activity in peritoneal macrophages from autoimmune-prone MRL/Mp-Ipr/Ipr mice. Int Arch Allergy Appl Immunol 87:113–119

Rollins TE, Springer MS (1985) Identification of the polymorphonuclear leukocyte C5a receptor. J Biol Chem 260:7157–7160

Romeo D, Zabucchi G, Soranzo MR, Rossi F (1971) Macrophage metabolism: activation of NADPH oxidase by phagocytosis. Biochem Biophys Res Commun 45:1056–1062

Romeo D, Zabucchi G, Rossi F (1973) Reversible metabolic stimulation of polymorphonuclear leukocytes and macrophages by concanavalin A. Nature [New Biol] 243:111–112

Romeo D, Jug M, Zabucchi G, Rossi F (1974) Perturbation of leukocyte metabolism by nonphagocytosable concanavalin A-coupled beads. FEBS Lett 42:90–93

Romeo D, Zabucchi G, Miani N, Rossi F (1975) Ion movement across leukocyte plasma membrane and excitation of their metabolism. Nature 253:542–544

Rooney TA, Hager R, Rubin E, Thomas AP (1989) Short chain alcohols activate guanine nucleotide-dependent phosphoinositidase C in turkey erythrocyte membranes. J Biol Chem 264:6817–6822

Roos D (1980) The metabolic response to phagocytosis. In: Glynn LE, Houck JC, Weissmann G (eds) The cell biology of inflammation. Elsevier/North-Holland, Amsterdam, pp 337–385 (Handbook of inflammation, vol 2)

Roos D, Weening RS, Voetman AA, van Schalk MLJ, Bot AAM, Meerhof LJ, Loos JA (1979) Protection of phagocytic leukocytes by endogenous glutathione: studies in a family with glutathione reductase deficiency. Blood 53:851–866

Roos D, Bot AAM, van Schaik LJ, de Boer M, Daha MR (1981) Interaction between human neutrophils and zymosan particles: the role of opsonins and divalent cations. J Immunol 126:433–440

Roos D, Voetman AA, Meerhof LJ (1983) Functional activity of enucleated human polymorphonuclear leukocytes. J Cell Biol 97:368–377

Roos D, Eckmann CM, Yazdanbakhsh M, Hamers MN, de Boer M (1984) Excretion of superoxide by phagocytes measured with cytochrome c entrapped in resealed erythrocyte ghosts. J Biol Chem 259:1770–1775

Root RK, Metcalf J, Oshino N, Chance B (1975) H_2O_2 release from human granulocytes during phagocytosis. I. Documentation, quantitation, and some regulating factors. J Clin Invest 55:945–955

Rose FR, Hirschhorn R, Weissmann G, Cronstein BN (1988) Adenosine promotes neutrophil chemotaxis. J Exp Med 167:1186–1194

Rosen A, Nairn AC, Greengard P, Cohn ZA, Aderem A (1989) Bacterial lipopolysaccharide regulates the phosphorylation of the 68K protein kinase C substrate in macrophages. J Biol Chem 264:9118–9121

Rosen GM, Freeman BA (1984) Detection of superoxide generated by endothelial cells. Proc Natl Acad Sci USA 81:7269–7273

Rosen H, Klebanoff SJ (1976) Chemiluminescence and superoxide production by myeloperoxidase-deficient leukocytes. J Clin Invest 58:50–60

Rosenthal W, Koesling D, Rudolph U, Kleuss C, Pallast M, Yajima M, Schultz G (1986) Identification and characterization of the 35-kDa β subunit of guanine-nucleotide-binding proteins by an antiserum raised against transducin. Eur J Biochem 158:255–263

Rosenthal W, Binder T, Schultz G (1987) NADP efficiently inhibits endogenous but not pertussis toxin-catalyzed covalent modification of membrane proteins incubated with NAD. FEBS Lett 211:137–143

Rosoff PM, Cantley LC (1985) Lipopolysaccharide and phorbol esters induce differentiation but have opposite effects on phosphatidylinositol turnover and Ca^{2+} mobilization in 70Z/3 pre-B-lymphocytes. J Biol Chem 260:9209–9215

Rossi F (1986) The O_2^--forming NADPH oxidase of the phagocytes: nature, mechanisms of activation and function. Biochim Biophys Acta 853:65–89

Rossi F, Zatti M, Patriarca P (1969) H_2O_2 production during NADPH oxidation by the granule fraction of phagocytosing polymorphonuclear leukocytes. Biochim Biophys Acta 184:201–203

Rossi F, Romeo D, Patriarca P (1972) Mechanism of phagocytosis-associated oxidative metabolism in polymorphonuclear leukocytes and macrophages. J Reticuloendothel Soc 12:127–149

Rossi F, Della Bianca V, Davoli A (1981a) A new way for inducing a respiratory burst in guinea pig neutrophils. Change in the Na^+, K^+ concentration of the medium. FEBS Lett 132:273–277

Rossi F, Della Bianca V, Bellavite P (1981b) Inhibition of the respiratory burst and of phagocytosis by nordihydroguairetic acid in neutrophils. FEBS Lett 127:183–187

Rossi F, De Togni P, Bellavite P, Della Bianca V, Grzeskowiak M (1983) Relationship between the binding of N-formylmethionylleucylphenylalanine and the respiratory response in human neutrophils. Biochim Biophys Acta 758:168–175

Rossi F, Della Bianca V, Grzeskowiak M, De Togni P, Cabrini G (1985) Relationships between phosphoinositide metabolism, Ca^{2+} changes and respiratory burst in formyl-methionyl-leucyl-phenylalanine-stimulated human neutrophils. The breakdown of phosphoinositides is not involved in the rise of cytosolic free Ca^{2+}. FEBS Lett 181:253–258

Rossi F, Grzeskowiak M, Della Bianca V (1986) Double stimulation with FMLP and Con A restores the activation of the respiratory burst but not of the phosphoinositide turnover in Ca^{2+}-depleted human neutrophils. A further example of dissociation between stimulation of the NADPH oxidase and phosphoinositide turnover. Biochem Biophys Res Commun 140:1–11

Rossi F, Della Bianca V, Grzeskowiak M, Bazzoni F (1989) Studies on molecular regulation of phagocytosis in neutrophils. Con A-mediated ingestion and associated respiratory burst independent of phosphoinositide turnover, rise in $[Ca^{2+}]_i$, and arachidonic acid release. J Immunol 142:1652–1660

Rossi F, Grzeskowiak M, Della Bianca V, Calzetti F, Ganini G (1990) Phosphatidic acid and not diacylglycerol generated by phospholipase D is functionally linked to the activation of the NADPH oxidase by FMLP in human neutrophils. Biochem Biophys Res Commun 168:320–327

Rot A, Henderson LE, Copeland TD, Leonard EJ (1987) A series of six ligands for the human formyl peptide receptor: tetrapeptides with high chemotactic potency and efficacy. Proc Natl Acad Sci USA 84:7967–7971

Rotrosen D, Gallin JI, Spiegel AM, Malech HL (1988) Subcellular localization of $G_{i\alpha}$ in human neutrophils. J Biol Chem 263:10958–10964

Rotrosen D, Kleinberg ME, Nunoi H, Leto T, Gallin JI, Malech HL (1990) Evidence for a functional cytoplasmic domain of phagocyte oxidase cytochrome b_{558}. J Biol Chem 265:8745–8750

Rouis M, Nigon F, Chapman MJ (1988) Platelet activating factor is a potent stimulant of the production of active oxygen species by human monocyte-derived macrophages. Biochem Biophys Res Commun 156:1293–1301

Roux-Lombard P, Cruchaud A, Dayer JM (1986) Effect of interferon-γ and $1\alpha,25$-dihydroxyvitamin D_3 on superoxide anion, prostaglandins E_2, and mononuclear cell factor production by U937 cells. Cell Immunol 97:286–296

Rouzer CA, Scott WA, Cohn ZA, Blackburn P, Manning JM (1980) Mouse peritoneal macrophages release leukotriene C in response to a phagocytic stimulus. Proc Natl Acad Sci USA 77:4928–4932

Rovin BH, Wurst E, Kohan DE (1990) Production of reactive oxygen species by tubular epithelial cells in culture. Kidney Int 37:1509–1514

Royer-Pokora B, Kunkel LM, Monaco AP, Goff SC, Newburger PE, Baehner RL, Cole FS, Curnutte JT, Orkin SH (1986) Cloning the gene for an inherited human disorder-chronic granulomatous disease-on the basis of its chromosomal location. Nature 322:32–38

Rubin BY, Anderson SL, Lunn RM, Smith LJ (1989) Induction of proteins in interferon-α- and interferon-γ-treated polymorphonuclear leukocytes. J Leukocyte Biol 45:396–400

Rubin R, Hoek JB (1988) Alcohol-induced stimulation of phospholipase C in human platelets requires G-protein activation. Biochem J 254:147–153

Rudolph U, Koesling D, Hinsch KD, Seifert R, Bigalke M, Schultz G, Rosenthal W (1989a) G-protein α-subunits in cytosolic and membranous fractions of human neutrophils. Mol Cell Endocrinol 63:143–153

Rudolph U, Schultz G, Rosenthal W (1989b) The cytosolic G-protein α-subunit in human neutrophils responds to treatment with guanine nucleotides and magnesium. FEBS Lett 251:137–142

Rüegg UT, Burgess GM (1989) Staurosporine, K-252 and UCN-01: potent but non-specific inhibitors of protein kinases. Trends Pharmacol Sci 10:218–220

Saad M, Strnad CF, Wong K (1987) Dual regulation of neutrophil adenylate cyclase by fluoride and its relationship to cellular activation. Br J Pharmacol 91:715–719

Sachs CW, Christensen RH, Pratt PC, Lynn WS (1990) Neutrophil elastase activity and superoxide production are diminished in neutrophils of alcoholics. Am Rev Respir Dis 141:1249–1255

Sager G, Bang BE, Pedersen M, Aarbakke J (1988) The human promyelocytic leukemia cell (HL-60 cell) β-adrenergic receptor. J Leukocyte Biol 44:41–45

Sagone AL Jr, King GW, Metz EN (1976) A comparison of the metabolic response to phagocytosis in human granulocytes and monocytes. J Clin Invest 57:1352–1358

Sakane F, Takahashi K, Koyama J (1983) Separation of three menadione-dependent, O_2^--generating, pyridine nucleotide-oxidizing enzymes in guinea pig polymorphonuclear leukocytes. J Biochem (Tokyo) 94:931–936

Sakane F, Takahashi K, Koyama J (1984) Purification and characterization of a membrane-bound NADPH-cytochrome c reductase capable of catalyzing menadione-dependent O_2^- formation in guinea pig polymorphonuclear leukocytes. J Biochem (Tokyo) 96:671–678

Sakane F, Takahashi K, Takayama H, Koyama J (1987a) Stabilizing effect of glutaraldehyde on the respiratory burst NADPH oxidase of guinea pig polymorphonuclear leukocytes. J Biochem (Tokyo) 102:247–253

Sakane F, Kojima H, Takahashi K, Koyama J (1987b) Porcine polymorphonuclear leukocytes NADPH-cytochrome c reductase generates superoxide in the presence of cytochrome b_{559} and phospholipid. Biochem Biophys Res Commun 147:71–77

Sakata A, Ida E, Tominaga M, Onoue K (1987a) Calmodulin inhibitors, W-7 and TFP, block the calmodulin-independent activation of NADPH oxidase by arachidonate in a cell-free system. Biochem Biophys Res Commun 148:112–119

Sakata A, Ida E, Tominaga M, Onoue K (1987b) Arachidonic acid acts as an intracellular activator of NADPH-oxidase in Fcγ receptor-mediated superoxide gneration in macrophages. J Immunol 138:4353–4359

Salmon JA, Higgs GA (1987) Prostaglandins and leukotrienes as inflammatory mediators. Br Med Bull 43:285–296

Salmon JE, Cronstein BN (1990) Fcγ receptor-mediated functions in neutrophils are modulated by adenosine receptor occupancy. A1 receptors are stimulatory and A2 receptors are inhibitory. J Immunol 145:2235–2240

Salvemini D, de Nucci G, Gryglewski RJ, Vane JR (1989) Human neutrophils and mononuclear cells inhibit platelet aggregation by releasing a nitric oxide-like factor. Proc Natl Acad Sci USA 86:6328–6332

Salvemini D, Masini E, Anggard E, Mannaioni PF, Vane J (1990) Synthesis of a nitric oxide-like factor from L-arginine by rat serosal mast cells: stimulation of guanylate cyclase and inhibition of platelet aggregation. Biochem Biophys Res Commun 169:596–601

Salzer W, Gerard C, McCall C (1987) Effect of an inhibitor of protein kinase C on human polymorphonuclear leukocyte degranulation. Biochem Biophys Res Commun 148:747–754

Samuelsson B, Dahlén SE, Lindgren JA, Rouzer CA, Serhan CN (1987) Leukotrienes and lipoxins: structures, biosynthesis and biological effects. Science 237:1171–1176

Sandborg RR, Smolen JE (1988) Biology of disease. Early biochemical events in leukocyte activation. Lab Invest 59:300–320

Sandler JA, Clyman RI, Manganiello VC, Vaughan M (1975a) The effect of serotonin (5-hydroxytryptamine) and derivatives on guanosine 3′,5′-monophosphate in human monocytes. J Clin Invest 55:431–435

Sandler JA, Gallin JI, Vaughan M (1975b) Effects of serotonin, carbamylcholine, and ascorbic acid on leukocyte cyclic GMP and chemotaxis. J Cell Biol 67:480–484

Sasada M, Johnston RB Jr (1980) Macrophage microbicidal activity. Correlation between phagocytosis-associated oxidative metabolisms and the killing of Candida by macrophages. J Exp Med 152:85–98

Sasada M, Pabst MJ, Johnston RB Jr (1983) Activation of mouse peritoneal macrophages by lipopolysaccharide alters the kinetic parameters of the superoxide-producing NADPH oxidase. J Biol Chem 258:9631–9635

Sasada M, Kubo A, Nishimura T, Kakita T, Moriguchi T, Yamamoto K, Uchino H (1987) Candidacidal activity of monocyte-derived human macrophages: relationship between Candida killing and oxygen radical generation by human macrophages. J Leukocyte Biol 41:289–294

Sasagawa S, Suzuki K, Sakatani T, Fujikura T (1985) Effects of nicotine on the functions of human polymorphonuclear leukocytes in vitro. J Leukocyte Biol 37:493–502

Sato M, Nakamura T, Koyama J (1987) Different abilities of two distinct Fcγ receptors on guinea pig polymorphonuclear leukocytes to trigger the arachidonic acid metabolic cascade. FEBS Lett 224:29–32

Satoh M, Nunri H, Takeshige K, Minakami S (1985) Pertussis toxin inhibits intracellular pH changes in human neutrophils stimulated by N-formyl-methionyl-leucyl-phenylalanine. Biochem Biophys Res Commun 131:64–69

Saussy DL Jr, Sarau HM, Foley JJ, Mong S, Crooke ST (1989) Mechanisms of leukotriene E_4 partial agonist activity at leukotriene D_4 receptors in differentiated U-937 cells. J Biol Chem 264:19845–19855

Sawyer DW, Sullivan JA, Mandell GL (1985) Intracellular free calcium localization in neutrophils during phagocytosis. Science 230:663–666

Sbarra AJ, Karnovsky ML (1959) The biochemical basis of phagocytosis. I. Metabolic changes during the ingestion of particles by polymorphonuclear leukocytes. J Biol Chem 234:1355–1362

Schächtele C, Seifert R, Osswald H (1988) Stimulus-dependent inhibition of platelet aggregation by the protein kinase C inhibitors polymyxin B, H-7 and staurosporine. Biochem Biophys Res Commun 151:542–547

Schächtele C, Wagner B, Rudolph C (1989) Effect of Ca^{2+} entry blockers on myosin light-chain kinase and protein kinase C. Eur J Pharmacol 163:151–155

Schainberg H, Borish L, King M, Rocklin RE, Rosenwasser LJ (1988) Leukocyte inhibitory factor stimulates neutrophil-endothelial cell adhesion. J Immunol 141:3055–3060

Schatzmann RC, Raynor RL, Kuo JF (1983) N-(6-Aminohexyl)-5-chloro-1-naphthalene-sulfonamide (W-7), a calmodulin antagonist, also inhibits phospholipid-sensitive calcium-dependent protein kinase. Biochim Biophys Acta 755:144–147

Schell-Frederick E (1974) Stimulation of the oxidative metabolism of polymorphonuclear leukocytes by the calcium ionophore A23187. FEBS Lett 48:37–40

Schell-Frederick E (1984) A comparison of the effects of soluble stimuli on free cytoplasmic and membrane bound calcium in human neutrophils. Cell Calcium 5:237–251

Schiffmann E, Corcoran BA, Wahl SM (1975) N-Formylmethionyl peptides as chemoattractants for leucocytes. Proc Natl Acad Sci USA 72:1059–1062

Schiffmann E, Aswanikumar S, Venkatasubramanian K, Corcoran BA, Pert CB, Brown J, Gross E, Day AR, Freer RJ, Showell AH, Becker EL (1980) Some characteristics of the neutrophil receptor for chemotactic peptides. FEBS Lett 117:1–7

Schinetti ML, Lazzarino G (1986) Inhibition of phorbol ester-stimulated chemiluminescence and superoxide production in human neutrophils by fructose 1,6-diphosphate. Biochem Pharmacol 35:1762–1764

Schinetti ML, Mazzini A, Greco R, Bertelli A (1984) Inhibiting effects of levamisole on superoxide production from rat mast cells. Biochem Biophys Res Commun 16:101–107

Schleimer RP, Freeland HS, Peters SP, Brown KE, Derse CP (1989) An assessment of the effects of glucocorticoids on degranulation, chemotaxis, binding to vascular endothelium and formation of leukotriene B4 by purified human neutrophils. J Pharmacol Exp Ther 250:598–605

Schlesinger MJ, Aliperti G, Kelley PM (1982) The response of cells to heat shock. Trends Biochem Sci 7:222–225

Schmeichel CJ, Thomas LL (1987) Methylxanthine bronchodilators potentiate multiple human neutrophil functions. J Immunol 138:1896–1903

Schmidt HHHW, Nau H, Wittfoht W, Gerlach J, Prescher KE, Klein MM, Niroomand F, Böhme E (1988) Arginine is a physiological precursor of endothelium-derived nitric oxide. Eur J Pharmacol 154:213–216

Schmidt HHHW, Seifert R, Böhme E (1989) Formation and release of nitric oxide from human neutrophils and HL-60 cells induced by a chemotactic peptide, platelet activating factor and leukotriene B4. FEBS Lett 244:357–360

Schneider C, Zanetti M, Romeo D (1981) Surface-reactive stimuli selectively increase protein phosphorylation in human neutrophils. FEBS Lett 127:4–8

Schneider GB (1982) An evaluation of the phagocytic cells in ia (osteopetrotic) rat: oxidative metabolism in monocytes and macrophages. J Reticuloendothel Soc 31:225–232

Schnur RA, Newman SL (1990) The respiratory burst to Histoplasma capsulatum by human neutrophils. Evidence for intracellular trapping of superoxide anion. J Immunol 144:4765–4772

Schoch P, Sargent DF (1980) Quantitative analysis of the binding of mellitin to planar lipid bilayers allowing for the discrete-charge effect. Biochim Biophys Acta 602:234–247

Schopf RE, Lemmel EM (1983) Control of the production of oxygen intermediates of human polymorphonuclear leukocytes and monocytes by β-adrenergic receptors. J Immunopharmacol 5:203–216

Schopf RE, Mattar J, Meyenburg W, Scheiner O, Hammann KP, Lemmel EM (1984) Measurement of the respiratory burst in human monocytes and polymorphonuclear leukocytes by Nitro Blue tetrazolium reduction and chemiluminescence. J Immunol Methods 67:109–117

Schrier DJ, Imre KM (1986) The effects of adenosine agonists on human neutrophil-function. J Immunol 137:3284–3289

Schröder H, Ney P, Woditsch I, Schrör K (1989) Cyclic GMP mediates SIN-1-induced inhibition of human polymorphonuclear leukocytes. Eur J Pharmacol 182:211–218

Schubert T, Müller WE (1989) N-Formyl-methionyl-leucyl-phenylalanine induced accumulation of inositol phosphates indicates the presence of oligopeptide chemoattractant receptors on circulating human lymphocytes. FEBS Lett 257:174–176

Schultz RM, Nanda SKW, Altom MG (1985) Effects of various inhibitors of arachidonic acid oxygenation on macrophage superoxide release and tumoricidal activity. J Immunol 135:2040–2044

Schwartz RH, Bianco AR, Handwerger BS, Kahn CR (1975) Demonstration that monocytes rather than lymphocytes are the insulin-binding cells in preparations of human peripheral blood mononuclear leukocytes: implications for studies of insulin-resistant states in man. Proc Natl Acad Sci USA 72:474–478

Scully SP, Segel GB, Lichtman MA (1986) Relationship of superoxide production to cytoplasmic free clacium in human monocytes. J Clin Invest 77:1349–1356

Sechler JMG, Malech HL, White CJ, Gallin JI (1988) Recombinant human interferon-γ reconstitutes defective phagocyte function in patients with chronic granulomatous disease of childhood. Proc Natl Acad Sci USA 85:4874–4878

Seeds MC, Parce JW, Szejda P, Bass DA (1985) Independent stimulation of membrane potential changes and the oxidative metabolic burst in polymorphonuclear leukocytes. Blood 65:233–240

Segal AW (1987) Absence of both cytochrome b_{-245} subunits from neutrophils in X-linked chronic granulomatous disease. Nature 326:88–91

Segal AW (1989a) The electron transport chain of the microbicidal oxidase of phagocytic cells and its involvement in the molecular pathology of chronic granulomatous disease. Biochem Soc Trans 117:427–434

Segal AW (1989b) The electron transport chain of the microbicidal oxidase of phagocytic cells and its involvement in the molecular pathology of chronic granulomatous disease. J Clin Invest 83:1785–1793

Segal AW, Coade SB (1978) Kinetics of oxygen consumption by phagocytosing human neutrophils. Biochem Biophys Res Commun 84:611–617

Segal AW, Jones OTG (1978) Novel cytochrome b system in phagocytic vacuoles of human granulocytes. Nature 276:515–517

Segal AW, Jones OTG (1979) The subcellular distribution and some properties of the cytochrome b component of the microbicidal oxidase system of human neutrophils. Biochem J 182:181–188

Segal AW, Jones OTG (1980) Rapid incorporation of the human neutrophil plasma membrane cytochrome b into phagocytic vacuoles. Biochem Biophys Res Commun 92:710–715

Segal AW, Garcia RC, Goldstone AH, Cross AR, Jones OTG (1981) Cytochrome b_{-245} of neutrophils is also present in human monocytes, macrophages and eosinophils. Biochem J 196:363–367

Segal AW, Cross AR, Garcia RC, Borregaard N, Valerius NH, Soothill JF, Jones OTG (1983) Absence of cytochrome b_{-245} in chronic granulomatous disease. A multicenter European evaluation of its incidence and relevance. N Engl J Med 308:245–251

Segal AW, Heyworth PG, Cockcroft S, Barrowman MM (1985) Stimulated neutrophils from patients with autosomal recessive chronic granulomatous disease fail to phosphorylate a M_r-44.000 protein. Nature 316:547–549

Seifert R (1988) Regulation of NADPH oxidase activity in cell-free systems of HL-60 leukemic cells. Naunyn Schmiedebergs Arch Pharmacol [Suppl] 337:R39

Seifert R, Schächtele C (1988) Studies with protein kinase C inhibitors presently available cannot elucidate the role of protein kinase C in the activation of NADPH oxidase. Biochem Biophys Res Commun 152:585–592

Seifert R, Schultz G (1987a) Fatty acid-induced activation of NADPH oxidase in plasma membranes of human neutrophils depends on neutrophil cytosol and is potentiated by stable guanine nucleotides. Eur J Biochem 162:563–569

Seifert R, Schultz G (1987b) Reversible activation of NADPH oxidase in membranes of HL-60 human leukemic cells. Biochem Biophys Res Commun 146:1296–1302

Seifert R, Schultz G (1989) Involvement of pyrimidinoceptors in the regulation of cell functions by uridine and by uracil nucleotides. Trends Pharmacol Sci 10:365–369

Seifert R, Rosenthal W, Schultz G (1986) Guanine nucleotides stimulate NADPH oxidase in membranes of human neutrophils. FEBS Lett 205:161–165

Seifert R, Schächtele C, Schultz G (1987) Activation of protein kinase C by *cis*- and *trans*-octadecadienoic acids in intact human platelets and its potentiation by diacylglycerol. Biochem Biophys Res Commun 149:762–768

Seifert R, Rosenthal W, Schächtele C, Schultz G (1988a) Arachidonic acid-induced activation of NADPH oxidase in membranes of human neutrophils and HL-60 leukemic cells is regulated by guanine nucleotide-binding proteins and is independent of Ca^{2+} and protein kinase C. In: Nigam S, McBrien DCH, Slater TF (eds) Eicosanoids, lipid peroxidation and cancer. Springer, Berlin Heidelberg New York, pp 169–179

Seifert R, RosenthalW, Schultz G, Wieland T, Gierschik P, Jakobs KH (1988b) The role of nucleoside-diphosphate kinase reactions in G protein activation of NADPH oxidase by guanine and adenine nucleotides. Eur J Biochem 175:51–55

Seifert R, Schächtele C, Rosenthal W, Schultz G (1988c) Activation of protein kinase C by *cis*- and *trans*-fatty acids and its potentiation by diacylglycerol. Biochem Biophys Res Commun 154:20–26

Seifert R, Burde R, Schultz G (1989a) Lack of effect of opioid peptides, morphine and naloxone on superoxide formation in human neutrophils and HL-60 leukemic cells. Naunyn Schmiedebergs Arch Pharmacol 340:101–106

Seifert R, Burde R, Schultz G (1989b) Activation of NADPH oxidase by purine and pyrimidine nucleotides involve G proteins and is potentiated by chemotactic peptides. Biochem J 259:813–819

Seifert R, Jungblut P, Schultz G (1989c) Differential expression of cytosolic activation factors for NADPH oxidase in HL-60 leukemic cells. Biochem Biophys Res Commun 161:1109–1117

Seifert R, Wenzel K, Eckstein F, Schultz G (1989d) Purine and pyrimidine nucleotides potentiate activation of NADPH oxidase and degranulation by chemotactic peptides and induce aggregation of human neutrophils via G proteins. Eur J Biochem 181:277–285

Seifert R, Schultz G, Richter-Freund M, Metzger J, Wiesmüller KH, Jung G, Bessler WG, Hauschildt S (1990b) Activation of superoxide formation and lysozyme release in human neutrophils by the synthetic lipopeptide, Pam₃Cys-Ser-(Lys₄): involvement of G-proteins and synergism with chemotactic peptides. Biochem J 267:795–802

Seifert R, Hilgenstock G, Fassbender M, Distler A (1991) Regulation of the superoxide-forming NADPH oxidase is not altered in essential hypertension. J Hypertension 9:147–153

Seifter JL, Aronson PS (1986) Properties and physiologic roles of the plasma membrane sodium-hydrogen exchanger. J Clin Invest 78:859–864

Sekiguchi K, Tsukuda M, Ogita K, Kikkawa U, Nishizuka Y (1987) Three distinct forms of rat brain protein kinase C: differential response to unsaturated fatty acids. Biochem Biophys Res Commun 145:797–802

Seligmann BE, Gallin JI (1980) Use of lipophilic probes of membrane potential to assess human neutrophil activation. J Clin Invest 66:493–503

Seligmann BE, Fletcher MP, Gallin JI (1982) Adaptation of human neutrophil responsiveness to the chemoattractant N-formylmethionylleucylphenylalanine. Heterogeneity and/or negative cooperative interaction of receptors. J Biol Chem 257:6280–6286

Seligmann BE, Fletcher MP, Gallin JI (1983) Histamine modulation of human neutrophil oxidative metabolism, locomotion, degranulation, and membrane potential changes. J Immunol 130:1902–1909

Serhan CN, Broekman MJ, Korchak HM, Smolen JE, Marcus AJ, Weissmann G (1983) Changes in phosphatidylinositol and phosphatidic acid in stimulated human neutrophils. Relationship to calcium mobilization, aggregation and superoxide radical generation. Biochim Biophys Acta 762:420–428

Serhan CN, Hamberg M, Samuelsson B (1984) Lipoxins: novel series of biologically active compounds formed from arachidonic acid in human leukocytes. Proc Natl Acad Sci USA 81:5335–5339

Serra MC, Bellavite P, Davoli A, Bannister JV, Rossi F (1984) Isolation from neutrophil membranes of a complex containing active NADPH oxidase and cytochrome b_{-245}. Biochim Biophys Acta 788:138–146

Serra MC, Bazzoni F, Della Bianca V, Greszkowiak M, Rossi F (1988) Activation of human neutrophils by substance P. Effect on oxidative metabolism, exocytosis, cytosolic Ca^{2+} concentration and inositol phosphate formation. J Immunol 141:2118–2124

Sha'afi RI (1989) Some effects of phorbol esters are not mediated by protein kinase C. Biochem J 261:688

Sha'afi RI, Molski TFP (1987) Signalling for increased cytoskeletal actin in neutrophils. Biochem Biophys Res Commun 145:934–941

Sha'afi RI, Molski TFP (1988) Activation of the neutrophil. Prog Allergy 42:1–64

Sha'afi RI, Williams K, Wacholtz MC, Becker EL (1978) Binding of the chemotactic synthetic peptide [^{3}H]formyl-Nor-Leu-Leu-Phe to plasma membrane of rabbit neutrophils. FEBS Lett 91:305–309

Sha'afi RI, White JR, Molski TFP, Shefcyk J, Volpi M, Naccache PH, Feinstein MB (1983) Phorbol 12-myristate 13-acetate activates rabbit neutrophils without an apparent rise in the level of intracellular free calcium. Biochem Biophys Res Commun 114:638–645

Sha'afi RI, Molski TFP, Huang CK, Naccache PH (1986) The inhibition of neutrophil responsiveness caused by phorbol esters is blocked by the protein kinase C inhibitor H7. Biochem Biophys Res Commun 137:50–60

Sha'afi RI, Molski TFP, Gomez-Cambronero J, Huang CK (1988) Dissociation of the 47-kilodalton protein phosphorylation from degranulation and superoxide production in neutrophils. J Leukocyte Biol 43:18–27

Sha'afi RI, Gomez-Cambronero J, Yamazaki M, Durstin M, Huang CK (1989) Activation of human neutrophils by granulocyte-macrophage-colony-stimulating factor: role of guanine-nucleotide binding proteins. In: Biology of cellular transducing signals '89. Vanderhoek JY (ed) Washington. Ninth International Washington Spring Symposium. p 44

Sha'ag D, Pick E (1988) Macrophage-derived superoxide-generating NADPH oxidase in an amphiphile-activated, cell-free system; partial purification of the cytosolic component and evidence that it may contain the NADPH binding site. Biochim Biophys Acta 952:213–219

Sha'ag D, Pick E (1990) Nucleotide binding properties of cytosolic components required for expression of activity of the superoxide generating NADPH oxidase. Biochim Biophys Acta 1037:405–412

Shah SV, Naum-Bedigian S (1981) Light emission by isolated rat glomeruli in response to phorbol myristate acetate. J Lab Clin Med 98:46–57

Shah SV, Wallin JD, Eilen SD (1983) Chemiluminescence and superoxide anion production by leukocytes from diabetic patients. J Clin Endocrinol Metab 57:402–409

Shak S, Perez HD, Goldstein IM (1983) A novel dioxygenation product of arachidonic acid possesses potent chemotactic activity for human polymorphonuclear leukocytes. J Biol Chem 258:14948–14953

Shalaby MR, Aggarwal BB, Rinderknecht E, Svedersky LP, Finkle BS, Palladino MA Jr (1985) Activation of human polymorphonuclear neutrophil functions by interferon-γ and tumor necrosis factors. J Immunol 135:2069–2073

Shalaby MR, Palladino MA Jr, Hirabayashi SE, Eessalu TE, Lewis GD, Shepard HM, Aggarwal BB (1987) Receptor binding and activation of polymorphonuclear neutrophils by tumor necrosis factor-alpha. J Leukocyte Biol 41:196–204

Shalit M, Von Allmen C, Atkins PC, Zweiman B (1988) Platelet activating factor increases expression of complement receptors on human neutrophils. J Leukocyte Biol 44:212–217

Shapiro BM, Schackmann RW, Gabel CA (1981) Molecular approaches to the study of fertilization. Annu Rev Biochem 50:815–843

Shappell SB, Toman C, Anderson DC, Taylor AA, Entman ML, Smith CW (1990) Mac-1 (CD11b/CD18) mediates adherence-dependent hydrogen peroxide production by human and canine neutrophils. J Immunol 144:2702–2711

Sharp BM, Keane WF, Suh HJ, Gekker G, Tsukayama D, Peterson PK (1985) Opioid peptides rapidly stimulate superoxide production by human polymorphonuclear leukocytes and macrophages. Endocrinology 117:793–795

Sharp BM, Tsukayama DT, Gekker G, Keane WF, Peterson PK (1987) β-Endorphin stimulates human polymorphonuclear leukocyte superoxide production via a stereoselective opiate receptor. J Pharmacol Exp Ther 242:579–582

Shaw JO, Pinckard RN, Ferrigni KS, McManus LM, Hanahan DJ (1981) Activation of human neutrophils with 1-O-hexadecyl/octadecyl-2-acetyl-sn-glyceryl-3-phosphorylcholine (platelet activating factor). J Immunol 127:1250-1255

Shearman MS, Naor Z, Sekiguchi K, Kishimoto A, Nishizuka Y (1989) Selective activation of the γ-subspecies of protein kinase C from bovine cerebellum by arachidonic acid and its lipoxygenase metabolites. FEBS Lett 243:177–182

Shefcyk J, Yassin R, Volpi M, Molski TFP, Naccache PH, Munoz JJ, Becker EL, Feinstein MB, Sha'afi RI (1985) Pertussis but not cholera toxin inhibits the stimulated increase in actin association with the cytoskeleton in rabbit neutrophils: role of the "G proteins" in stimulus-response coupling. Biochem Biophys Res Commun 126:1174–1181

Shelly J, Hoff SF (1989) Effects of non-steroidal anti-inflammatory drugs on isolated polymorphonuclear leukocytes (PMN): chemotaxis, superoxide production, degranulation and N-formyl-L-methionyl-L-leucyl-L-phenylalanine (FMLP) receptor binding. Gen Pharmacol 20:329-334

Shen L, Collins J (1989) Monocyte superoxide secretion triggered by human IgA. Immunology 68:491–496

Shepherd VL, Campbell EJ, Senior RM, Stahl PD (1982) Characterization of the mannose/fucose receptor on human mononuclear phagocytes. J Reticuloendothel Soc 32:423–431

Sherman JW, Goetzl EJ, Koo CH (1988) Selective modulation by guanine nucleotides of the high affinity subset of plasma membrane receptors for leukotriene B₄ on human polymorphonuclear leukocytes. J Immunol 140:3900–3904

Shibanuma M, Kuroki T, Nose K (1987) Effects of the protein kinase C inhibitor H-7 and calmodulin antagonist W-7 on superoxide production in growing and resting human histiocytic leukemia cells (U937). Biochem Biophys Res Commun 144:1317–1323

Shigeoka AO, Charette RP, Wyman ML, Hill HR (1981) Defective oxidative metabolic responses of neutrophils of neutrophils from stressed neonates. J Pediatr 98:392–398

Shirasawa Y, White RP, Robertson JT (1983) Mechanisms of the contractile effect induced by uridine 5-triphosphate in canine cerebral arteries. Stroke 14:347–355

Shirato M, Takahashi K, Nagasawa S, Koyama J (1988) Different sensitivities of the responses of human neutrophils stimulated with immune complex and C5a anaphylatoxin to pertussis toxin. FEBS Lett 234:231–234

Showell HJ, Freer RJ, Zigmond SH, Schiffmann E, Aswanikumar S, Corcoran B, Becker EL (1976) The structure-activity relations of synthetic peptides as chemotactic factors and inducers of lysosomal enzyme secretion for neutrophils. J Exp Med 143:1154–1169

Showell HJ, Mackin WM, Becker EL, Muthukumarasway N, Day AR, Freer R (1981) Nc-formyl-norleucyl-leucyl-phenylalanyl chloromethylketone. Mol Cell Biochem 41:19–25

Shpungin S, Dotan I, Abo A, Pick E (1989) Activation of the superoxide forming NADPH oxidase in a cell-free system by sodium dodecyl sulfate. Absolute lipid dependency of the solubilized enzyme. J Biol Chem 264:9195–9203

Shult PA, Graziano FM, Wallow IH, Busse WW (1985) Comparison of superoxide generation and luminol-dependent chemiluminescence with eosinophils and neutrophils from normal individuals. J Lab Clin Med 106:638–645

Shurin SB, Cohen HJ, Whitin JC, Newburger PE (1983) Impaired granulocyte superoxide production and prolongation of the respiratory burst due to a low-affinity NADPH-dependent oxidase. Blood 62:564–571

Sibinga NES, Goldstein A (1988) Opioid peptides and opioid receptors in cells of the immune system. Annu Rev Immunol 6:219–249

Sibley DR, Benovic JL, Caron MG, Lefkowitz RJ (1987) Regulation of transmembrane signaling by receptor phosphorylation. Cell 48:913–922

Silverman DHS, Wu H, Karnovsky ML (1985) Muramyl peptides and serotonin interact at specific binding sites on macrophages and enhance superoxide release. Biochem Biophys Res Commun 131:1160–1167

Silverman DHS, Krueger JM, Karnovsky ML (1986) Specific binding sites for muramyl peptides on murine macrophages. J Immunol 136:2195–2201

Silverstein SC, Steinman RM, Cohn ZA (1977) Endocytosis. Annu Rev Biochem 46:669–722

Simchowitz L (1985a) Intracellular pH modulates the generation of superoxide radicals by human neutrophils. J Clin Invest 76:1079–1089

Simchowitz L (1985b) Chemotactic peptide-induced activation of Na^+/H^+ exchange in human neutrophils. I. Sodium fluxes. J Biol Chem 260:13237–13247

Simchowitz L, Spilberg I (1979) Chemotactic factor-induced generation of superoxide radicals by human neutrophils: evidence for the role of sodium. J Immunol 123:2428–2435

Simchowitz L, Atkinson JP, Spilberg I (1980a) Stimulus-specific deactivation of chemotactic factor-induced cyclic AMP response and superoxide generation by human neutrophils. J Clin Invest 66:736–747

Simchowitz L, Fischbein LC, Spilberg I, Atkinson JP (1980b) Induction of a transient elevation in intracellular levels of adenosine-3',5'-cyclic monophosphate by chemotactic factors: an early event in human neutrophil activation. J Immunol 124:1482–1491

Simchowitz L, Atkinson JP, Spilberg I (1982) Stimulation of the respiratory burst in human neutrophils by crystal phagocytosis. Arthritis Rheum 25:181–188

Simchowitz L, Spilberg I, Atkinson JP (1983) Evidence that the functional responses of human neutrophils occur independently of transient elevations in cyclic AMP levels. J Cyclic Nucleotide Protein Phosphor Res 9:35–47

Simchowitz L, Foy MA, Cragoe EJ Jr (1990) A role for a Na^+/Ca^{2+} exchange in the generation of superoxide radicals by human neutrophils. J. Biol Chem 265:13449–13456

Simons JM, 'T Hart BA, Ip Vai Ching TRAM, van Dijk H, Labadie RP (1990) Metabolic activation of natural phenols into selective oxidative burst agonists by activated human neutrophils. Free Radic Biol Med 8:251–258

Simpkins CO, Ives N, Tate E, Johnson M (1985) Naloxone inhibits superoxide release from human neutrophils. Life Sci 37:1381–1386

Simpkins CO, Alailima ST, Tate EA, Johnson M (1986) The effect of enkephalins and prostaglandins on O_2^- release by neutrophils. J Surg Res 41:645–652

Sinico-Durieux I, Gougerot-Pocidalo MA, Perianin A, Hakim J (1986) Effect of doxycycline on oxygen-dependent killing mechanism of human neutrophils. Biochem Pharmacol 35:1801–1804

Sklar LA, Oades ZG (1985) Signal transduction and ligand-receptor dynamics in the neutrophil. Ca^{2+} modulation and restoration. J Biol Chem 260:11468–11475

Sklar LA, Jesaitis AJ, Painter RG, Cochrane CG (1981a) The kinetics of neutrophil activation. The response to chemotactic peptides depends upon whether ligand-receptor interaction is rate-limiting. J Biol Chem 256:9909–9914

Sklar LA, Oades ZG, Jesaitis AJ, Painter RG, Cochrane CG (1981b) Fluoresceinated chemotactic peptide and high-affinity antifluorescein antibody as a probe of the temporal characteristics of neutrophil stimulation. Proc Natl Acad Sci USA 78:7540–7544

Sklar LA, Finney DA, Oades ZG, Jesaitis AJ, Painter RG, Cochrane CG (1984) The dynamics of ligand-receptor interactions. Real-time analyses of association, dissociation, and internalization of an N-formyl peptide and its receptors on the human neutrophil. J Biol Chem 259:5661–5669

Sklar LA, Omann GM, Painter RG (1985a) Relationship of actin polymerization and depolymerization to light scattering in human neutrophils: dependence on receptor occupancy and intracellular Ca^{++}. J Cell Biol 101:1161–1166

Sklar LA, Hyslop PA, Oades ZG, Omann GM, Jesaitis AJ, Painter RG, Cochrane CG (1985b) Signal transduction and ligand-receptor dynamics in the human neutrophil. Transient responses and occupancy-response relations at the formyl peptide receptor. J Biol Chem 260:11461–11467

Sklar LA, Bokoch GM, Button D, Smolen JE (1987) Regulation of ligand-receptor dynamics by guanine nucleotides. Real-time analysis of interconverting states for the neutrophil formyl peptide receptor. J Biol Chem 262:135–139

Skubitz KM, Hammerschmidt DE (1986) Effects of ibuprofen on chemotactic peptide-receptor binding and granulocyte response. Biochem Pharmacol 35:3349–3354

Slater TF (1984) Free-radical mechanisms in tissue injury. Biochem J 222:1–15

Smail EH, Melnick DA, Ruggeri R, Diamond RD (1988) A novel natural inhibitor from Candida albicans hyphae causing dissociation of the neutrophil respiratory burst response to chemotactic peptides from other post-activation events. J Immunol 140:3893–3899

Smith CD, Lane BC, Kusaka I, Verghese MW, Snyderman R (1985) Chemoattractant receptor-induced hydrolysis of phosphatidylinositol 4,5-biphosphate in human polymorphonuclear leukocyte membranes. Requirement for a guanine nucleotide regulatory protein. J Biol Chem 260:5875–5878

Smith CD, Cox CC, Snyderman R (1986) Receptor-coupled activation of phosphoinositide-specific phospholipase C by an N protein. Science 232:97–100

Smith CD, Uhing RJ, Snyderman R (1987) Nucleotide regulatory protein-mediated activation of phospholipase C in human polymorphonuclear leukocytes is disrupted by phorbol esters. J Biol Chem 262:6121–6127

Smith CD, Glickman JF, Chang J (1988) The antiproliferative effects of staurosporine are not exclusively mediated by inhibition of protein kinase C. Biochem Biophys Res Commun 156:1250–1256

Smith RJ, Iden SS (1981) Modulation of human neutrophil superoxide anion generation by the calcium antagonist 8-(N,N-diethylamino)-octyl-(3,4,5-trimethoxy) benzoate hydrochloride. J Reticuloendothel Soc 29:215–225

Smith RJ, Ignarro LJ (1975) Bioregulation of lysosomal enzyme secretion from human neutrophils: roles of guanosine 3':5'-monophosphate and calcium in stimulus-secretion coupling. Proc Natl Acad Sci USA 72:108–112

Smith RJ, Bowman BJ, Iden SS (1981) Effects of trifluoperazine on human neutrophil function. Immunology 44:677–684

Smith RJ, Bowman BJ, Iden SS (1984a) Effects of an anion channel blocker, 4,4'-diisothiocyano-2,2'-disulfonic acid stilbene (DIDS), on human neutrophil function. Biochem Biophys Res Commun 120:964–972

Smith RJ, Iden SS, Bowman BJ (1984b) Activation of the human neutrophil respiratory burst with zymosan-activated serum. Biochem Biophys Res Commun 121:695–701

Smith RJ, Speziale SC, Bowman BJ (1985) Properties of interleukin-1 as a complete secretagogue for human neutrophils. Biochem Biophys Res Commun 130:1233–1240

Smith RJ, Epps DE, Justen JM, Sam LM, Wynalda MA, Fitzpatrick FA, Yein FS (1987) Human neutrophil activation with interleukin-1. A role for intracellular calcium and arachidonic acid lipoxygenation. Biochem Pharmacol 36:3851–3858

Smith RJ, Sam LM, Justen JM (1988a) Diacylglycerols modulate human polymorphonuclear neutrophil responsiveness: effects on intracellular calcium mobilization, granule exocytosis, and superoxide anion production. J Leukocyte Biol 43:411–419

Smith RJ, Justen JM, Sam LM (1988b) Effects of a protein kinase C inhibitor, K-252a, on human polymorphonuclear neutrophil responsiveness. Biochem Biophys Res Commun 152:1497–1503

Smith RJ, Justen JM, Sam LM (1988c) Functions and stimulus-specific effects of phorbol 12-myristate 13-acetate on human polymorphonuclear neutrophils: autoregulatory role for protein kinase C in signal transduction. Inflammation 12:597–611

Smith RJ, Sam LM, Justen JM, Bundy GL, Bala GA, Bleasdale JE (1990) Receptor-coupled signal transduction in human polymorphonuclear neutrophils: effects of a novel inhibitor of phospholipase C-dependent processes on cell responsiveness. J Pharmacol Exp Ther 253:688–697

Smith RL, Hunt NH, Merritt JE, Evans T, Weidemann MJ (1980) Cyclic nucleotide metabolism and reactive oxygen production by macrophages. Biochem Biophys Res Commun 96:1079–1087

Smith RM, Curnutte JT, Babior BM (1989a) Affinity labeling of the cytosolic and membrane components of the respiratory burst oxidase by the 2',3'-dialdehyde derivative of NADPH. Evidence for a cytosolic location of the nucleotide-binding site in the resting cell. J Biol Chem 264:1958–1962

Smith RM, Curnutte JT, Mayo LA, Babior BM (1989b) Use of an affinity label to probe the function of the NADPH binding component of the respiratory burst oxidase of human neutrophils. J Biol Chem 264:12243–12248

Smolen JE (1984) Lag period for superoxide anion generation and lysosomal enzyme release from human neutrophils: effects of calcium antagonists and anion channel blockers. J Lab Clin Med 104:1–10

Smolen JE, Weissmann G (1980) Effects of indomethacin, 5,8,11,14-eicosatetraynoic acid, and p-bromophenacyl bromide on lysosomal enzyme release and superoxide anion generation by human polymorphonuclear leukocytes. Biochem Pharmacol 39:533–538

Smolen JE, Korchak HM, Weissmann G (1980) Increased levels of cyclic adenosine-3′,5′-monophosphate in human polymorphonuclear leukocytes after surface stimulation. J Clin Invest 65:1077–1085

Smolen JE, Korchak HM, Weissmann G (1981) The roles of extracellular and intracellular calcium in lysosomal enzyme release and superoxide anion generation by human neutrophils. Biochim Biophys Acta 677:512–520

Smolen JE, Korchak HM, Weissmann G (1982) Stimulus-response coupling in neutrophils. Trends Biochem Sci 3:483–485

Snyder RM, Mirabelli CK, Crooke ST (1986) Cellular association, intracellular distribution, and efflux of auranofin via sequential ligand exchange reactions. Biochem Pharmacol 35:923–932

Snyder RM, Mirabelli CK, Crooke ST (1987) Cellular interactions of auranofin and a related gold complex with RAW 264.7 macrophages. Biochem Pharmacol 36:647–654

Snyderman R (1984) Regulatory mechanisms of a chemoattractant receptor on leukocytes. Fed Proc 43:2743–2748

Snyderman R, Pike MC, Edge S, Lane B (1984) A chemoattractant receptor on macrophages exists in two affinity states regulated by guanine nucleotides. J Cell Biol 98:444–448

Solomkin JS, Brodt JK, Zemlan FP (1986) Degranulation inhibition. A potential mechanism for control of neutrophil superoxide production in sepsis. Arch Surg 121:77–80

Sonderer B, Wild P, Wyler R, Fontana A, Peterhans E, Schwyzer M (1987) Murine glia cells in culture can be stimulated to generate reactive oxygen. J Leukocyte Biol 42:463–473

Sottrup-Jensen L (1989) α-Macroglobulins: structure, shape, and mechanism of proteinase complex formation. J Biol Chem 264:11539–11542

Southwick FS, Dairi GA, Paschetto M, Zigmond SH (1989) Polymorphonuclear leukocyte adherence induces actin polymerization by a transient pathway which differs from that used by chemoattractants. J Cell Biol 109:1561–1569

Souza LM, Boone TC, Gabrilove J, Lai PH, Zsebo KM, Murdock DC, Chazin VR, Bruszewski J, Lu H, Chen KK, Barendt J, Platzer E, Moore MAS, Mertelsmann R, Welte K (1986) Recombinant human granulocyte colony-stimulating factor: effects on normal and leukemic myeloid cells. Science 232:61–65

Spangrude GJ, Sacchi F, Hill HR, Van Epps DE, Daynes RA (1985) Inhibition of lymphocyte and neutrophil chemotaxis by pertussis toxin. J Immunol 135:4135–4143

Speer CP, Pabst MJ, Hedegaard HB, Rest RF, Johnston RB Jr (1984) Enhanced release of oxygen metabolites by monocyte-derived macrophages exposed to proteolytic enzymes: Activity of neutrophil elastase and cathepsin G. J Immunol 133:2151–2156

Spirer Z, Zakuth V, Golander A, Bogair N, Fridkin M (1975) The effect of tuftsin on the nitrous blue tetrazolium reduction of normal human polymorphonuclear leukocytes. J Clin Invest 55:198–200

Spitalny GL (1980) Suppression of bactericidal activity of macrophages in ascites tumors. J Reticuloendothel Soc 28:223–235

Stabinsky Y, Gottlieb P, Zakuth V, Spirer Z, Fridkin M (1978) Specific binding sites for the phagocytosis stimulating peptide tuftsin on human polymorphonuclear leukocytes and monocytes. Biochem Biophys Res Commun 83:599–606

Stabinsky Y, Bar-Shavit Z, Fridkin M, Goldman R (1980) On the mechanism of action of the phagocytosis-stimulating peptide tuftsin. Mol Cell Biochem 30:71–77

Stadel JM, Crooke ST (1989) Fluoride interaction with G-proteins. Biochem J 258:932–933

Stähelin H, Karnovsky ML, Farnham AE, Suter E (1957) Studies on the interaction between phagocytes and tubercle bacilli. III. Some metabolic effects in guinea pigs associated with infection with tubercle bacilli. J Exp Med 105:265–277

Stahl PD, Rodman JS, Miller MJ, Schlesinger PH (1978) Evidence for receptor-mediated binding of glycoproteins, glycoconjugates, and lysosomal glycosidases by alveolar macrophages. Proc Natl Acad Sci USA 75:1399–1403

Stahl PD, Schlesinger PH, Sigardson E, Rodman JS, Lee YC (1980) Receptor-mediated pinocytosis of mannose glycoconjugates by macrophages: characterization and evidence for receptor recycling. Cell 19:207–215

Starke (1987) Presynaptic α-autoreceptors. Rev Physiol Biochem Pharmacol 107:74–146

Stasia MJ, Dianoux AC, Vignais PV (1989) A 23 kDa protein as a substrate for protein kinase C in bovine neutrophils. Purification and partial characterization. Biochemistry 28:9659–9667

Stauber GB, Aggarwal BB (1989) Characterization and affinity cross-linking of receptors for human recombinant lymphotoxin (tumor necrosis factor-β) on a human histiocytic lymphoma cell line, U-937. J Biol Chem 264:3573–3576

Steadman R, Knowlden J, Lichodziejewska M, Williams J (1990) The influence of net surface charge on the interaction of uropathogenic *Escherichia coli* with human neutrophils. Biochim Biophys Acta 1053:37–42

Steffens U, Bessler W, Hauschildt S (1989) B cell activation by synthetic lipopeptide analogues of bacterial lipoprotein bypassing phosphatidylinositol metabolism and proteinkinase C translocation. Mol Immunol 26:897–904

Steinbeck MJ, Roth JA, Kaeberle ML (1986) Activation of bovine neutrophils by recombinant interferon-γ. Cell Immunol 98:137–144

Steinberg TH, Silverstein SC (1987) Extracellular ATP^{4-} promotes cation fluxes in the J774 mouse macrophage cell line. J Biol Chem 262:3118–3122

Stendal O, Coble BI, Dahlgren C, Hed J, Molin L (1984) Myeloperoxidase modulates the phagocytic activity of polymorphonuclear neutrophil leukocytes. Studies with cells from a myeloperoxidase-deficient patient. J Clin Invest 73:366–373

Stenson WF, Mehta J, Spilberg I (1984) Sulfasalazine inhibition of binding of N-formyl-methionyl-leucyl-phenylalanine (FMLP) to its receptor on human neutrophils. Biochem Pharmacol 33:407–412

Sternweis PC, Gilman AG (1982) Aluminum: a requirement for activation of the regulatory component of adenylate cyclase by fluoride. Proc Natl Acad Sci USA 79:4888–4891

Sternweis PC, Northup JK, Smigel MD, Gilman AG (1981) The regulatory component of adenylate cyclase. Purification and properties. J Biol Chem 256:11517–11526

Stewart AG, Dubbin PN, Harris T, Dusting GJ (1990) Platelet-activating factor may act as a second messenger in the release of icosanoids and superoxide anions from leukocytes and endothelial cells. Proc Natl Acad Sci USA 87:3215–3219

Stickle DF, Daniele RP, Holian A (1984) Cytosolic calcium, calcium fluxes, and regulation of alveolar macrophage superoxide anion production. J Cell Physiol 121:458–466

Stock JL, Coderre JA (1982) Calcitonin and parathyroid hormone inhibit accumulation of cyclic AMP in stimulated human mononuclear cells. Biochem Biophys Res Commun 109:935–942

Stock JL, Coderre JA (1987) Pertussis toxin blocks the inhibitory effects of calcitonin on cyclic AMP accumulation in stimulated cultured human monocytes. J Leukocyte Biol 42:504–509

Stocker R, Richter C (1982) Involvement of calcium, calmodulin and phospholipase A in the alteration of membrane dynamics and superoxide production of human neutrophils stimulated by phorbol myristate acetate. FEBS Lett 147:243–246

Stocker R, Winterhalter KH, Richter C (1982) Increased fluorescence polarization of 1,6-diphenyl-1,3,5-hexatriene in the phorbol myristate acetate-stimulated plasma membrane of human neutrophils. FEBS Lett 144:199–203

Stoehr SJ, Smolen JE (1990) Human neutrophils contain a protein kinase C-like enzyme that utilizes guanosine triphosphate as a phosphate donor. Cofactor requirements, kinetics, and endogenous acceptor proteins. Blood 75:479–487

Stolc V (1977) Mechanism of regulation of adenylate cyclase activity in human polymorphonuclear leukocytes by calcium, guanosyl nucleotides, and positive effectors. J Biol Chem 252:1901–1907

Storch J, Ferber E, Munder PG (1988) Influence of platelet activating factor and a nonmetabolizable analogue on superoxide production by bone marrow derived macrophages. J Leukocyte Biol 44:385–390

Strathmann M, Wilkie TM, Simon MI (1989) Diversity of the G-protein family: sequences from five additional α subunits in the mouse. Proc Natl Acad Sci USA 86:7407–7409

Strausz J, Müller-Quernheim J, Steppling H, Ferlinz R (1990) Oxygen radical production by alveolar inflammatory cells in idiopathic pulmonary fibrosis. Am Rev Respir Dis 141:124–128

Streb H, Irvine RF, Berridge MJ, Schulz I (1983) Release of Ca^{2+} from a nonmitochondrial intracellular store in pancreatic acinar cells by inositol-1,4,5-triphosphate. Nature 306:67–69

Strnad CF, Wong C (1985a) Effect of the calcium ionophore A23187 on superoxide generation in phorbol ester stimulated human neutrophils. Can J Physiol Pharmacol 63:1543–1546

Strnad CF, Wong K (1985b) Calcium mobilization in fluoride activated human neutrophils. Biochem Biophys Res Commun 133:161–167

Strnad CF, Parente JE, Wong K (1986) Use of fluoride ion as a probe for the guanine nucleotide-binding protein involved in the phosphoinositide-dependent neutrophil transduction pathway. FEBS Lett 206:20–24

Stryer L, Bourne HR (1986) G proteins: a family of signal transducers. Annu Rev Cell Biol 2:391–419

Stuehr DJ, Nathan CF (1989) Nitric oxide. A macrophage product responsible for cytostasis and respiratory inhibition in tumor target cells. J Exp Med 169:1543–1555

Stuehr DJ, Gross SS, Sakuma I, Levi R, Nathan CF (1989) Activated murine macrophages secrete a metabolite of arginine with the bioactivity of endothelium-derived relaxing factor and the chemical reactivity of nitric oxide. J Exp Med 169:1011–1020

Stutchfield J, Cockcroft S (1988) Guanine nucleotides stimulate polyphosphoinositide phosphodiesterase and exocytic secretion from HL60 cells permeabilized with streptolysin O. Biochem J 250:375–382

Styrt B (1989) Species variation in neutrophil biochemistry and function. J Leukocyte Biol 46:63–74

Styrt B, Walker RD, White JC (1989) Neutrophil oxidative metabolism after exposure to bacterial phospholipase C. J Lab Clin Med 114:51–57

Subjeck JR, Shyy TT (1986) Stress protein systems of mammalian cells. Am J Physiol 250:C1–C17

Sugar AM, Field KG (1988) Characterization of murine bronchoalveolar macrophage respiratory burst: comparison of soluble and particulate stimuli. J Leukocyte Biol 44:500–507

Sugimoto M, Ando M, Senba H, Tokuomi H (1980) Lung defenses in neonates: effects of bronchial lavage fluids from adult and neonatal rabbits on superoxide production by their alveolar macrophages. J Reticuloendothel Soc 27:595–606

Sugimoto M, Higuchi S, Ando M, Horio S, Tokuomi H (1982) The effect of cytochalasin B on the superoxide production by alveolar macrophages obtained from normal rabbit lungs. J Reticuloendothel Soc 31:117–130

Suki WN, Abramowitz J, Mattera R, Codina J, Birnbaumer L (1987) The human genome encodes at least three non-allellic G proteins with α_i-type subunits. FEBS Lett 220:187–192

Sullivan GW, Donowitz GR, Sullivan JA, Mandell GL (1984) Interrelationships of polymorphonuclear neutrophil membrane-bound calcium, membrane potential, and chemiluminescence: studies in single living cells. Blood 64:1184–1192

Sullivan GW, Carper HT, Sullivan JA, Murata T, Mandell GL (1989) Both recombinant interleukin-1 (beta) and purified human monocyte interleukin-1 prime human neutrophils for increased oxidative activity and promote neutrophil spreading. J Leukocyte Biol 45:389–395

Sullivan R, Griffin JD, Simons ER, Schafer AI, Meshulam T, Fredette JP, Maas AK, Gadenne AS, Leavitt JL, Melnick DA (1987) Effects of recombinant human granulocyte and macrophage colony-stimulating factors on signal transduction pathways in human granulocytes. J Immunol 139:3422–3430

Sullivan R, Griffin JD, Wright J, Melnick DA, Leavitt JL, Fredette JP, Horne JH Jr, Lyman CA, Lazzari KG, Simons ER (1988) Effects of recombinant human granulocyte-macrophage colony-stimulating factor on intracellular pH in mature granulocytes. Blood 72:1665–1673

Sullivan R, Fredette JP, Griffin JD, Leavitt JL, Simons ER, Melnick DA (1989a) An elevation in the concentration of free cytosolic calcium is sufficient to activate the respiratory burst of granulocytes primed with recombinant human granulocyte-macrophage colony-stimulating factor. J Biol Chem 264:6302–6309

Sullivan R, Fredette JP, Leavitt JL, Gadenne AS, Griffin JD, Simons ER (1989b) Effects of recombinant human granulocyte-macrophage colony-stimulating factor (GM-CSF$_{rh}$) on transmembrane electrical potentials in granulocytes: relationship between enhancement of ligand-mediated depolarization and augmentation of superoxide (O_2^-) production. J Cell Physiol 139:361–369

Sumimoto H, Takeshige K, Minakami S (1984) Superoxide production of human polymorphonuclear leukocytes stimulated by leukotriene B_4. Biochim Biophys Acta 803:271–277

Sun FF, McGuire JC (1984) Metabolism of arachidonic acid by human neutrophils. Characterization of the enzymatic reactions that lead to the synthesis of leukotriene B_4. Biochim Biophys Acta 794:56–64

Sung CP, Mirabelli K, Badger AM (1984) Effect of gold compounds on phorbol myristic acetate (PMA) activated superoxide (O_2^-) production by mouse peritoneal macrophages. J Rheumatol 11:153–157

Sung SSJ, Young JDE, Origlio AM, Heiple JM, Kaback HR, Silverstein SC (1985) Extracellular ATP perturbs transmembrane ion fluxes, elevates cytosolic [Ca^{2+}], and inhibits phagocytosis in mouse macrophages. J Biol Chem 260:13442–13449

Sutherland C, Walsh MP (1989) Activation of protein kinase C by the dihydropyridine calcium channel blocker, felodipine. Biochem Pharmacol 38:1263–1270

Sutherland CA, Amin D (1982) Relative activities of rat and dog platelet phospholipase A_2 and diglyceride lipase. Selective inhibition of diglyceride lipase by RHC 80267. J Biol Chem 257:14006–16010

Sutherland MW, Nelson J, Harrison G, Forman HJ (1985) Effects of t-butyl hydroperoxide on NADPH, glutathione, and the respiratory burst of rat alveolar macropages. Arch Biochem Biophys 243:325–331

Suzuki H, Kurita T, Kakinuma K (1982) Effects of neuraminidase on O_2 consumption and release of O_2^- and H_2O_2 from phagocytosing human polymorphonuclear leukocytes. Blood 60:446–453

Suzuki H, Pabst MJ, Johnston RB (1985) Enhancement by Ca^{2+} or Mg^{2+} of catalytic activity of the superoxide-producing NADPH oxidase in membrane fractions of human neutrophils and monocytes. J Biol Chem 260:3635–3639

Suzuki Y, Furuta H (1988) Stimulation of guinea pig neutrophil superoxide anion-producing system with thymol. Inflammation 12:575–584

Suzuki Y, Lehrer RI (1980) NAD(P)H oxidase activity in human neutrophils stimulated by phorbol myristate acetate. J Clin Invest 66:1409–1418

Suzuki Y, Nakamura S, Sugiyama K, Furuta H (1987) Differences of superoxide production in blood leukocytes stimulated with thymol between human and nonhuman primates. Life Sci 41:1659–1664

Szefler SJ, Norton CE, Ball B, Gross JM, Aida Y, Pabst MJ (1989) IFN-γ and LPS overcome glucucorticoid inhibition of priming for superoxide release in human monocytes. Evidence that secretion of IL-1 and tumor necrosis factor-α is not essential for monocyte priming. J Immunol 142:3985–3992

Szuro-Sudol A, Nathan CF (1982) Suppression of macrophage oxidative metabolism by products of malignant and nonmalignant cells. J Exp Med 156:945–961

Taffet SM, Greenfield ARL, Haddox MK (1983) Retinal inhibits TPA activated, calcium-dependent, phospholipid-dependent protein kinase ("C" kinase). Biochem Biophys Res Commun 114:1194–1199

Takahashi A, Totsune-Nakano H, Nakano M, Mashiko S, Suzuki N, Ohma C, Inaba H (1989) Generation of O_2^- and tyrosine cation-mediated chemiluminescence during the fertilization of sea urchin eggs. FEBS Lett 246:117–119

Takasugi S, Ishida K, Takeshige K, Minakami S (1989) Effect of 2',3'-dialdehyde NADPH on activation of superoxide-producing NADPH oxidase in a cell-free system of pig neutrophils. J Biochem 105:155–157

Takayama H, Iwaki S, Tamoto K, Koyama J (1984) Inhibition of the O_2^--generating NADPH oxidase of guinea-pig polymorphonuclear leukocytes by rabbit antibody to homologous liver NADPH-cytochrome c (P-450) reductase. Biochim Biophys Acta 799:151–157

Takenawa T, Nagai Y (1981) Purification of phosphatidylinositol-specific phospholipase C from rat liver. J Biol Chem 256:6769–6775

Takenawa T, Ishitoya J, Homma Y, Kato M, Nagai Y (1985) Role of enhanced inositol phospholipid metabolism in neutrophil activation. Biochem Pharmacol 34:1931–1935

Takenawa T, Ishitoya J, Nagai Y (1986) Inhibitory effect of prostaglandin E_2, forskolin, and dibutyryl cAMP on arachidonic acid release and inositol phospholipid metabolism in guinea pig neutrophils. J Biol Chem 261:1092–1098

Takeshige K, Minakami S (1981) Involvement of calmodulin in phagocytotic respiratory burst of leukocytes. Biochem Biophys Res Commun 99:484–490

Takeshige K, Nabi ZF, Tatschek B, Minakami S (1980) Release of calcium from membranes and its relation to phagocytic metabolic changes: a fluorescence study on leukocytes loaded with chlortetracycline. Biochem Biophys Res Commun 95:410–415

Tamaoki T, Nomoto H, Takahashi I, Kato Y, Morimoto M, Tomita F (1986) Staurosporine, a potent inhibitor of phospholipid/Ca^{++} dependent protein kinase. Biochem Biophys Res Commun 135:397–402

Tamoto K, Hazeki H, Mori Y, Koyama J (1989) Phosphorylation of NADPH-cytochrome c reductase in guinea pig peritoneal macrophages stimulated with phorbol myristate acetate. FEBS Lett 244:159–162

Tamura M, Tamura T, Tyagi SR, Lambeth JD (1988) The superoxide-generating respiratory burst oxidase of human neutrophil plasma membrane. Phosphatidylserine as an effector of the activated enzyme. J Biol Chem 263:17621–17626

Tanaka T, Ohmura T, Yamakado T, Hidaka H (1982) Two types of calcium-dependent protein phosphorylations modulated by calmodulin antagonists. Naphthalene-sulfonamide derivatives. Mol Pharmacol 22:408–412

Tanaka T, Kanegasaki S, Makino R, Iizuka T, Ishimura Y (1987) Saturated and *trans*-unsaturated fatty acids elicit high levels of superoxide generation in intact and in cell-free preparations of neutrophils. Biochem Biophys Res Commun 144:606–612

Tanaka T, Makino R, Iizuka T, Ishimura Y, Kanegasaki S (1988) Activation by saturated and monounsaturated fatty acids of the O_2^--generating system in a cell-free preparation from neutrophils. J Biol Chem 263:13670–13676

Tanaka Y, Kiyotaki C, Tanowitz H, Bloom BR (1982) Reconstitution of a variant macrophage cell line defective in oxygen metabolism with a H_2O_2-generating system. Proc Natl Acad Sci USA 79:2584–2888

Tanaka T, Imajoh-Ohmi S, Kanagasaki S, Takagi Y, Makino R, Ishimura Y (1990) A 63 kilodalton cytosolic polypeptide involved in superoxide generation in porcine neutrophils. Purification and characterization. J Biol Chem 265:18717–18720

Taniguchi K, Takanaka K (1984) Inhibitory effects of various drugs on phorbol myristate acetate and *n*-formyl methionyl leucyl phenylalanine induced O_2^- production in polymorphonuclear leukocytes. Biochem Pharmacol 33:3165–3169

Tao W, Molsi TFP, Sha'afi RI (1989) Arachidonic acid release in rabbit neutrophils. Biochem J 257:633–637

Tarsi-Tsuk D, Levy R (1990) Stimulation of the respiratory burst in peripheral blood monocytes by lipoteichoic acid. The involvement of calcium ions and phospholipase A_2. J Immunol 144:2665–2670

Tauber AI (1981) Current views of neutrophil dysfunction. An integrated clinical perspective. Am J Med 70:1237–1246

Tauber AI (1987) Protein kinase C and the activation of human neutrophil NADPH oxidase. Blood 69:711–720

Tauber AI, Goetzl EJ (1979) Structural and catalytic properties of the solubilized superoxide-generating activity of human polymorphonuclear leukocytes. Solubilization, stabilization in solution, and partial characterization. Biochemistry 18:5576–5584

Tauber AI, Goetzl EJ (1981) Inhibition of complement-mediated functions of human neutrophils by impermeant stilbene disolfonic acids. J Immunol 126:1786–1789

Tauber AI, Simons ER (1983) Dissociation of human neutrophil membrane depolarization, respiratory burst stimulation and phospholipid metabolism by quinacrine. FEBS Lett 156:161–164

Tauber AI, Brettler DB, Kennington EA, Blumberg PM (1982) Relation of human neutrophil phorbol ester receptor occupancy and NADPH-oxidase activity. Blood 60:333–339

Tauber AI, Borregaard N, Simons E, Wright J (1983) Chronic granulomatous disease: a syndrome of phagocyte oxidase deficiencies. Medicine (Baltimore) 62:286–309

Tauber AI, Fay JR, Marletta MA (1984) Flavonoid inhibition of the human neutrophil NADPH oxidase. Biochem Pharmacol 33:1367–1369

Tauber AI, Cox JA, Curnutte JT, Carrol PM, Nakakuma H, Warren B, Gilbert H, Blumberg PM (1989a) Activation of human neutrophil NADPH-oxidase in vitro by the catalytic fragment of protein kinase-C. Biochem Biophys Res Commun 158:884–890

Tauber AI, Pavlotsky N, Shin Lin J, Rice PA (1989b) Inhibition of human neutrophil NADPH oxidase by *Chlamydia* serovars E, K, and L_2. Infect Immun 57:1108–1112

Taylor R (1986) Insulin receptors and the clinician. Br Med J 292:919–922

Taylor R, Agius L (1988) The biochemistry of diabetes. Biochem J 250:625–640

Teahan CG, Rowe P, Parker P, Totty N, Segal AW (1987) The X-linked chronic granulomatous disease gene codes for β-chain of cytochrome b_{-245}. Nature 327:720–721

Teahan CG, Totty N, Casimir CM, Segal AW (1990) Purification of the 47 kDa phosphoprotein associated with the NADPH oxidase of human neutrophils. Biochem J 267:485–489

Tecoma ES, Motulsky HJ, Traynor AE, Omann GM, Muller H, Sklar LA (1986) Transient catecholamine modulation of neutrophil activation: Kinetic and intracellular aspects of isoproterenol action. J Leukocyte Biol 40:629–644

Tennenberg SD, Zemlan FP, Solomkin JS (1988) Characterization of N-formyl-methionyl-leucyl-phenylalanine receptors on human neutrophils. Effects of isolation and temperature on receptor expression and functional activity. J Immunol 141:3937–3944

Tenner TE Jr, Earley KJ, Young JA (1989) Characterization of β-adrenoceptor subtypes in rabbit mononuclear leukocytes. Eur J Pharmacol 160:291–293

Terkeltaub RA, Sklar LA, Mueller H (1990) Neutrophil activation by inflammatory microcrystals of monosodium urate monohydrate utilizes pertussis toxin-insensitive and -sensitive pathways. J Immunol 144:2719–2724

Terranova VP, DiFlorio H, Hujanen ES, Lyall RM, Liotta LA, Thorgeirsson U, Siegal GP, Schiffmann E (1986) Laminin promotes rabbit neutrophil motility and attachment. J Clin Invest 77:1180–1186

'T Hart BA, Simons JM, Rijkers GT, Hoogvliet JC, van Dijk H, Labadie RP (1990) Reaction products of 1-naphthol with reactive oxygen species prevent NADPH oxidase activation in activated human neutrophils, but leave phagocytosis intact. Free Radic Biol Med 8:241–249

Thelen M, Baggiolini M (1990) Reconstitution of cell-free NADPH-oxidase from human monocytes and comparison with neutrophils. Blood 75:2223–2228

Thelen M, Wolf M, Baggiolini M (1988a) Activation of monocytes by interferon-gamma has no effect on the level or affinity of the nicotinamide adenine dinucleotide-phosphate oxidase and on agonist-dependent superoxide formation. J Clin Invest 81:1889–1895

Thelen M, Peveri P, Kernen P, von Tscharner V, Walz A, Baggiolini M (1988b) Mechanism of neutrophil activation by NAF, a novel monocyte-derived peptide agonist. FASEB J 2:2702–2706

Thelen M, Rosen A, Nairn AC, Aderem A (1990) Tumor necrosis factor α modifies agonist-dependent responses in human neutrophils by inducing the synthesis and myristoylation of a specific protein kinase C substrate. Proc Natl Acad Sci USA 87:5603–5607

Therrien S, Naccache PH (1989) Guanine nucleotide-induced polymerization of actin in electropermeabilized human neutrophils. J Cell Biol 109:1125–1132

Thomassen MJ, Barna BP, Wiedemann HP, Farmer M, Ahmad M (1988) Human alveolar macrophage function: differences between smokers and nonsmokers. J Leukocyte Biol 44:313–318

Thompson BY, Sivam G, Britigan BE, Rosen GM, Cohen MS (1988) Oxygen metabolism of the HL-60 cell line: Comparison of the effects of monocytoid and neutrophilic differentiation. J Leukocyte Biol 43:140–147

Thomsen MK, Ahnfelt-Ronne IA (1989) Inhibition by the LTD$_4$ antagonist, SR2640, of effects of LTD$_4$ on canine polymorphonuclear leukocyte functions. Biochem Pharmacol 38:2291–2295

Tiku ML, Liesch JB, Robertson FM (1990) Production of hydrogen peroxide by rabbit articular chrondrocytes. Enhancement by cytokines. J Immunol 145:690–696

Tomlinson S, MacNeil S, Walker SW, Ollis CA, Merritt JE, Brown BL (1984) Calmodulin and cell function. Clin Sci 66:497–508

Toper R, Aviram A, Aviram I (1987) Fluoride-mediated activation of guinea pig neutrophils. Biochim Biophys Acta 931:262–266

Torres M, Coates TD (1984) Neutrophil cytoplasts: relationships of superoxide release and calcium pools. Blood 64:891–895

Tosi MF, Berger M (1988) Functional differences between the 40 kDa and 50 to 70 kDa IgG Fc receptors on human neutrophils revealed by elastase treatment and antireceptor antibodies. J Immunol 141:2097–2103

Tosk J, Lau BHS, Lui P, Myers RC, Torrey RR (1989) Chemiluminescence in a macrophage cell line modulated by biological response modifiers. J Leukocyte Biol 46:103–108

Totsuka Y, Nielsen TB, Field JB (1982) Roles of GTP and GDP in the regulation of the thyroid adenylate cyclase system. Biochim Biophys Acta 718:135–143

Totti N III, McCusker KT, Campbell EJ, Griffin GL, Senior RM (1984) Nicotine is chemotactic for neutrophils and enhances neutrophil responsiveness to chemotactic peptides. Science 223:169–171

Traynor AE, Scott PJ, Harris AL, Badwey JA, Sklar LA, Babior BM, Curnutte JT (1989) Respiratory burst oxidase activation can be dissociated from phosphatidylinositol bisphosphate degradation in a cell-free system from human neutrophils. Blood 73:296–300

Traynor-Kaplan AE, Thompson BL, Harris AL, Taylor P, Omann GM, Sklar LA (1989) Transient increase in phosphatidylinositol 3,4-bisphosphate and phosphatidylinositol trisphosphate during activation of human neutrophils. J Biol Chem 264:15668–15673

Tremblay J, Gerzer R, Hamet P (1988) Cyclic GMP in cell function. Adv Scd Mess Phosphoprot Res 22:319–383

Treves S, di Virgilio F, Vaselli GM, Pozzan T (1987) Effect of cytochalasins on cytosolic-free calcium concentration and phosphoinositide metabolism in leukocytes. Exp Cell Res 168:285–298

Trinchieri G, Kobayashi M, Rosen M, Loudon R, Murphy M, Perussia B (1986) Tumor necrosis factor and lymphotoxin induce differentiation of human myeloid cell lines in synergy with interferon. J Exp Med 164:1206–1225

Tritsch GL, Niswander PW (1982) Positive correlation between superoxide release and intracellular adenosine deaminase activity during macrophage membrane perturbation regardless of nature or magnitude of stimulus. Mol Cell Biochem 49:49–51

Tritsch GL, Niswander PW (1983) Modulation of macrophage superoxide release by purine metabolism. Life Sci 32:1359–1362

Trudel S, Downey GP, Grinstein S, Paquet MR (1990) Activation of permeabilized HL60 cells by vanadate. Evidence for divergent signalling pathways. Biochem J 269:127–131

Truett AP III, Verghese MW, Dillon SB, Snyderman R (1988) Calcium influx stimulates a second pathway for sustained diacylglycerol production in leukocytes activated by chemoattractants. Proc Natl Acad Sci USA 85:1549–1553

Truett AP III, Snyderman R, Murray JJ (1989) Stimulation of phosphorylcholine turnover and diacylglycerol production in human polymorphonuclear leukocytes. Novel assay for phosphorylcholine. Biochem J 260:909–913

Trush MA, Wilson ME, Van Dyke K (1978) The generation of chemiluminescence (CL) by phagocytic cells. Methods Enzymol 57:462–494

Tsan MF (1983) Inhibition of neutrophil sulfhydryl groups by chloromethyl ketones. A mechanism for their inhibition of superoxide production. Biochem Biophys Res Commun 112:671–677

Tsan MF, Newman B, McIntyre PA (1976) Surface sulfhydryl groups and phagocytosis-associated oxidative metabolic changes in human polymorphonuclear leukocytes. Br J Haematol 33:189–204

Tsuchiya M, Okimasu E, Ueda W, Hirakawa M, Utsumi K (1988) Halothane, an inhalation anesthetic, activates protein kinase C and superoxide generation by neutrophils. FEBS Lett 242:101–105

Tsujimoto M, Yokota S, Vilcek J, Weissmann G (1986) Tumor necrosis factor provokes superoxide anion generation from neutrophils. Biochem Biophys Res Commun 137:1094–1100

Tsunawaki S, Nathan CF (1984) Enzymatic basis of macrophage activation. Kinetic analysis of superoxide production in lysates of resident and activated mouse peritoneal macrophages and granulocytes. J Biol Chem 259:4305–4312

Tsunawaki S, Nathan CF (1986) Release of arachidonate and reduction of oxygen. J Biol Chem 261:11563–11570

Tsunawaki S, Kaneda M, Kakinuma K (1983) Activation of guinea pig polymorphonuclear leukocytes with soluble stimulators leads to nonrandom distribution of NADPH oxidase in the plasma membrane. J Biochem 94:655–664

Tsung PK, Sakamoto T, Weissmann G (1975) Protein kinase and phosphatases from human polymorphonuclear leukocytes. Biochem J 145:437–448

Tsung PK, Kegeles SW, Becker EL (1978) The evidence for the existence of chymotrypsin-like esterase activity in the plasma membranes of rabbit neutrophils and the specific chemotactic peptide binding activity of the subcellular fractions. Biochim Biophys Acta 541:150–160

Tsusaki BE, Kanda S, Huang L (1986) Stimulation of superoxide release in neutrophils by 1-oleoyl-2-acetylglycerol incorporated into pH-sensitive liposomes. Biochem Biophys Res Commun 136:242–246

Turner E, Somers CE, Shapiro BM (1985) The relationship between a novel NAD(P)H oxidase activity of ovoperoxidase and the CN⁻-resistant respiratory burst that follows fertilization of sea urchin eggs. J Biol Chem 260:13163–13171

Turrens JF, McCord JM (1988) How relevant is the reoxidation of ferrocytochrome c by hydrogen peroxide when determining superoxide anion production? FEBS Lett 227:43–46

Tyagi SR, Tamura M, Burnham DN, Lambeth JD (1988) Phorbol myristate acetate (PMA) augments chemoattractant-induced diglyceride generation in human neutrophils but inhibits phosphoinositide hydrolysis. Implications for the mechanism of PMA priming of the respiratory burst. J Biol Chem 263:13191–13198

Tyagi SR, Burnham DN, Lambeth JD (1989a) On the biological occurrence and regulation of 1-acyl and 1-O-alkyl-diradylglycerols in human neutrophils. J Biol Chem 264:12977–12982

Tyagi SR, Winton EF, Lambeth JD (1989b) Granulocyte/macrophage colony-stimulating factor primes human neutrophils for increased diacylglycerol generation in response to chemoattractant. FEBS Lett 257:188–190

Tzehoval E, Segal S, Stabinsky Y, Fridkin M, Spirer Z, Feldman M (1978) Tuftsin (an Ig-associated tetrapeptide) triggers the immunogenic function of macrophages: Implications for activation of programmed cells. Proc Natl Acad Sci USA 75:3400–3404

Tzeng DY, Deuel TF, Huang JS, Senior RM, Boxer LA, Baehner RL (1984) Platelet-derived growth factor promotes polymorphonuclear leukocyte activation. Blood 64:1123–1128

Uesaka H, Yokoyama M, Ohtsuki K (1987) Physiological correlation between nucleoside-diphosphate kinase and the enzyme-associated guanine nucleotide binding proteins. Biochem Biophys Res Commun 143:552–559

Uhing RJ, Polakis PG, Snyderman R (1987) Isolation of GTP-binding proteins from myeloid HL-60 cells. Identification of two pertussis toxin substrates. J Biol Chem 262:15575–15579

Uhing RJ, Prpic V, Hollenbach PW, Adams DO (1989) Involvement of protein kinase C in platelet-activating factor-stimulated diacylglycerol accumulation in murine peritoneal macrophages. J Biol Chem 264:9224–9230

Ulevitch RJ, Tobias PS, Mathison JC (1984) Regulation of the host response to bacterial lipopolysaccharides. Fed Proc 43:2755–2759

Umei T, Takeshige K, Minakami S (1986) NADPH binding component of neutrophil superoxide-generating oxidase. J Biol Chem 261:5229–5232

Umei T, Takeshige K, Minakami S (1987) NADPH-binding component of the super-oxide-generating oxidase in unstimulated neutrophils and in neutrophils from the patients with chronic granulomatous disease. Biochem J 243:467–472

Umeki S (1990) Human neutrophil cytosolic activation factor of the NADPH oxidase. Characterization of activation kinetics. J Biol Chem 265:5049–5054

Umeki S, Soejima R (1990) Hydrocortisone inhibits the respiratory burst oxidase from human neutrophils in whole-cell and cell-free systems. Biochim Biophys Acta 1052:211–215

Unkeless JC, Scigliano E, Freedman VH (1988) Structure and function of human and murine receptors for IgG. Annu Rev Immunol 6:251–281

Urban JL, Shepard HM, Rothstein JL, Sugarman BJ, Schreiber H (1986) Tumor necrosis factor: a potent effector molecule for tumor cell killing by activated macrophages. Proc Natl Acad Sci USA 83:5233–5237

Valentino M, Governa M, Fiorini R, Curatola G (1986) Changes of membrane fluidity in chemotactic peptide-stimulated polymorphonuclear leukocytes. Biochem Biophys Res Commun 141:1151–1156

Valletta EA, Berton G (1987) Desensitization of macrophage oxygen metabolism on immobilized ligands: different effect of immunoglobulin G and complement. J Immunol 138:4366–4373

Van Epps DE, Bender JG, Simpson SJ, Chenoweth DE (1990) Relationship of chemotactic receptors for formyl peptide and C5a to CR1, CR3, and Fc receptors on human neutrophils. J Leukocyte Biol 47:519–527

Varecka L, Peterajová E, Pogády J (1987) Polymyxin B, a novel inhibitor of red cell Ca^{2+}-activated K^+ channel. FEBS Lett 225:173–177

Varga Z, Jacob MP, Robert L, Fülöp T Jr (1989) Identification and signal transduction mechanism of elastin peptide receptor in human leukocytes. FEBS Lett 258:5–8

Vargaftig BB, Braquet PG (1987) PAF-acether today — relevance for acute experimental anaphylaxis. Br Med Bull 43:312–335

Vasconcelles MJ, Weitzman SA, Nam Lee S, Prachard S, Gordon LI (1990) Inhibition of human polymorphonuclear leukocyte respiratory burst activity and aggregation by 6-ketocholestanol. Free Radic Res Commun 8:185–193

Vercellotti GM, Wickham NWR, Gustafson KS, Yin HQ, Hebert M, Jacob HS (1989) Thrombin-treated endothelium primes neutrophil functions: inhibition by platelet-activating factor receptor antagonists. J Leukocyte Biol 45:483–490

Verghese MW, Snyderman R (1983) Hormonal activation of adenylate cyclase in macrophage membranes is regulated by guanine nucleotides. J Immunol 130:869–873

Verghese MW, Smith CD, Snyderman R (1985a) Potential role for a guanine nucleotide regulatory protein in chemoattractant receptor mediated polyphosphoinositide metabolism, Ca^{++} mobilization and cellular responses by leukocytes. Biochem Biophys Res Commun 127:450–457

Verghese MW, Fox K, McPhail LC, Snyderman R (1985b) Chemoattractant-elicited alterations of cAMP levels in human polymorphonuclear leukocytes require a Ca^{2+}-dependent mechanism which is independent of transmembrane activation of adenylate cyclase. J Biol Chem 260:6769–6775

Verghese MW, Smith CD, Snyderman R (1986a) Role of guanine nucleotide regulatory protein in polyphosphoinositide degradation and activation of phagocytic leukocytes by chemoattractants. J Cell Biochem 32:59–69

Verghese MW, Smith CD, Charles LA, Jakoi L, Snyderman R (1986b) A guanine nucleotide regulatory protein controls polyphosphoinositide metabolism, Ca^{2+}

mobilization, and cellular responses to chemoattractants in human monocytes. J Immunol 137:271–275

Verghese MW, Uhing RJ, Snyderman R (1986c) A pertussis/choleratoxin-sensitive N protein may mediate chemoattractant receptor signal transduction. Biochem Biophys Res Commun 138:887–894

Verghese MW, Charles LA, Jakoi L, Dillon SB, Snyderman R (1987) Role of a guanine nucleotide regulatory protein in the activation of phospholipase C by different chemoattractants. J Immunol 138:4374–4380

Verhagen J, Bruynzeel PLB, Koedam JA, Wassink GA, de Boer M, Terpstra GK, Kreukniet J, Veldink GA, Vliegenhart FG (1984) Specific leukotriene formation by purified human eosinophils and neutrophils. FEBS Lett 168:23–28

Verhoeven AJ, Bolscher GJM, Meerhof LJ, van Zwieten R, Keijer J, Weening RS, Roos D (1989) Characterization of two monoclonal antibodies against cytochrome b_{558} of human neutrophils. Blood 73:1686–1694

Verkest V, McArthur M, Hamilton S (1988) Fatty acid activation of protein kinase C: dependence on diacylglycerol. Biochem Biophys Res Commun 152:825–829

Verspaget HW, Miermet-Ooms MAC, Weterman IT, Pena AS (1984) Partial defect of neutrophil oxidative metabolism in Crohn's disease. Gut 25:849–853

Verspaget HW, Elmgreen J, Weterman IT, Pena AS, Riis P, Lamers CBHW (1986) Impaired Activation of the Neutrophil Oxidative Metabolism in Chronic Inflammatory Bowel Disease. Scand J Gastroenterol 21:1124–1130

Vianello A, Macri F (1989) NAD(P)H oxidation elicits anion superoxide formation in radish plasmalemma vesicles. Biochim Biophys Acta 980:202–208

Volkman DJ, Buescher ES, Gallin JI, Fauci AS (1984) B cell lines as models for inherited phagocytic diseases: abnormal superoxide generation in chronic granulomatous disease and giant granules in Chediak-Higashi syndrome. J Immunol 133:3006–3009

Volpe P, Krause KH, Hashimoto S, Zorzato F, Pozzan T, Meldolesi J, Lew DP (1988) "Calciosome," a cytoplasmic organelle: the inositol 1,4,5-triphosphate-sensitive Ca^{2+} store of nonmuscle cells? Proc Natl Acad Sci USA 85:1091–1095

Volpi M, Yassin R, Tao W, Molski TFP, Naccache PH, Sha'afi RI (1984) Leukotriene B_4 mobilizes calcium without the breakdown of polyphosphoinositides and the production of phosphatidic acid in rabbit neutrophils. Proc Natl Acad Sci USA 81:5966–5969

Volpi M, Naccache PH, Molski TFP, Shefcyk J, Huang CK, Marsh ML, Munoz J, Becker EL, Sha'afi RI (1985) Pertussis toxin inhibits fMet-Leu-Phe- but not phorbol ester-stimulated changes in rabbit neutrophils: role of G proteins in excitation response coupling. Proc Natl Acad Sci USA 82:2708–2712

Volpp BD, Nauseef WM, Clark RA (1988) Two cytosolic neutrophil oxidase components absent in autosomal chronic granulomatous disease. Science 242:1295–1297

Volpp BD, Nauseef WM, Clark RA (1989a) Subcellular distribution and membrane association of human neutrophil substrates for ADP-ribosylation by pertussis toxin and cholera toxin. J Immunol 142:3206–3212

Volpp BD, Nauseef WM, Donelson JE, Moser DR, Clark RA (1989b) Cloning of the cDNA and functional expression of the 47-kilodalton cytosolic component of human neutrophil respiratory burst oxidase. Proc Natl Acad Sci USA 86:7195–7199

Von der Mark K, Kühl U (1985) Laminin and its receptor. Biochim Biophys Acta 823:147–160

Von Tscharner V, Deranleau DA, Baggiolini M (1986a) Calcium fluxes and calcium buffering in human neutrophils. J Biol Chem 261:10163–10168

Von Tscharner V, Prod'hom B, Baggiolini M, Reuter H (1986b) Ion channels in human neutrophils activated by a rise in free cytosolic calcium concentration. Nature 324:369–372

Vostal JG, Reid DM, Jones CE, Shulman NR (1989) Anion channel blockers cause apparent inhibition of exocytosis by reacting with agonist or secretory product, not with cell. Proc Natl Acad Sci USA 86:5839–5843

Waite M, DeChatelet LR, King L, Shirley PS (1979) Phagocytosis-induced release of arachidonic acid from human neutrophils. Biochem Biophys Res Commun 90:984–992

Wakeyama H, Takeshige K, Takayanagi R, Minakami S (1982) Superoxide-forming NADPH oxidase preparation of pig polymorphonuclear leucocytes. Biochem J 205:593–601

Waldman SA, Murad F (1987) Cyclic GMP synthesis and function. Pharmacol Rev 39:163–196

Walker BAM, Cunningham TW, Freyer DR, Todd RF III, Johnson KJ, Ward PA (1989) Regulation of superoxide responses of human neutrophils by adenine compounds. Independence of requirement for cytoplasmic granules. Lab Invest 61:515–521

Walker F, Burgess AW (1985) Specific binding of radioiodinated granulocyte-macrophage colony-stimulating factor to hemopoietic cells. EMBO J 4:933–939

Walker RJ, Lazzaro VA, Duggin GG, Horvath JS, Tiller DJ (1989) Cyclosporin A inhibits protein kinase C activity: a contributing mechanism in the development of nephrotoxicity? Biochem Biophys Res Commun 160:409–415

Wallaert B, Lassalle P, Fortin F, Aerts C, Bart F, Fournier E, Voisin C (1990) Superoxide anion generation by alveolar inflammatory cells in simple pneumoconiosis and in progressive massive fibrosis of nonsmoking coal workers. Am Rev Respir Dis 141:129–133

Walsh MP, Valentine KA, Ngai PK, Carruthers CA, Hollenberg MD (1984) Ca^{2+}-dependent hydrophobic-interaction chromatography. Biochem J 224:117–127

Walter RJ, Marasco WA (1987) Direct visualization of formylpeptide receptor binding on rounded and polarized human neutrophils. Cellular and receptor heterogeneity. J Leukocyte Biol 41:377–391

Walz A, Peveri P, Aschauer H, Baggiolini M (1987) Purification and amino acid sequencing of NAF, a novel neutrophil-activating factor produced by monocytes. Biochem Biophys Res Commun 149:755–761

Ward PA, Cunningham TW, McCulloch KK, Phan SH, Powell J, Johnson KJ (1988a) Platelet enhancement of O_2^- responses in stimulated human neutrophils. Identification of platelet factor as adenine nucleotide. Lab Invest 58:37–47

Ward PA, Cunningham TW, Walker BAM, Johnson KJ (1988b) Differing calcium requirements for regulatory effects of ATP, ATPγS and adenosine on O_2^- responses of human neutrophils. Biochem Biophys Res Commun 154:746–751

Ward PA, Cunningham TW, McCulloch KK, Johnson KJ (1988c) Regulatory effects of adenosine and adenine nucleotides on oxygen radical responses of neutrophils. Lab Invest 58:438–447

Warner JA, Yancey KB, MacGlashan DW Jr (1987) The effect of pertussis toxin on mediator release from human basophils. J Immunol 139:161–165

Warr GA (1980) A macrophage receptor for (mannose/glucosamine)-glycoproteins of potential importance in phagocytic activity. Biochem Biophys Res Commun 93:737–745

Washida N, Sagawa A, Tamoto K, Koyama J (1980) Comparative studies on superoxide anion production by polymorphonuclear leukocytes stimulated with various agents. Biochim Biophys Acta 631:371–379

Watson SP (1984) Are the proposed substance P receptor sub-types, substance P receptors? Life Sci 25:797–808

Watson SP (1987) Multiple receptors for substance P and other mammalian tachykinins. In: Grimwade AM (editor and publisher) Philadelphia Pharmacology. pp 82–85 (ISI atlas of science)

Weber H, Heilmann P, Meyer S, Maier KL (1990) Effect of canine surfactant protein (SP-A) on the respiratory burst of phagocytic cells. FEBS Lett 270:90–94

Weening RS, Corbeel L, de Boer M, Lutter R, van Zwieten R, Hamers MN, Roos D (1985) Cytochrome *b* deficiency in an autosomal recessive form of chronic granulomatous disease. A third form of chronic granulomatous disease recognized by monocyte hybridization. J Clin Invest 75:915–920

Wei EP, Kontos HA, Christman CW, DeWitt DS, Povlishock JT (1985) Superoxide generation and reversal of acetylcholine-induced cerebral arteriolar dilation after acute hypertension. Circ Res 57:781–787

Weidman PJ, Kay ES, Shapiro BM (1985) Assembly of the sea urchin fertilization membrane: isolation of proteoliaisin, a calcium-dependent ovoperoxidase binding protein. J Cell Biol 100:938–946

Weigt C, Just I, Wegener A, Aktories K (1989) Nonmuscle actin ADP-ribosylated by botulinum C2 toxin caps actin filaments. FEBS Lett 246:181–184

Weinberg JB, Misukonis MA (1983) Phorbol diester-induced H_2O_2 production by peritoneal macrophages. Different H_2O_2 production by macrophages from normal and BCG-infected mice despite comparable phorbol diester receptors. Cell Immunol 80:405–415

Weisbart RH, Golde DW, Clark SC, Wong GG, Gasson JC (1985) Human granulocyte-macrophage colony-stimulating factor is a neutrophil activator. Nature 314:361–363

Weisbart RH, Golde DW, Gasson JC (1986) Biosynthetic human GM-CSF modulates the number and affinity of neutrophil fMet-Leu-Phe receptors. J Immunol 137:1584–3587

Weisbart RH, Kwan L, Golde DW, Gasson JC (1987) Human GM-CSF primes neutrophils for enhanced oxidative metabolism in response to the major physiological chemoattractants. Blood 69:18–21

Weiss SJ (1989) Tissue destruction by neutrophils. N Engl J Med 320:365–376

Weiss SJ, LoBuglio AF, Kessler HB (1980) Oxidative mechanisms of monocyte-mediated cytotoxicity. Proc Natl Acad Sci USA 77:584–587

Weissmann G (1987) Pathogenesis of inflammation: effects of the pharmacological manipulation of arachidonic acid metabolism on the cytological response to inflammatory stimuli. Drugs [Suppl 1] 33:28–37

Weissmann G, Goldstein I, Hoffstein S, Tsung PK (1975) Reciprocal effects of cAMP and cGMP on microtubule-dependent release of lysosomal enzymes. Ann NY Acad Sci 253:750–762

Weissmann G, Smolen JE, Korchak H (1980) Prostaglandins and inflammation: receptor/cyclase coupling as an explanation of why PGEs and PGI_2 inhibit functions of inflammatory cells. Adv Prostaglandin Thromboxane Res 8:1637–1653

Weitzman SA, Weitberg AB, Clark EP, Stossel TP (1985) Phagocytes as carcinogens: malignant transformation produced by human neutrophils. Science 227:1231–1233

Welch WD (1984) Effect of enflurane, isoflurane, and nitrous oxide on the microbicidal activity of human polymorphonuclear leukocytes. Anesthesiology 61:188–192

Welch WD, Zaccari J (1982) Effect of halothane and N_2O on the oxidative activity of human neutrophils. Anesthesiology 57:172–176

Welch WD, Graham CW, Zaccari J, Thrupp LD (1980) Analysis and comparison of the luminol-dependent chemiluminescence responses of alveolar macrophages and neutrophils. J Reticuloendothel Soc 28:275–283

Welte K, Platzer E, Lu L, Gabrilove JL, Levi E, Mertelsmann R, Moore MAS (1985) Purification and biochemical characterization of human pluripotent hematopoietic colony-stimulating factor. Proc Natl Acad Sci USA 82:1526–1530

Wender PA, Cribbs CM, Koehler KF, Sharkey NA, Herald CL, Kamano Y, Pettit GR, Blumberg PM (1988) Modeling of the bryostatins to the phorbol ester pharmacophore on protein kinase C. Proc Natl Acad Sci USA 85:7197–7201

Wenzel-Seifert K, Seifert R (1990) Nucleotide-, chemotactic peptide- and phorbol ester-induced β-glucuronidase release in HL-60 leukemic cells. Immunobiology 181:298–316

Werns SW, Lucchesi BR (1987) Inflammation and myocardial infarction. Br Med Bull 43:460–471

Werns SW, Lucchesi BR (1988) Leukocytes, oxygen radicals, and myocardial injury due to ischemia and reperfusion. Free Radic Biol Med 4:31–37

Whetton AD, Huang SJ, Monk PN (1988) Adenosine triphosphate can maintain multi-potent haemopoietic stem cells in the absence of interleukin 3 via a membrane permeabilization mechanism. Biochem Biophys Res Commun 152:1173–1178

White JR, Naccache PH, Sha'afi RI (1983a) Stimulation by chemotactic factor of actin association with the cytoskeleton in rabbit neutrophils. Effects of calcium and cytochalasin B. J Biol Chem 258:14041–14047

White JR, Naccache PH, Molski TFP, Borgeat P, Sha'afi RI (1983b) Direct demonstration of increased intracellular concentration of free calcium in rabbit and human neutrophils following stimulation by chemotactic factor. Biochem Biophys Res Commun 113:44–50

Whitin JC, Chapman CE, Simons ER, Chovaniec ME, Cohen HJ (1980) Correlation between membrane potential changes and superoxide production in human granulocytes stimulated by phorbol myristate acetate. Evidence for defective activation in chronic granulomatous disease. J Biol Chem 255:1874–1878

Wieland T, Jakobs KH (1989) Receptor-regulated formation of GTP[γS] with subsequent persistent G_s-protein activation in membranes of human platelets. FEBS Lett 245:189–193

Wierusz-Wysocka B, Wysocki H, Siekierka H, Wykretowicz A, Szczepanik A, Klimas R (1987) Evidence of polymorphonuclear neutrophils (PMN) activation in patients with insulin-dependent diabetes mellitus. J Leukocyte Biol 42:519–523

Wightman PD, Raetz CRH (1984) The activation of protein kinase C by biologically active lipid moities of lipopolysaccharide. J Biol Chem 259:10048–10052

Wilde MW, Carlson KE, Manning DR, Zigmond SH (1989) Chemoattractant-stimulated GTPase activity is decreased on membranes from polymorphonuclear leukocytes incubated in chemoattractant. J Biol Chem 264:190–196

Williams AJ, Cole PJ (1981) Polymorphonuclear leukocyte membrane-stimulated oxidative metabolic activity — the effect of divalent cations and cytochalasins. Immunology 44:847–858

Williams LT, Snyderman R, Pike MC, Lefkowitz RJ (1977) Specific receptor sites for chemotactic peptides on human polymorphonuclear leukocytes. Proc Natl Acad Sci USA 74:1204–1208

Williams M (1987) Purine receptors in mammalian tissues: pharmacology and functional significance. Annu Rev Pharmacol Toxicol 27:315–345

Williams WR, Kagamimori S, Davies BH, Watanabe M (1986) Formation of extracellular cGMP in blood mononuclear and platelet cell preparations. J Clin Lab Immunol 20:89–92

Williamson K, Dickey BT, Yung Pyun H, Navarro J (1988) Solubilization and reconstitution of the formylmethionylleucylphenylalanine receptor coupled to guanine nucleotide regulatory protein. Biochemistry 27:5371–5377

Willis HE, Browder B, Feister AJ, Mohanakumar T, Ruddy S (1988) Monoclonal antibody to human IgG Fc receptors. Cross-linking of receptors induces lysosomal enzyme release and superoxide generation by neutrophils. J Immunol 140:234–239

Wilson CB, Tsai V, Remington JS (1980) Failure to trigger the oxidative metabolic burst by normal macrophages. Possible mechanism for survival of intracellular pathogens. J Exp Med 151:328–346

Wilson E, Olcott MC, Bell RM, Merrill AH Jr, Lambeth JD (1986) Inhibition of the oxidative burst in human neutrophils by sphingoid long-chain bases. Role of protein kinase C in activation of the burst. J Biol Chem 261:12616–12623

Wilson E, Laster SM, Gooding LR, Lambeth JD (1987) Platelet-derived growth factor stimulates phagocytosis and blocks agonist-induced activation of the neutrophil oxidative burst: a possible cellular mechanism to protect against oxygen radical damage. Proc Natl Acad Sci USA 84:2213–2217

Wilson ME, Jones DP, Munkenbeck P, Morrison DC (1982) Serum-dependent and -independent effects of bacterial lipopolysaccharides on human neutrophil oxidative capacity in vitro. J Reticuloendothel Soc 31:43–57

Wing EJ, Ampel NM, Waheed A, Shadduck RK (1985) Macrophage colony-stimulating factor (M-CSF) enhances the capacity of murine macrophages to secrete oxygen reduction products. J Immunol 135:2052–2056

Wirthmueller U, de Weck AL, Dahinden CA (1989) Platelet-activating factor production in human neutrophils by sequential stimulation with granulocyte-macrophage colony-stimulating factor and the chemotactic factors C5a or formyl-methionyl-leucyl-phenylalanine. J Immunol 142:3213–3218

Wise BC, Glass DB, Chou CHJ, Raynor RL, Katoh N, Schatzman RC, Turner RS, Kibler RF, Kuo JF (1982) Phospholipid-sensitive Ca^{2+}-dependent protein kinase from heart. II. Substrate specificity and inhibition by various agents. J Biol Chem 257:8489–8495

Witz G, Goldstein BD, Amoruso M, Stone DS, Troll W (1980) Retinoid inhibition of superoxide anion radical production by human polymorphonuclear leukocytes stimulated with tumor promoters. Biochem Biophys Res Commun 97:883–888

Witz G, Lawrie NJ, Amoruso MA, Goldstein BD (1987) Inhibition by reactive aldehydes or superoxide anion radical production from stimulated polymorphonuclear leukocytes and pulmonary alveolar macrophages. Effects on cellular sulfhydryl groups and NADPH oxidase activity. Biochem Pharmacol 36:721–726

Wolf JE, Abegg AL, Travis SJ, Kobayashi GS, Little Jr (1989) effects of *Histoplasma capsulatum* on murine macrophage functions: inhibition of macrophage priming, oxidative burst, and antifungal activities. Infect Immun 57:513–519

Wolfson M, McPhail LC, Nasrallah VN, Snyderman R (1985) Phorbol myristate acetate mediates redistribution of protein kinase C in human neutrophils: potential role in the activation of the respiratory burst enzyme. J Immunol 135:2057–2062

Wong GG, Witek JS, Temple PA, Wilkens KM, Leary AC, Luxenberg DP, Jones SS, Brown EL, Kay RM, Orr EC, Shoemaker C, Golde DW, Kaufman RJ, Hewick RM, Wang EA, Clark SC (1985) Human GM-CSF: molecular cloning of the complementary DNA and purification of the natural and recombinant proteins. Science 228:810–815

Wong K (1983) The interactive effects of fluoride and N-formyl-L-methionyl-L-leucyl-L-phenylalanine on superoxide production and cAMP levels in human neutrophils. Can J Biochem Cell Biol 61:569–578

Wong K, Chew C (1984) The respiratory burst of human neutrophils treated with various stimulators in vitro is dampened by exogenous unsaturated fatty acids. J Cell Physiol 119:89–95

Wong K, Chew C (1986) Comparison of 1-oleoyl-2-acetyl-glycerol and phorbol myristate acetate as secretagogues for human neutrophils. Can J Physiol Pharmacol 64:1149–1152

Wong RCK, Remold-O'Donnell E, Vercelli D, Sancho J, Terhorst C, Rosen F, Geha R, Chatila T (1990) Signal transduction via leukocyte antigen CD43 (sialophorin). Feedback regulation by protein kinase C. J Immunol 144:1455–1460

Woodroofe MN, Hayes GM, Cuzner ML (1989) Fc receptor density, MHC antigen expression and superoxide production are increased in interferon-gamma-treated microglia isolated from adult rat brain. Immunology 68:421-426

Woolf AD, Dieppe PA (1987) Mediators of crystal-induced inflammation in the joint. Br Med Bull 43:429–444

Worthen GS, Seccombe JF, Clay KL, Guthrie LA, Johnston RB Jr (1988) The priming of neutrophils by lipopolysaccharide for production of intracellular platelet-activating factor. Potential role in mediation of enhanced superoxide secretion. J Immunol 140:3553–3559

Wozniak A, McLennan G, Betts WH, Murphy GA, Scicchitano R (1989) Activation of human neutrophils by substance P: effect of FMLP-stimulated oxidative and arachidonic acid metabolism and on antibody-dependent cell-mediated cytotoxicity. Immunology 68:359–364

Wright CD, Hoffman MD (1986) The protein kinase C inhibitors H-7 and H-9 fail to inhibit human neutrophil activation. Biochem Biophys Res Commun 135:749–755

Wright CD, Hoffman MD (1987) Comparison of the roles of calmodulin and protein kinase C in activation of the human neutrophil respiratory burst. Biochem Biophys Res Commun 142:53–62

Wright CD, Hoffman MD, Thueson DO, Conroy MC (1987a) Inhibition of human neutrophil activation by the allergic mediator release inhibitor, CI-922: differential inhibition of responses to a variety of stimuli. J Leukocyte Biol 42:30–35

Wright CD, Hoffman MD, Thueson DO, Conroy MC (1987b) Inhibition of human neutrophil activation by the allergic mediator release inhibitor, CI-922: mechanism of inhibitory activity. Biochem Biophys Res Commun 148:1110–1117

Wright CD, Mülsch A, Busse R, Osswald H (1989) Generation of nitric oxide by human neutrophils. Biochem Biophys Res Commun 160:813–819

Wright GG, Mandell GL (1986) Anthrax toxin blocks priming of neutrophils by lipopolysaccharide and by muramyl dipeptide. J Exp Med 164:1700–1709

Wright J, Schwartz JH, Olson R, Kosowsky JM, Tauber AI (1986) Proton secretion by the sodium/hydrogen ion antiporter in the human neutrophil. J Clin Invest 77:782–788

Wright J, Maridonneau-Parini I, Cragoe EJ Jr, Schwartz JH, Tauber AI (1988) The role of the Na^+/H^+ antiporter in human neutrophil NADPH oxidase activation. J Leukocyte Biol 43:183–186

Wright SD, Silverstein SC (1983) Receptors for C3b and C3bi promote phagocytosis but not the release of toxic oxygen from human phagocytes. J Exp Med 158:2016–2023

Wymann MP, von Tscharner V, Deranleau DA, Baggiolini M (1987a) Chemiluminescence detection of H_2O_2 produced by human neutrophils during the respiratory burst. Anal Biochem 163:371–378

Wymann MP, von Tscharner V, Deranleau DA, Baggiolini M (1987b) The onset of the respiratory burst in human neutrophils. Real-time studies of H_2O_2 formation reveal a rapid agonist-induced transduction process. J Biol Chem 262:12048–12053

Wymann MP, Kernen P, Deranleau DA, Baggiolini M (1989) Respiratory burst oscillations in human neutrophils and their correlation with fluctuations in apparent cell shape. J Biol Chem 264:15829–15834

Yagawa K, Itoh T, Tomoda A (1983) Effect of transmethylation-reaction and increased levels of cAMP on superoxide generation of guinea-pig macrophages induced with wheat germ agglutinin and phorbor myristate. FEBS Lett 154:383–386

Yamaguchi T, Kakinuma K (1982) Inhibitory effect of Cibacron blue F3GA on the O_2^- generating enzyme of guinea pig polymorphonuclear leukocytes. Biochem Biophys Res Commun 104:200–206

Yamaguchi T, Kaneda M, Kakinuma K (1983) Essential requirement of magnesium ion for optimal activity of the NADPH oxidase of guinea pig polymorphonuclear leukocytes. Biochem Biophys Res Commun 115:261–267

Yamaguchi T, Kaneda M, Kakinuma K (1986) Effect of saturated and unsaturated fatty acids on the oxidative metabolism of human neutrophils. The role of calcium ion in the extracellular medium. Biochim Biophys Acta 861:440–446

Yamaguchi T, Hayakawa T, Kaneda M, Kakinuma K, Yoshikawa A (1989) Purification and some properties of the small subunit of cytochrome b_{558} from human neutrophils. J Biol Chem 264:112–118

Yamamoto K, Johnston RB Jr (1984) Dissociation of phagocytosis from stimulation of the oxidative metabolic burst in macrophages. J Exp Med 159':405–416

Yamashita T (1983) Effect of maleimide derivatives, sulfhydryl reagents, on stimulation of neutrophil superoxide anion generation with concanavalin A. FEBS Lett 164:267–271

Yamashita T, Someya A, Tsutakawa-Kido Y (1984) Effect of maleimide derivatives on superoxide-generating system of guinea-pig neutrophils stimulated by different soluble stimuli. Eur J Biochem 145:71–76

Yamashita T, Someya A, Hara E (1985) Response of superoxide anion production by guinea pig eosinophils to various soluble stimuli: comparison to neutrophils. Arch Biochem Biophys 241:447–452

Yamazaki A, Bitensky MW, Garcia-Sainz JA (1987) The GTP-binding protein of rod outer segments. II. An essential role for Mg^{2+} signal amplification. J Biol Chem 262:9324–9331

Yamazaki M, Gomez-Combranero J, Durstin M, Molski TFP, Becker EL, Sha'afi RI (1989) Phorbol 12-myristate 13-acetate inhibits binding of leukotriene B$_4$ and platelet-activating factor and the responses they induce in neutrophils: site of action. Proc Natl Acad Sci USA 86:5791–5794

Yasaka T, Boxer LA, Baehner RL (1982) Monocyte aggregation and superoxide anion release in response to formyl-methionyl-leucyl-phenylalanine (FMLP) and platelet-activating factor (PAF) J Immunol 128:1939–1944

Yazdanbakhsh M, Eckmann CM, Roos D (1985) Characterization of the interaction of human eosinophils and neutrophils with opsonized particles. J Immunol 135:1378–1384

Yazdanbakhsh M, Eckmann CM, De Boer M, Roos D (1987a) Purification of eosinophils from normal human blood, preparation of eosinoplasts and characterization of their functional response to various stimuli. Immunology 60:123–129

Yazdanbakhsh M, Eckmann CM, Koenderman L, Verhoeven AJ, Roos D (1987b) Eosinophils do respond to fMLP. Blood 70:379–383

Yea CM, Cross AR, Jones OTG (1990) Purification and some properties of the 45 kDa diphenylene iodonium-binding flavoprotein of neutrophil NADPH oxidase. Biochem J 265:95–100

Yoon PS, Boxer LA, Mayo LA, Yang AY, Wicha MS (1987) Human neutrophil laminin receptors: activation-dependent receptor expression. J Immunol 138:259–265

Yoshie O, Majima T, Saito H (1989) Membrane oxidative metabolism of human eosinophilic cell line EoL-1 in response to phorbol diester and formyl peptide: synergistic augmentation by interferon-γ and tumor necrosis factor. J Leukocyte Biol 45:10–20

Young JDE, Ko SS, Cohn ZA (1984) The increase in intracellular free calcium associated with IgGγ2b/γ1 Fc receptor-ligand interactions: role in phagocytosis. Proc Natl Acad Sci USA 81:5430–5434

Yufu Y, Nishimura J, Takahira H, Ideguchi H, Nawata H (1989) Down-regulation of a M_r 90,00 heat shock cognate protein during granulocytic differentiation in HL-60 human leukemia cells. Cancer Res 49:2405–2408

Yukawa T, Kroegel C, Chanez P, Dent G, Ukena D, Chung KF, Barnes PJ (1989) Effect of theophylline and adenosine on eosinophil function. Am Rev Respir Dis 140:327–333

Yuli I, Snyderman R (1986) Extensive hydrolysis of N-formyl-L-methionyl-L-leucyl-L-[^{3}H]phenylalanine by human polymorphonuclear leukocytes. A potential mechanism for modulation of the chemoattractant signal. J Biol Chem 261:4902–4908

Yuli I, Tomonaga A, Snyderman R (1982) Chemoattractant receptor functions in human polymorphonuclear leukocytes are divergently altered by membrane fluidizers. Proc Natl Acad Sci USA 79:5906–5910

Yuo A, Kitagawa S, Okabe T, Urabe A, Komatsu Y, Itoh S, Takaku F (1987) Recombinant human granulocyte colony-stimulating factor repairs the abnormalities of neutrophils in patients with myelodysplastic syndromes and chronic myelogenous leukemia. Blood 70:404–411

Yuo A, Kitagawa S, Suzuki I, Urabe A, Okabe T, Saito M, Takaku F (1989a) Tumor necrosis factor as an activator of human granulocytes. Potentiation of the metabolism triggered by the Ca^{2+}-mobilizing agonists. J Immunol 142:1678–1684

Yuo A, Kitagawa S, Ohsaka A, Ohta M, Miyazono K, Okabe T, Urabe A, Saito M, Takaku F (1989b) Recombinant human granulocyte colony-stimulating factor as an activator of human granulocytes: potentiation of responses triggered by receptor-mediated agonists and stimulation of C3bi receptor expression and adherence. Blood 74:2144–2149

Zabrenetzky V, Gallin EK (1988) Inositol 1,4,5-trisphosphate concentrations increase after adherence in the macrophage-like cell line J774.1. Biochem J 255:1037–1043

Zakhireh B, Root RK (1979) Development of oxidase activity by human bone marrow granulocytes. Blood 54:429–439

Zatti M, Rossi F (1967) Relationship between glycolysis and respiration in surfactant-treated leucocytes. Biochem Biophys Acta 148:553–555

Zeller JM, Caliendo J, Lint TF, Nelson DJ (1988) Changes in respiratory burst activity during human monocyte differentiation in suspension culture. Inflammation 12:585–594

Zigmond SH, Tranquillo AW (1986) Chemotactic peptide binding by rabbit polymorphonuclear leukocytes. Presence of two compartments having similar affinities but different kinetics. J Biol Chem 261:5283–5288

Zimmerli W, Lew PD, Waldvogel FA (1984) Pathogenesis of foreign body infection. Evidence for a local granulocyte defect. J Clin Invest 73:1191–1200

Zimmerli W, Seligmann B, Gallin JI (1986) Exudation primes human and guinea pig neutrophils for subsequent responsiveness to the chemotactic peptide N-formyl-methionylleucylphenylalanine and increases complement component C3bi receptor expression. J Clin Invest 77:925–933

Zimmerman GA, Renzetti AD, Hill HR (1983) Functional and metabolic activity of granulocytes from patients with adult respiratory distress syndrome. Am Rev Respir Dis 127:290–300

Zimmerman GA, McIntyre TM, Prescott SM (1985) Thrombin stimulates the adherence of neutrophils to human endothelial cells in vitro. J Clin Invest 76:2235–2246

Zimmerman JJ, Shelhamer JH, Parrillo JE (1985) Examination of the enzyme assay for NADPH oxidoreductase; application to polymorphonuclear leukocyte superoxide anion generation. Crit Care Med 13:197–203

Zimmerman JJ, Zuk SM, Millard JR (1989) In vitro modulation of human neutrophil superoxide anion generation by various calcium channel antagonists used in ischemia-reperfusion resuscitation. Biochem Pharmacol 38:3601–3610

Zylber-Katz E, Glazer RI (1985) Phospholipid- and Ca^{2+}-dependent protein kinase activity and protein phosphorylation patterns in the differentiation of human promyelocytic leukemia cell line HL-60. Cancer Res 45:5159–5164

Subject Index

acetyl salicylate 169
actin 61
– polymerization 61, 62, 63, 107, 109, 132
activation factor, purified 196, 197, 198
– factors, cytosolic, NADPH oxidase 189ff
adenosine 78
– desaminase 74
– respiratory burst 148, 162, 163, 165
adenylyl cyclase 171
β-adrenergic agonists 148, 159, 160, 161, 165
β-adrenoreceptors 82
AIDS 100
alcohols, respiratory burst 122
alkylacylglycerol 21, 24
alkylamines 31, 157
allergy, respiratory burst 208, 212, 213
AMG-C_{16} 48, 51, 52
amiloride 128, 129, 157
aminoglycosides 155
amphotericin B 83
anthracyclines 155
anthrax toxin 119, 152
aprotinin 70
arachidonic acid 180, 183, 184
– –, Ca^{2+}, mobilization 60
– – release 59, 60
arginine, R-NO 127
asbestos 79, 125
auranofin 150, 170, 171

benoxaprofen 21, 28
bile salts 80
bleomycin 122
botulinum C2 toxin 63, 74
bromophenacylbromide 151
bryostatin 22, 27

C-reactive protein 154, 173, 174
Ca^{2+} channel blockers 48, 56, 57
– ionophores 24, 25, 79
– –, respiratory burst 56
– mobilization 71, 84, 86
– –, phosphoinositol metabolism 55
– processes, respiratory burst 47ff
calmodulin, NADPH oxidase 182
calpain 29
cAMP, chemotactic peptides 64
– decreasing agents, respiratory burst 166, 167
– increasing agents, respiratory burst 147ff
candida albicans 153, 172
carotid body, NADPH oxidase 135, 143
Chediak-Higashi syndrome 207, 210
chemotaxis 127
chlamydia trachomatis 153, 172
chloroquine 150, 170
chlorpromazine 47, 51
cholera toxin 32, 149, 164
chondrocytes, H_2O_2 143
cibacron blue 154
clindamycin 155
clofazimine 81
colchicine 63, 64
collagen 75, 101
colony-stimulating factors 73, 95–99, 100
complement components, respiratory burst 72, 102, 103
concanavalin A 78, 115, 116
cross-linking agents 176
cyclic nucleotides, respiratory burst 64–69
– – see also cAMP, cGMP
cyclosporin A, respiratory burst 150, 168
cytochalasins, superoxide formation 61, 62, 63, 79

cytochrome b_{-245} 9, 10, 15−17
− −, defect   203, 204
− − translocation   18
cytokines, respiratory burst   155, 175
−, NADPH oxidase   72, 73, 91, 100
−, R-NO synthesis   126
cytoplasts   15
cytoskeleton   105, 109
−, formyl peptide receptors   83
−, superoxide   61−64
cytosolic activation factors, NADPH oxidase   189 ff

desensitization, heterologous   165
−, homologous   84
deuterium oxide   64
dexamethasone   167, 168
diabetes mellitus, respiratory burst   209, 214, 215
diacylglycerol   19, 201
− kinase   22
− − inhibitors, respiratory burst   53, 54
− lipase   54
−, NADPH oxidase   25, 26
− release   116
diazepam   157
diclofenac   151, 169
digitonin, respiratory burst   80, 120, 121
doxycyclin   155

elastin peptides   71, 102
endothelial cells, superoxide   143, 144
endothelin-1   91
eosinophil granule major basic protein   91
eosinophils, respiratory burst   133, 134, 137
esculetin   60
ethacrynic acid   157, 176

famotidine   161
fat cells, H_2O_2 143
fatty acids   80
− −, NADPH-oxidase   29−31, 179 ff
− −, protein kinase C   180−182
− −, respiratory burst   156
fibrinogen   154
fibroblasts, respiratory burst   135, 143
flavine adenine dinucleotide   9, 13
flavonoids, respiratory burst   48, 52
flavoprotein   10, 13
−, translocation   17
fMet-Leu-Phe   68, 83, 84, 85, 100

−, membrane potential   129, 130
−, R-NO   126
formyl peptide receptors   82, 83, 84, 98, 100, 117, 138
− − −, internalization   62
forskolin   149, 163

G-proteins   17, 32, 61, 62, 68, 87, 90, 97, 104, 112, 188, 189
−, low molecular mass   39
−, membrane receptors, interaction   32 ff
−, NADPH oxidase   31 ff, 184 ff
−, NaF   39−41
glia, respiratory burst   142
glucocorticoids   60
−, respiratory burst   150, 167, 168
glucose-6-phosphate dehydrogenase deficiency   207, 210
glutathione synthetase deficiency   207, 210
− peroxidase deficiency   207, 210
glycogen storage disease   217
gold compounds, NADPH oxidase   150, 170, 171
gramicidin, formyl peptide receptors   71, 87
granulomatous disease, chronic   203−206
guanine nucleotides, superoxide formation   114
guanylyl cyclase   65

halothane, NADPH oxidase   22, 27
haptoglobin   155, 174
heat-shock proteins   154, 173
hematological disorders, respiratory burst   206, 207
hemodialysis, respiratory burst   215, 216
hemoglobinuria, nocturnal   120, 207
5-HETE   111
hexachlorocyclohexanes, respiratory burst   80, 121
histamine   77, 148, 157, 159, 161, 162
histoplasma capsulatum   153, 172
HL-60 cells, NADPH oxidase   179
− −, NADPH oxidase activation factors   191, 193, 195, 196
− −, respiratory burst   134, 138, 139
H_2O_2 158
hypertension, essential, NADPH oxidase   208, 213, 214

ibuprofen   151, 169

immune complexes, respiratory burst 72, 103 ff
impromidine 161
indomethacin 22, 54, 151, 169
infections, respiratory burst 207, 208, 211, 212
influenza virus, respiratory burst 211
insulin 143
interferon 73, 92−95, 100
interleukin-1 73, 91, 92
interleukin-6 73, 99
interleukin-8 72, 99
iodonium compounds 14
isoproterenol 35

J774 cells, respiratory burst 135, 140

kapras 120 cells 134
kupffer's cells, respiratory burst 135, 139

laminin 75, 101
lassa fever 212
latex particles 79, 125
lectins, NADPH oxidase 78, 115, 116
leishmania donovani 153, 172
leukemia, myeloid 100
leukocyte-inhibitory factor 73, 99, 100
leukotrienes, respiratory burst 76, 109, 110
levamisole 156
lipocortin I 150, 168
lipopeptides, NADPH oxidase 71, 89, 90
lipopolysaccharides 116
−, respiratory burst 76, 116−118
lipoxins 76, 111
lipoxygenase pathway 109−111
local anesthetics 156
LTB$_4$ 32, 34, 42, 54, 58
lupus erythematosus-like 209
B-lymphocytes, respiratory burst 135, 140, 141
lymphotoxin 94
lysophosphatides 80

macrophages, activation factors 194
−, priming 7, 8
−, respiratory burst generation 20, 25
mannose 156
mastoparan 90
mellitin 184
mepacrine 150, 170
mesangial cells, respiratory burst 135, 141, 142, 168

mezerein 22, 27
microtubules, disruption 63
muramyl peptides 75, 119
muscarinic agonists 65, 66
myeloperoxidase 206, 210
myotonic dystrophy 208, 214

NADPH cytochrome c reductase 14
− dialdehyde 154
− diaphorase activity 5
− oxidase, K_m value 5, 8
− − activation 195
− − −, arachidonic acid release 59, 60
− − −, Ca^{2+}-dependent 56−58
− − −, fatty acids 29−31
− − −, formyl peptides 71, 82 ff
− − −, general mechanisms 19 ff
− − −, miscellaneous aspects 126 ff
− − −, Na^+/Ca^{2+} exchange 57
− − −, proteases 69, 70
− − −, receptor agonists, mechanisms 31 ff
− − −, various stimuli 70 ff
− − activity, age-related alterations 219
− − −, phorbol esters 200
− − −, phosphatidic acid 201, 202
− −, adenine nucleotides 7
− −, adherent neutrophils 131
− −, anionic amphiphiles 182, 183
− −, anti-inflammatory agents 150, 151, 167 ff
− −, antibodies 12, 13, 16, 189
− −, cAMP 64, 65, 66, 67
− −, catalytic properties 5, 6, 7, 8
− −, cell-free systems 177 ff
− −, cellular localization 17, 18
− −, cGMP 65, 66, 67
− −, chemoattractants 8
− −, cofactor requirements 6, 7
− −, cytochrome b_{-245} 16, 17
− −, cytosolic activation factors 189 ff
− −, − − −, defect 205, 206
− −, diacylglycerols 25, 26
− −, divalent cations 6, 7
− −, electrogenic 130
− −, fatty acids 178 ff
− −, G-proteins 184 ff
− −, hexachlorocyclohexanes 80, 121
− −, high-activity 191
− −, immunoglobulin 72, 103 ff
− −, inhibition 147 ff
− −, molecular mass 12
− −, Na^+/H^+ exchange 128

NADPH (cont.)
– –, NADPH-binding component 9, 12
– –, NaF 39–41
– –, pathology 203 ff
– –, phagocytosis 1
– –, phorbol esters 20–25
– –, phospholipases 42–45
– –, protein kinase C 19 ff, 41 ff, 200, 201
– –, – – inhibitors 46 ff
– –, R-NO, synthesis 127
– –, reconstitution 177 ff
– –, retinoids 80, 120
– – see also respiratory burst or superoxide
– –, solubilized 6, 12
– –, structural components 9–17
– –, translocation 17, 18
–, purified 9, 11, 13
NaF 37, 78, 149, 165
–, NADPH oxidase 39–41, 187
NEM, NADPH oxidase 157, 175
neuraminidase 74, 123
neutrophil functions, anti-inflammatory drugs 169, 170
– granules, NADPH oxidase 179
neutrophils, NADPH oxidase activation factors 192, 194
–, respiratory burst 216
–, resting potential changes 129, 130
nicotine, phagocytes 208, 213
nitric oxide 126
nucleoside diphosphate kinase 187, 188
nucleotides, respiratory burst 78, 111 ff

opioids 75, 88
opsonophagocytosis 106
osteopetrosis 209, 216
oxygen intermediate, role 1, 2
– radical formation, methods 2
– radicals, biochemistry 2

paraquat 156
peptides, chemotactic 152, 172
pertussis dependent activation 34–37
– toxin 32, 38, 40, 90, 95, 97, 104, 108, 109, 112, 113, 115, 147, 152, 186, 211
phagocyte functions 162, 163
phagocytes, activation by receptor agonists 33–37
–, cell-free systems 178, 179
–, IgG receptors 104, 105
–, NADPH oxidase defect 203

–, respiratory burst 132, 133–137, 215, 219
–, signal transduction 2
phagocytosis 1
–, deficiency 211
phagosomes 17
phalloidin 156
phenylbutazone 151, 169
phorbol esters 19, 54, 180
– –, NADPH oxidase 20–25, 26
phosphatidic acid 21, 22, 42
– –, respiratory burst 201, 202
phosphatidylserine 200, 201
phosphodiesterase inhibitors 159, 160, 161, 164
phosphodiesterases, cAMP 64, 65
phosphoinositide pathway 19, 33, 41, 42
phosphoinositol metabolism 53, 71, 78, 80, 84, 86, 103, 106, 107, 109, 111, 115, 117, 121, 140, 160, 165, 219
phospholipase 74, 169
phospholipase A_2 184
– –, activation 59
– C 71, 86, 162, 165, 166, 200
– D 201
phospholipases, NADPH oxidase 42–45
phospholipids, NADPH oxidase 6
phosphotyrosine phosphatase 67, 68
plant cells, NADPH oxidase 145
platelet aggregation 127
– – activating factor 76, 107, 108
– – derived growth factor 68, 69
polymyxin B 47, 49
propionic acid 81
propranolol 22
prostacyclin 160
prostaglandin E_1 35, 65
– –, respiratory burst 26
prostaglandins 117, 147, 148, 159, 160
proteases 74
–, superoxide formation 69, 70
protein kinase C 71, 85
– – –, Ca^{2+}/phospholipid-independent 28, 29
– – – inhibitors, NADPH oxidase 46 ff
– – –, inhibitory role, respiratory burst 54, 55
– – – isoenzymes 19, 20
– – –, NADPH oxidase 19 ff, 41 ff
– – –, priming by activators 45, 46
– – cAMP-dependent 64, 166
– –, cGMP-dependent 65
– tyrosine kinases 67, 68

pseudorabies virus 212
pulmonary disorder, respiratory burst 208, 212, 213
purine nucleotides 78, 111, 112, 113
purinoceptors 111, 113
pyrimidine 154
− nucleotides 78, 112, 113

quartz 79, 125
quinones 9, 10, 14, 15

radish, NAD(P)H oxidase 136
renal failure, respiratory burst 209, 215, 216
− tubules, respiratory burst 135, 142
resiniferatoxin 28
respiratory burst 13
− −, activation by various agents 20ff
− −, adherent neutrophils 131, 132
− −, alcohols 122
− −, anion transport blocker 131
− −, anti-inflammatory drugs 150, 151, 167ff
− −, arachidonic acid 59, 60
− −, L-arginine 127
− −, botulinum C2 toxin 63
− −, Ca^{2+} ionophores 56
− −, complement 72, 102, 103
− −, cytokines 72, 73, 91, 94ff, 100
− −, cytoplasmic Ca^{2+} 57
− −, definition 1
− −, desensitization 165
− −, digitonin 80, 120, 121
− −, fMet-Leu-Phe 58, 68
− −, formyl peptides 82, 98
− −, guanine nucleotides 114
− −, infections 207, 208, 211, 212
− −, inhibition 46ff, 147−176
− −, interferon 73, 92−95, 100
− −, intracellular pH 128, 129
− −, latex beads 79, 125
− −, lectins 78, 115, 116
− −, lipoxins 76, 111
− −, LTB$_4$ 109, 110
− −, mechanism of inhibition 164−166
− −, membrane potential changes 129, 130
− −, methods 2
− −, miscellaneous inhibitory agents 171ff
− −, Na$^+$/H$^+$ exchange 128
− −, NaF 40
− −, neonatal animals 219

− −, nucleotides 78, 111, 112, 113
− −, opioids 75, 88
− −, opsonized particles 106, 107
− −, PAF 76, 107, 108
− −, particulate stimuli 79, 123ff
− −, parturition 219
− −, pathology 203, 204, 206ff
− −, PK inhibitors 47−53
− −, priming 7, 38, 44, 87
− −, −, protein kinase C activators 45, 46
− −, proteases 69, 70
− −, quinones 14
− −, reactive NO 126, 127
− −, see also NADPH oxidase or superoxide
− −, serotonin 67
− −, sex related 220
− −, somatotropin 75, 88, 89
− −, specialized cell types 133−145
− −, substance P 86, 87
− −, termination 6, 84
− −, tyrosine residues and phosphatases 67−69
− −, wasp venom 72, 90
retinoids, NADPH oxidase 80, 120
−, respiratory burst 52

sarcoidosis 208, 213
sea urchin eggs, respiratory burst 135, 144
sensitization, homologous 85
serotonin 119
SH reagents, NADPH oxidase 175, 176
sodium nitroprusside 67
somatotropin 75, 88
sphingosine 47, 49
staurosporine 47, 49, 50, 120
sterols 156
stilbene 157
substance P 35
− −, NADPH oxidase 86
sulfasalazine 150, 170
superoxide dismutase 127
− formation 83−86
− −, cAMP 65
− −, cytochalasins 61, 62
− −, enzymatic 5, 6
− −, lipopolysaccharides 76, 116−118
− −, proteases 69, 70
− − see also NADPH oxidase or respiratory burst suramin 48, 52, 53

thymol 81, 122
thyroid hormones, synthesis 142
tiazofurine 157
trifluoperazine 47, 51
triphenyltin 156
tuftsin 75, 87, 88
tumor necrosis factor 72, 94, 95
tyrosine phosphorylation 67, 68

U-937 cells, respiratory burst 134, 138

urate, NADPH oxidase 79, 126

vinblastine 63
vitamin D$_3$ 81, 123
vitamin E 48, 52

wasp venom 72, 90

zymosan 124
zymosan, respiratory burst 106